DENNIS FAST www.dennisfast.smugmug.com

Library and Archives Canada Cataloguing in Publication

Beattie, Heather, 1980-
Wild West : nature living on the edge / Heather Beattie and Barbara Huck.

ISBN 978-1-896150-35-2

1. Endangered species--North America. 2. Biotic communities--North America. I. Huck, Barbara II. Title.

QH77.N56B44 2009 591.68097 C2009-906910-5

ENVIRONMENTAL BENEFITS STATEMENT

Heartland Associates Inc saved the following resources by printing the pages of this book on chlorine free paper made with 10% post-consumer waste.

TREES	WATER	SOLID WASTE	GREENHOUSE GASES
8	3,777	[illegible]	784
FULLY GROWN	GALLONS	POUNDS	POUNDS

Calculations based on research by Environmental Defense and the Paper Task Force.
Manufactured at Friesens Corporation

Wild West

Nature Living on the Edge

By Heather Beattie
and
Barbara Huck

Heartland Associates, Inc.
Winnipeg, Canada

Printed in Manitoba, Canada

Credits

Editorial assistance
Theresa Harvey Pruden, Doug Whiteway, Peter St. John

Maps, design and layout
Dawn Huck

Image research
Jane Huck, Dennis and Frieda Fast,
Jack and Nancy Most

Printing and binding
Friesens Corporation, Manitoba, Canada

Cover Image
Source: Getty Images

This young ferret is part of the Black-footed Ferret Recovery Program in Colorado.

COURTESY OF THE NATIONAL BLACK-FOOTED FERRET RECOVERY PROGRAM / US FISH AND WILDLIFE SERVICE

ACKNOWLEDGEMENTS

WILD WEST: Nature Living on the Edge would not have been possible without the time and talents (and in some cases, the research capabilities and persuasive powers) of dozens of people. Many have dedicated their lives to creating a place for the animals that once swam our oceans, darkened our skies or filled our many and incredibly varied landscapes. Others have battled for decades to preserve our rivers, our forests and our disappearing grasslands. Every North American owes them a debt of gratitude. They, and in many cases their prescient predecessors, as well as those who are determined to follow in their footsteps, have – hopefully just in time – realized what we are losing, and devoted their lives to saving it.

Many of them took time from their busy schedules and personal lives – even, in one case, the birth of a baby – to assist us, to read sections on endangered species as varied as blue whales and prairie skinks, or landscapes from the Arctic to the southern deserts, or to provide images, many of them absolutely magnificent, of the more than 50 species in this book. In some cases, several people reviewed a particular section, making comments on various aspects of it and ensuring that our information was as up-to-date as possible.

We thank them all.

For our section on The Arctic, the species and ecosystem experts included Parks Canada ecologist and University of Manitoba professor Micheline Manseau, who reviewed our section on Peary caribou (not once, but twice), and allowed us to use one of her excellent photographs; U.S. Fish and Wildlife Service biologist Thomas Evans and media relations chief Bruce Woods of Anchorage, Alaska, who sorted us out over several controversial aspects of polar bears; biologist Craig George and regional deputy director Harry Brower, Jr., of North Slope Burough, Alaska, who reviewed our section on bowhead whales, and Mark Mallory, a seabird biologist with the Canadian Wildlife Service in Iqualuit, Nunavut, who made comments (some of them gratifyingly positive) on the Ross's gull entry. Others, who did not wish to be named, assisted with our chapter on The Arctic.

For the chapter on The Boreal Forest, Dale Seip, wildlife ecologist with the BC government's Ministry of Forests in Prince George, as well as Al Arsenault, senior ecologist with Saskatchewan's Ministry of Environment in Saskatoon both reviewed our section on the far-flung (but almost everywhere threatened) woodland caribou; Dale also appraised our section on the mountain caribou ecotype, found in the chapter on The Western Mountains. Brian Johns, wildlife biologist with the Canadian Wildlife Service in Saskatoon set us straight and up-to-date over the state of the poster child for endangered wildlife, the whooping crane, while Edmonton-based Canadian Wildlife Service biologist Greg Wilson offered suggestions and corrections to our section on wood bison.

We had a great deal of help for our lengthy Sea to Sky Region chapter, including information on the Kermode bear from Wayne McCrory, bear biologist with the Valhalla Wilderness Society, and information and corrections on the Vancouver Island marmot from wildlife biologist Don Doyle of the BC Ministry of the Environment, as well as on-site scouting from photographer Jack Most, who visited the marmot reintroduction site on Mount Washington not once but twice. For the section on sea otters, we had help from research wildlife biologist James Bodkin of the Alaska Science Center in Anchorage, while Lance Barrett-Lennard, head of the Marine Mammal Research Program at the Vancouver Aquarium's Marine Science Centre, reviewed our section on orcas. And John Ford, of Canada's Fisheries and Oceans Cetacean Research Program in Nanaimo, BC, was extremely helpful with two sections, the North Pacific right whale and the blue whale. The latter appears in The California Coast chapter.

But few were more exhaustively thorough than Vancouver-based Ken Morgan, of Fisheries and Oceans Canada, who had dozens of suggestions for our section on the short-tailed albatross.

For the chapter on The California Coast, we are grateful to Michael Woodbridge, public affairs specialist at the Hopper Mountain National Wildlife Refuge Complex in Ventura, California, who delivered not one but two edited versions of our section on the California condor, and to Scott Benson, research fish biologist with the National Oceanic and Atmospheric Administration in Moss Landing, California, who assisted with information and corrections to the section on the globally endangered leatherback turtle.

We had lots of help with our section on The Western Mountains, including Bill Pednault, director of the Sooke River Hatchery in Sooke, BC, who reviewed our sections on both chinook and sockeye salmon; Ron Ek, director of the BC Freshwater Fisheries Society's Kootenay Trout Hatchery, assisted with information on our white sturgeon entry, while Michael Chutter, provincial bird specialist with the Victoria-based BC Ministry of the Environment, set us straight over a number of aspects of the section on the northern spotted owl.

Our section on The Grasslands had input over several years from Gene Fortney and Cathy Shaluk of The Nature Conservancy of Canada, Manitoba chapter, and national board member Sheldon Bowles, as well as Manitoba educator, birding expert and wildlife photographer *par extraordinaire* Dennis Fast, whose many photographs grace this book. Ken De Smet, species at risk specialist with Manitoba Conservation, sent us his corrections and comments on both the burrowing owl and the loggerhead shrike just before our deadline, despite travelling and meetings.

Great dedication and expertise were provided by the U.S. Fish and Wildlife's Black-footed Ferret Recovery Project's staff and scientists, including Pete Gober, field supervisor and chief project coordinator, education specialist Sarah Bexell, who helped source many images and Paul Marinari, fish and wildlife biologist for the National Black-footed Ferret Conservation Center.

Glen McMaster, manager of the Habitat Assessment Unit of the Saskatchewan Watershed Authority in Regina, and his colleague, avian ecologist Corie White, both assisted with our section on piping plovers; Corie straightened us out on a number of details. And as for the section on prairie skinks, both University of Brandon biology professor Pamela Rutherford, who – with her students – is very involved in on-site research on Canada's very small population of skinks, and James Duncan, manager of Manitoba Conservation, both made helpful suggestions and added information.

For the final section, on The Western Deserts, many people obliged our increasingly tight timelines with remarkably quick responses and, in many cases, offers of excellent photographs and additional information. Reviewers for our section on the shrub-steppe ecosystem include Rex Crawford, national heritage ecologist for the Washington Department of Natural Resources in Olympia, and Denise Eastlick of British Columbia's Osoyoos Desert Society, while John and Mary Theberge of Oliver, BC, also commented on the section.

Erin Fernandez, the Mexico program coordinator with the U.S. Fish and Wildlife's Tucson-based Arizona Ecological Services reviewed both the jaguar and Sonoran pronghorn sections, giving us detailed and up-to-the minute information, as well as excellent background material. For our section on the Mexican wolf, Susan Dicks, a veterinarian with the U.S. Fish and Wildlife's Mexican Wolf Recovery Program in Albuquerque, New Mexico, was both informative and encouraging, while Lois Grunwald of the Sacramento Fish and Wildlife Office went over the rather complex section on the Sierra Nevada bighorn sheep with great care. We were delighted when biologist Jody Mays of the Laguna Atascosa National Wildlife Refuge in Alamo, Texas, agreed to review our entry on ocelots, for she heads the U.S. Fish and Wildlife's ocelot breeding program.

Wild West would not be the beautiful book it is, however, without the images – many of them spectacular – of species that are not only hard to find, but even harder to photograph. Many came from Dennis Fast, who has become almost intimate with polar and grizzly bears over the years, and has recorded the splendor of the tall grass prairie in every season; and Jack Most, who photographed spawning chinook salmon, orcas (with the help of Victoria, BC's Wild Cat Whale Watching) and Vancouver Island marmots. We're grateful, too, to his daughter Heather Most, also a talented photographer, who visited Victoria's Garry oak woodlands repeatedly over many months to record the changing seasons.

Some of the photographs have stories to go with them: Herb Segars rescued a leatherback turtle during his undersea photography session; Jerry Kautz broke an arm just before photographing the Okotoks glacial erratic; Ron Wolf shipped us an image we requested at 4 a.m. the day the book went to the printer and biologist Kara Dodge of the Large Pelagics Research Centre at the University of New Hampshire, not only sent us a series of photographs, but accompanied them with a clear explanation of the university's work tagging large leatherback turtles off the US East Coast.

But this book would not have been possible without the endless assistance of the dedicated scientists and researchers of the U.S. Fish and Wildlife Service and the National Oceans and Atmospheric Administration. They not only offered help and suggested contacts, but allowed us to use countless images of many endangered species. The staff responded to every query we made, often breathtakingly quickly.

On our end, those contacts would not have been made, or the sections vetted, without the ingenuity, dedication and perseverance of Jane Huck and Theresa Harvey Pruden. With a real passion for endangered species, Jane spent weeks sourcing photographs, paintings and drawings and, in the process got to know dozens of photographers and scientists across the continent. And with typical professionalism, Theresa Harvey Pruden coordinated the entire expert editing process, contacting biologists, botanists, ecologists and researchers from Alaska to Arizona and from Nunavut to New Mexico.

Thanks, too, to Heartland's wonderful partners Doug Whiteway and Peter St. John. Doug tackled several sections as time ran short, producing entries with typical finesse. And Peter read and reread the manuscript, weeding out errors, and also provided a number of lovely photographs.

We all owe a huge debt of gratitude to Dawn Huck, our talented designer, who not only created all the maps, but wove the information and images into a book that we hope is worthy of its subject.

And finally to the hundreds of scientists, staff and ordinary citizens who have spent years or decades working to bring these and so many other endangered and threatened species back from the brink, our heartfelt thanks.

Barbara Huck
November 2009

I would also like to express my most heartfelt gratitude to my parents, Jim and Della Beattie. Your love and support while I was working on this book, and every other day of my life, mean so much to me. Thank you for making anything seem possible.

Heather Beattie
November 2009

Table of Contents

Introduction

We **are newcomers** to this planet. Mammals have existed for perhaps 225 million of Earth's 4.65 billion years. Humanity, in the form of *Homo sapiens*, arose only 45,000 years ago. A breath. A blink of the eye.

Whether it was pure chance, or divine providence, the combination of geology, climate, evolution and time has created an almost unimaginable diversity of life on this planet. Geology gives the land its structure and soil, and climate determines what plant life will grow. The great diversity of environments that are found in North America has led to many different kinds of plants and animals establishing themselves on this continent. Western North America, with its tundra, Pacific and Arctic coasts, its forests, mountains, grasslands and deserts, is home to some of the most beautiful, majestic and increasingly rare wildlife and eco-systems in existence.

This is the story of some of those species. It spans hundreds of thousands of years, and provides a glimpse into their pasts, presents and possible futures. It is intimately connected with the land, and begins with its formation.

The Mesozoic

About 90 million years ago, North America was on its way to becoming the continent we know today. It had been more than 100 million years since Pangaea, Earth's most recent supercontinent, had begun to break up. On the Pacific coast, collisions between island terranes riding on the Pacific Plate and the North American continent had added bits and pieces to what would one day be the West Coast. Farther inland, the new territory served as an enormous bulldozer, rumpling the sedimentary layers like tinfoil and creating the Columbia Mountains and, farther east, jackknifing the main ranges of the Rockies skyward about 140 million years ago.

Then, between 85 and 90 million years ago, North America collided head on with Wrangellia, a huge terrane that includes today's Vancouver Island and Haida Gwaii. This enormous collision melted the crust of the collision zone and created the upheaval of the largest mass of granite in the world along the Coast Belt. And far to the east, it caused the front ranges and foothills of the Rockies to rise (for more on mountain building, see page 194).

Volcanic explosions and earthquakes were common as the ages passed and the resulting sediment was carried east by wind and water to the great Mid-Continental Seaway. This warm, shallow sea had long before formed a giant Y across the length of the continent, connecting the Arctic waters to Hudson Bay and the Gulf of Mexico and creating a remarkably stable climate, with little temperature variation from day to day and season to season. Tropical and sub-tropical vegetation flourished in the rich volcanically fertilized soil. The waters teemed with fish and reptiles, there were at least two species of birds in the sky, and small mammals scurried about the undergrowth. But dinosaurs ruled the land.

The Age of Mammals

Then, about 65 million years ago, the Mesozoic era came to an abrupt end, ushering in the beginning of our Cenozoic era, the age of mammals. Most scientists now believe this was the result of an asteroid or a comet slamming into the Earth, creating a massive explosion that threw rock, dust and water vapor into the air, virtually blacking out the sun and eventually smothering most of the world's plant life. The Earth cooled and, their habitats and food sources irrevocably changed, most of the dinosaurs and all the other large land animals died. The foundation of the complex marine food pyramid was shattered because, without sunshine, much of the oceans' plankton population disappeared. According to scientists, this astronomical event led to the extinction of 65 per cent of the world's species. However, some species survived and, with their major predators gone, mammals flourished and diversified, as did the dinosaurs' only living descendants, birds.

Time passed, species evolved and changed, and then, about 1.8 million years ago, the world's climate began to cool dramatically, as it had many times before and as it will again. This was the beginning of the Pleistocene epoch, the glacial age that continues today and, scientists believe, may last for at least another eight million years.

DENNIS FAST www.dennisfast.smugmug.com

Glacial ages occur when the shifting of the earth's tectonic plates arranges the continents in a way that prevents the oceans from distributing the sun's heat freely around the globe. When a land barrier prevents warm tropical water from circulating with the north and south polar waters, as is the case today, the global climate becomes more extreme. Summers at high latitudes are often not warm enough to melt all of the snow that falls during the winter, so snow and ice accumulate.

Still, the global climate is not consistent during a glacial age. During periods of intense global cooling, commonly known as glaciations, enormous ice sheets cover much of the land at temperate latitudes; between the glaciations are interglacial periods, when the ice recedes to the polar regions. Glaciations and interglacials alternate on a regular basis; some climatologists believe there have been as many as 17 glaciations since the Pleistocene began. These variations occur because of astronomical phenomena known as the Milankovitch cycle, after its discoverer, Serbian scientist Militun Milankovitch. This 100,000-year cycle is dependent on three other cyclical factors: 105,000-year variation in the shape of the Earth's orbit around the sun; the 41,000-year cycle in the tilt of the Earth's axis; and the 21,000-year cycle in the time of year at which the Earth is closest to the sun.

The most recent glaciation of North America, the Wisconsin glaciation, reached its peak 18,000 years ago. Most of present-day Canada and the northern United States were covered by the Laurentide ice sheet, which joined the smaller Cordilleran ice sheet just east of the Rockies. Because so much of the Earth's water was in glacial form, ocean levels were much lower than they are today, creating a land bridge, known as Beringia, between eastern Asia and Alaska. This broad northern isthmus, which – rather remarkably – stayed largely ice free, provided a corridor for animals and ultimately humans to move from Asia to the Americas.

Conditions at the southern edge of the glaciers were harsh, with cold winds blowing off the ice sheets and a border of permafrost surrounding them.

Southeastern North America was a combination of forest and tundra, while to the west, on today's Great Plains, the climate was drier and there were grasslands and large areas of sand dunes. Many large herbivorous mammals lived south of the glaciers, though very few can be seen today. Most, including mastodons and mammoths, several species of horses, camels, ground sloths, shrub-oxen, giant beavers, tapirs and stag moose, are now confined to museums. Here, too, were ancient bison, rather late arrivals on the continent. Preying on these animals were dire wolves, sabre-tooth cats, American lions and cheetahs, as well as giant short-faced bears.

BARBARA ENDRES

Humans in America

Humans had evolved during the Pleistocene, along with many other large mammals, but when and how they first came to North America is still the subject of much controversy. Once, it was theorized that groups of early hunters followed their prey across the Bering land bridge from Asia about 12,000 years ago. Moving deep into North America by way of an ice age corridor that opened as the Laurentide and Cordilleran glaciers began to melt, early North Americans were believed to have spread quickly throughout the continent. Distinctive spear points, named after the site near Clovis, New Mexico, where they were first discovered, had been found from Alberta to South America, and their age seemed to indicate that the people who made them had, in the space of only 500 years, established themselves throughout the hemisphere.

Today, however, the Clovis theory has been abandoned. It not only raised too many unanswerable questions – Why would Arctic hunters have continued travelling south as quickly as suggested when there were abundant food sources everywhere? How would they have so quickly adapted to the tropical climates of Central and South America? – but, more importantly, it was contradicted by archaeological evidence of a 12,800-year old human settlement at Mount Verde, Chile, deep in South America, and even older evidence of the presence of humans in many other locations. Albert Goodyear, of the University of South Carolina, has found tools that have been radiocarbon dated to 50,000 years old, although there is some dispute as to whether they were truly made by humans.

The Monte Verde discovery, which was proven beyond dispute in 1997, gave rise to other theories of the peopling of North America, each with its proponents. The Greenberg hypothesis suggests that three waves of people came by foot from Asia, establishing three basic linguistic groups.

Anthropologist C. Loring Brace argues that there were two crossings, one on foot about 15,000 years ago, the other by water 10,000 years later. This theory relies on the growing realization that while the continent and the western mountains were glaciated, and while coastal glaciers undoubted calved directly into the Pacific, many coastal refugia – areas that were unglaciated thanks to protective mountain slopes or relatively warm ocean currents – also existed. And ocean levels were low, leading to hypotheses that people could have migrated down the West Coast.

We also know that people have been using boats for 50,000 years; British Columbia palaeontologist Knut Fladmark has suggested that early people could have entered North America by sea along the comparatively mild West Coast. It is not surprising that evidence of this theory is scarce, because anything left behind at coastal settlements would now be deep under the ocean.

However the first North Americans arrived, when the global climate warmed, Beringia flooded and humans in significant numbers joined the other predators in hunting the grazers and browsers in a vast and varied land. The Holocene, or modern age, had begun (see geological chart on page 357).

The Great Extinction

Soon after the end of the Wisconsin glaciation, relatively speaking, some 40 species of large North American mammals became extinct. There are two main explanations for this mass extinction. The overkill hypothesis, first proposed in the 1960s by American geoscientist Paul S. Martin, argues that the extinction of so many large mammals was directly related to the arrival of human hunters. Throughout the world, he argues, the arrival of humans has preceded mass extinctions.

"Near-time extinctions [those that occurred within the past 50,000 years] of large animals swept Australia over 40,000 years ago, peaked in America 10,000 to 13,000 years ago and ended historically with the settlement of remote Pacific islands," Martin responded in a *New Scientist* interview following the 2005 publication of his *Twilight of the Mammoths: Ice Age Extinctions and the Rewilding of America.*

Native fauna were easy prey for early Americans; unlike African animals that had evolved alongside humans, or Asian animals that had adapted over hundreds of thousands of years to the presence of hunters, North American animals had no experience with how lethal this physically unimposing predator could be. Martin also points to evidence that it was only large mammals – those that would have been targeted for hunting and more vulnerable to human destruction of their habitat – that were affected during this period. And animals that were hunted for their meat, such as mammoths, bison

Though compared to their prey they were unimposing, and their weapons were, at least to our eyes, primitive, the first North Americans (opposite, above) tipped the balance for many other species, including American lions.

and horses, affect many other species, from the carnivores that had long preyed on them to smaller herbivores that depended on the megafauna to transform tall grasslands into areas where shorter, more nutritious grasses could grow.

The other explanation for this mass extinction is based on the rapid climate change the world faced between 12,000 and 6,000 years ago. Other mammalian extinction events have been associated with climate change where human exploitation was definitely not a factor. Scientists who support this theory argue that the human population was too small at the time the majority of the extinctions occurred for hunting to have played a significant role. Shifts in weather and vegetation patterns, on the other hand, would have disrupted ecosystems, and it was large animals with low birth rates that would have had the most trouble adapting.

There's evidence that the range of many smaller mammals also decreased dramatically during this time period. Competition and predation from species that migrated across the land bridge from Asia and penetrated temperate North America for the first time could also have impacted the ability of native mammals to survive.

In fact, as British paleontologists Antony Hallam and Paul Wignall have written, it's likely that both climate change and the arrival of humans contributed to the demise of most of the large mammals that were living in North America at the end of the last ice age. The few survivors were mainly species that had migrated from Asia during the last 300,000 years of the Pleistocene, among them elk, bison and bighorn sheep, as well as grizzly and black bears. Species native to the Americas were virtually all extirpated, including horses and camels, which survive anywhere today only thanks to ancestors that migrated from North America to Asia and proliferated on the other side of the globe. The only large mammal indigenous to the Americas that is still here is the pronghorn (see page 336).

Evolution

Whether they evolved here or elsewhere on the globe, North America's plant and animal species are the products of tens of thousands of years of evolution. Evolution is not something that occurs through any conscious will of an organism, or results from adaptations that individuals are forced to make in response to their surroundings. It takes place simply because certain characteristics confer an advantage or disadvantage in a particular environment. Whenever the environment changes, natural selection favors those members of a population that are best suited to the new conditions.

Every time a pair of animals reproduces, the offspring receives genetic material and hereditary characteristics from each parent. In addition to unique genetic material, mutations sometimes occur that give individuals a survival advantage, or put them at a disadvantage, in which case they will die out. Members of a given population that possess a survival advantage are more likely than others to reach maturity, reproduce and pass their genes onto the next generation. Hereditary traits include both physical characteristics and behaviors. And behaviour patterns can be instinctual or learned; both are hereditary in that an animal's ability to perform the behavior is determined by its genetic make-up.

In rapidly changing environments evolution can have a directional influence, with a species developing traits that best enable members to survive under the new conditions. When there are several open ecological niches, evolution can also spur diversity, as members of a population evolve into differing forms specifically adapted to different types of living space. Over many generations, these processes have filled gaps left by the Great Extinction of native North American fauna.

P. RINDISBACHER / NATIONAL ARCHIVES OF CANADA/ ACC. NO. / e002291334

Prairie wolves were still common – though being trapped and killed in large numbers – when 15-year-old Peter Rindisbacher, already a talented artist, arrived in Red River (later Winnipeg, Manitoba) in 1821.

The Great Killing

Wild West: Nature Living on the Edge delves into the rapid devastation of many of Western North America's species and ecosystems over the past 150 years. This is essentially the period since Europeans, armed with modern weaponry, and bearing the fundamental attitude that Nature exists for the use and pleasure of humans, or even to fulfill the biblical command to subdue the Earth, entered the vast western half of the continent in significant numbers.

While the disappearance of species native to the Americas at the end of the last glaciation – the period discussed above and widely known as the Great Extinction – took perhaps 2,000 years to accomplish, it has taken modern North Americans less than a tenth of that time to bring dozens of species to the brink of the abyss. As a result, many are calling this modern destruction of the continent's bountiful environments and animals the Great Hunt or the Great Killing, for it has been accomplished with a much greater awareness of its consequences.

Nevertheless, almost from the beginning, there were those who could see the longterm consequences of mindless slaughter. Artist George Catlin was one of them. Sent in 1832 by William Clark, then superintendent of Indian Affairs for the Upper Louisiana Territory, deep into what is now the western US, he was charged with recording the landscape and its people, including the Pawnee, Blackfoot, Crow, Ojibwe and Commache. Travelling on the plains and in the foothills until 1836, Catlin was among the first to record the disappearing wildlife, particularly the plains bison. During a stay at a fur post, he learned of the butchery of 1,400 bison, simply for their tongues, which he witnessed being dumped on the ground.

"This profligate waste of the lives of these noble and useful animals," he wrote in his journal, "when, from all that I could learn, not a skin or a pound of the meat [except the tongues], was brought in fully supports me in the seemingly extravagant predictions that I have made as to their extinction, which I am certain is near at hand."

During the summer of 1833, while Catlin was touring the Western Plains, North America's most celebrated wildlife artist, John James Audubon, was touring and painting in Eastern Canada. Twenty years before, in the autumn of 1813, Audubon had personally witnessed a flock of passenger pigeons, writes David M. Lank in *Audubon's Wilderness Palette,* "… so dense that the birds literally eclipsed the noonday sun, and continued to pass in undiminished numbers for three days in succession. It's estimated that the flock contained more than one billion birds."

The birds' passage created a noise, Audubon wrote, like "a hard gale at sea passing through the rigging of a close-reefed vessel", or a "roar of distant thunder". More telling, however, was his statement that the sound of the migration was so loud, "that the firing of guns could not be heard even as the constant slaughter continued."

This annual hunt, combined with the destruction of the eastern forests where the pigeons nested, spelled the end of the species. The last great nesting, according to Lank, "took place in 1878 near Petrosky, Michigan," and the last passenger pigeon died in September 1914, at the Cincinnati Zoo.

The visibly declining flocks, not only of passenger pigeons, but of other species, prompted some to speak out against the devastation and the squandering of life. Quoting a witness to the killing of migrating Eskimo curlews (see page 66) in Nebraska, Robert McCracken Peck wrote in *Land of the Eagle,* "'The slaughter of these poor birds was appalling and almost unbelievable … Hunters would shoot the birds without mercy until they had literally slaughtered a wagonload of them … The compact flocks and tameness of the birds made the slaughter possible, and at each shot … dozens of birds would fall.'"

By the end of the 19th century, the wanton slaughter was becoming evident across Western North America. The plains bison, once numbering 30 million, had dwindled to fewer than 200 by the 1880s; taken into captivity, they spent the next century as "domestic" animals behind tall fences – though anyone who has ever met a bison bull will know that it can never be realistically termed "domesticated". Pronghorns, too, though once numbering in the millions, hovered on the brink of extinction, and in the oceans the world's largest creature – the blue whale – became the most hunted creature on Earth in the early 20th century. The blue whale slaughter reached a peak in 1931, when 29,649 were taken.

Still, it was widely accepted that Nature's bounty was endless. It wasn't until the 1950s, as the Earth's human population soared toward three billion (which it passed in 1960), that some began to question the idea, not just among friends, but openly, and loudly. Perhaps it was the whooping crane (see page 92), that forced North Americans to take a closer look in the mirror. Perhaps it was the sheer numbers of species in trouble. Or our fast-disappearing wild spaces.

Whatever it was, the reality is here, staring us all in the face if only we care to look or listen. Our newspapers, TV stations and podcasts are filled with the dire reports: millions of "missing" salmon; species that exist only because of captive rearing programs; an enormous and rapidly growing "island" of plastic in the middle of the Pacific Ocean; sea and land mammals with rates of cancer heretofore unseen, and the list goes on and on.

Nevertheless, despite – or perhaps because of – these dire pronouncements, despite a global human population that is soaring toward seven billion, there is reason for optimism. And that comes from the growing awareness, the rising concern, the emergent determination that we – and only we – can do something about our continental, indeed our global environmental crisis.

In apparent despair, Henry David Thoreau, the great American naturalist and

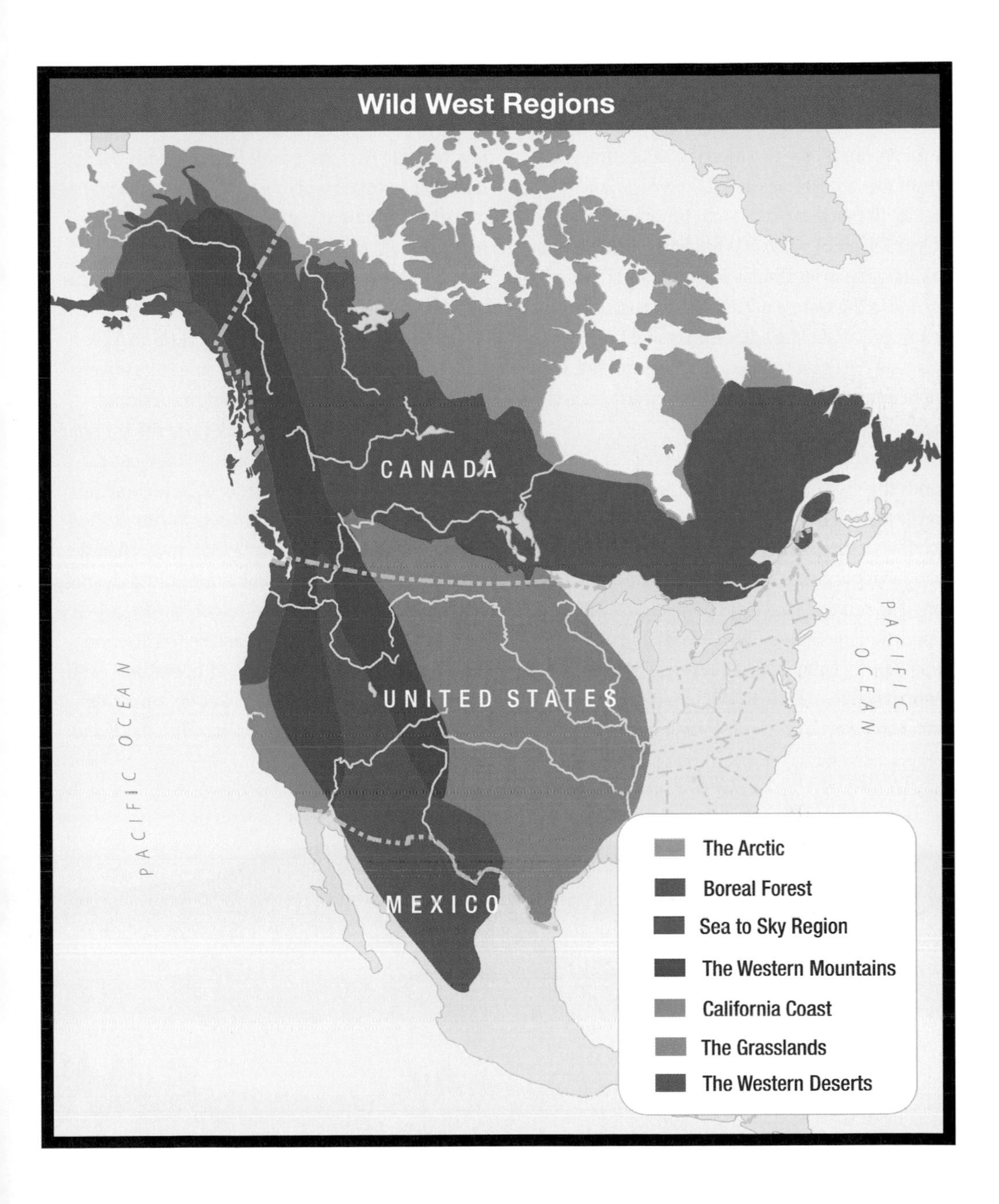

author, once described our national parks as "Little oases of wilderness in the desert of our civilization." Perhaps. But, as anyone lost in a desert will tell you, an oasis can be a life saver. And in many cases we still have time to turn those oases into corridors and the corridors into reserves. All it takes is human ingenuity and determination. And it's to assist this determination that *Wild West* was written.

Arctic Transformation

In Yukon, during the summer of 2007, Zelma Lake, one of the largest lakes in Old Crow Flats, an immense, flat section of the arctic tundra 835 kilometres or 520 miles north of Whitehorse, lost more than half its water in less than a month. Though a natural change in the water flow may have caused the water to abruptly drain into adjacent, smaller lakes, hydrologists suspect that shrinkage in the permafrost – the permanently frozen subsurface soil layer – is the more likely culprit.

Elsewhere, far to the northeast of Zelma Lake, ponds that had existed for more than a millennium on Ellesmere Island's Cape Herschel evaporated over a couple of summers in 2005 and 2006. Unlike the many shallow ponds that dot the Arctic in the tundra zone and serve as vital areas of biological activity, supporting micro-organisms, plants and animals, the Cape Herschel ponds were underlain by granite and seemed stable, making the water's disappearance even more startling. Scientists who had been studying the Cape Herschel ponds intermittently over a quarter of a century, blamed the changes on warmer and drier conditions that shifted the evaporation/precipitation ratio.

Nowhere is global climate change more apparent than in the Arctic. No place is likely to experience warming to a degree greater than the global average. And no place is more vulnerable to its effects.

Observed changes in the North, including warmer winters, earlier ice break-up in the spring, later ice formation in the fall and thinner ice year-round, prompted the international community to sponsor a four-year study of Arctic climate . Its findings, released in 2004, indicate that major changes are underway. Computer simulations using a variety of climate change scenarios support the theory that the effects of climate change will be felt first and most intensely in the Arctic. Air temperatures in the north have, on average, increased by more than 5° C or 11° F over the last 100 years. This has led to

DENNIS FAST www.dennisfast.smugmug.com

increased melting of sea ice; models suggest that if the warming continues, by 2080 the Arctic waters will be completely free of ice during the summer months.

The melting of snow and sea ice has a feedback effect on the climate; as white snow and ice melts to reveal darker colored land and water, less solar energy is reflected away from the Earth, and the climate becomes ever warmer. So far, satellite measurements have shown that Arctic sea ice has been decreasing at a rate of about three per cent per decade since the 1970s, and this rate will likely increase. Overall, the study found that in the northern Arctic, the total coverage of sea ice in summer had decreased by about 15 per cent, while in the Hudson Bay area to the south, this figure was close to 40 per cent.

There are other challenges to the arctic environment as well. The Canadian Arctic possesses 59 per cent of the nation's estimated oil resources, 48 per cent of potential gas resources, and considerable reserves of valuable metallic minerals. As a result, northern industrial activity includes major mining operations and oil and gas exploration and development.

Northern Canada has 99 known oil and natural gas fields. Some drilling began in the 1920s, but the number of wells drilled during the late 1970s and early 1980s increased significantly. Much of this activity has taken place in the mainland Northwest Territories, but increasingly, drilling is moving offshore to the Mackenzie Delta and Arctic islands. Northeastern Alaska is also rich in gas and oil reserves, which has led to an ongoing debate about allowing drilling on the coastal plain of the 19-million-acre or 7.7-million-hectare Arctic National Wildlife Refuge.

Though these developments are economically appealing to governments and industry, there are compelling environmental arguments against expanding oil and gas production in the Arctic. One is that it requires building significant infrastructure in and through ecologically intact areas, which can lead to habitat destruction, fragmentation of migration routes, erosion, gravel mining and draining freshwater resources to create ice roads. Underwater pipelines can cause significant damage to sea floor habitats and benthic organisms, such as corals.

The establishment of industrial infrastructure

Though the effects of climate change are being felt worldwide, perhaps nowhere are they as evident as they are in the Arctic.

such as roads and airstrips can also affect northern areas indirectly. Once access to remote areas becomes easier and less expensive, barriers to other forms of resource exploitation, such as logging, mining and commercial fishing and hunting are lowered.

Another strong argument against oil drilling in sensitive northern regions is the potential for oil spills from blowouts, pipeline leaks or shipping accidents. Because of the extremely low temperatures and limited or non-existent sunlight for much of the year, the productive season in the north is short. This means that it can take decades for northern ecosystems to recover from disturbances or disasters such as oil spills. The results of an oil spill can be particularly devastating for marine ecosystems. Marine mammals, birds, and some fish gather in large groups during certain seasons, and if an oil spill takes place in one of these places it can have very serious impacts on the population of the species. If there is ice in the water the situation will be even worse; there is no effective method for containing and cleaning up oil spills in icy waters.

Toxins are passed from prey to predator, and the levels become higher farther up the food chain, a process called biomagnification. Top predators, including many Inuit, are exposed to contaminants throughout their lives and can build up high levels of toxic chemicals in their bodies.

The climate of the Arctic raises another environmental issue, the seriousness of which is only now being recognized. Many industrial and agricultural chemicals have a tendency to move from warmer to colder climates. They are carried north in air and water currents and become trapped in the Arctic, in some cases in much higher concentrations than those found in the places where the chemicals were made or used. A class of chemicals known as persistent organic pollutants, or POPs, are often transported to the Arctic in the air, fall to the ground in rain or snow, and take many years to break down in the cold, dark, northern climate. These chemicals are incorporated into

vegetation through the air, soil and water and are ingested by animals. These toxins may not have a major effect on short-lived animals that are low on the Arctic food chain, but they accumulate in the fat, blood, and organs of longer-lived animals such as polar bears, whales, seals and humans. Toxins pass from prey to predator, and the levels become higher further up the food chain, a process called biomagnification. Top predators, including many Inuit, are exposed to contaminants throughout their lives and can build up high levels of toxic chemicals in their bodies. Further study of these chemicals is necessary to determine the health effects that they may have on humans and wildlife. They have the potential to cause reproductive problems and interfere with hormone function, development, and immunity.

Polar bear prints in the sand, left, and a bright orange sky are two signature sights of the far north.

BOTH IMAGES: DENNIS FAST www.dennisfast.smugmug.com

The consequences of these trends could be dire, for both the Arctic and the rest of the world. In the north, vegetation zones are expected to shift as the climate allows species to expand their growing range. The diversity, ranges and distribution of wildlife will change as well, with less adaptable species' populations suffering while more aggressive and invasive species thrive. Forest fires, insect infestations and invasion by non-native species will also become more common in northern ecosystems. An increasing number of extreme weather events will likely occur, and coastal communities in the Arctic and elsewhere will be exposed to more storms. As the ground thaws, roads and buildings erected on permafrost will be damaged, and major construction or repairs will become necessary. Reduction in sea ice will have an impact on the populations of many species, including seals that use the ice to give birth and raise their young. Any reduction in the populations of seals and other species will have a social and economic impact on Inuit communities. Worldwide, changes in the climate of the Arctic will lead to changes in the global heat balance and rising sea levels. Changes in ocean currents may affect climate both locally and globally by affecting the rate at which warm waters circulate between the tropics and the poles. The cause of these changes, and the rate at which they are taking place, will continue to be debated by scientists. What is undeniable is that climate change has the potential to cause major change in the north, with devastating effects for Arctic vegetation, wildlife, and cultures. Once these changes have started they may be impossible to halt or reverse. Indeed, many feel that global warming is simply a prelude to rapid, and ultimately catastrophic global cooling – in other words, another glaciation.

The brillant colors of the tundra surprise the eyes.

DENNIS FAST www.dennisfast.smugmug.com

The Arctic

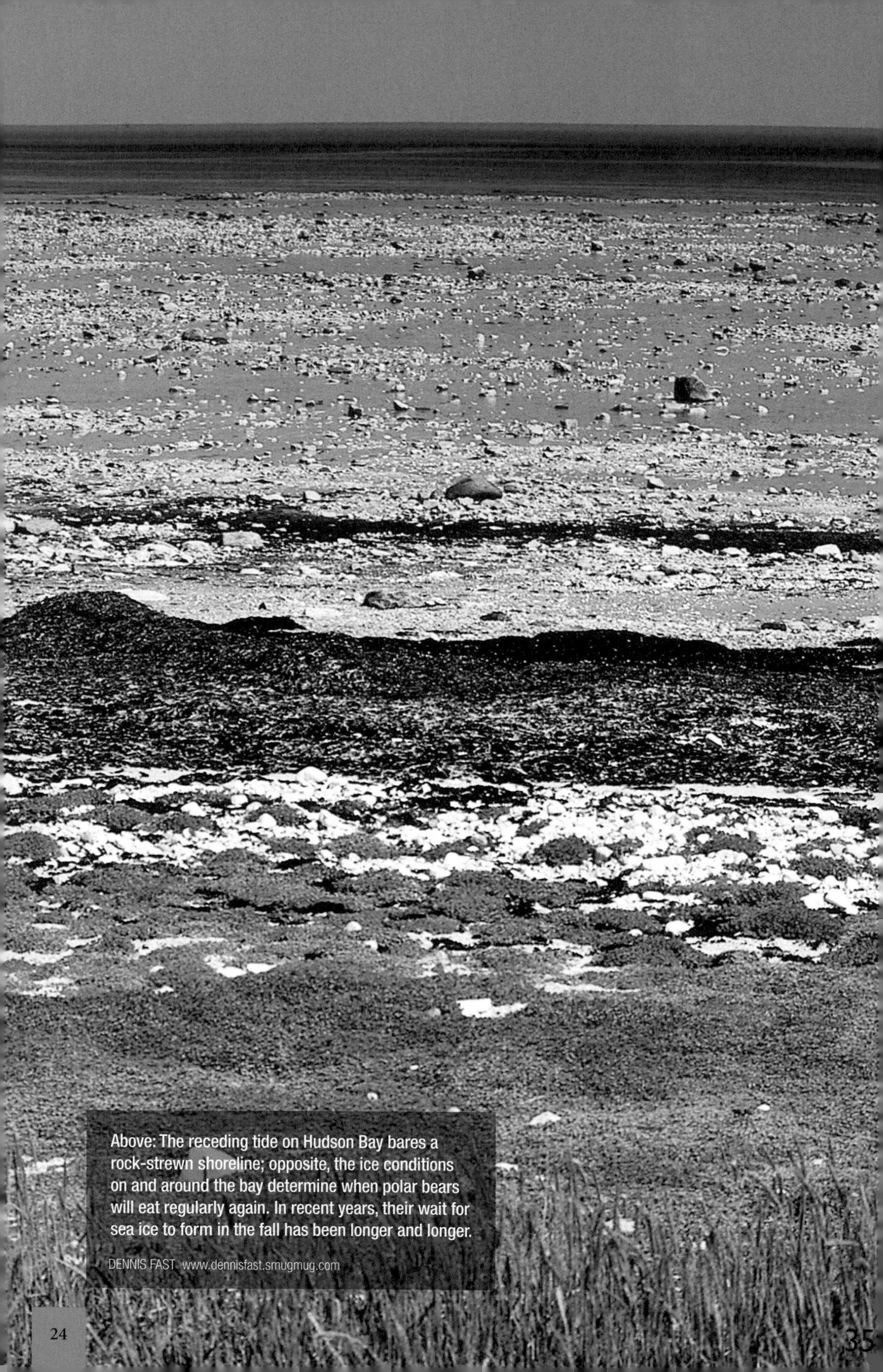

Above: The receding tide on Hudson Bay bares a rock-strewn shoreline; opposite, the ice conditions on and around the bay determine when polar bears will eat regularly again. In recent years, their wait for sea ice to form in the fall has been longer and longer.

DENNIS FAST www.dennisfast.smugmug.com

The Arctic

Around the top of the world, south of the Earth's shrinking ice cap, is a vast, largely treeless expanse known as the arctic tundra.

Occupying a significant portion of North America's land mass, this "ice desert" or "frozen prairie", as it's been called, gets its name from two, very different cultures. The word "arctic" comes from the Greek word for bear, *arktos*, a reference to the constellations – Ursa Major and Ursa Minor (which are, in turn, Latin for "great bear" and "little bear") – that dominate the skies of this enormous region for nearly six months of the year. The word "tundra" is derived from the Finnish word *tunturi*, meaning "barren" or "treeless land".

This is the coldest of all terrestrial ecosystems, with average annual temperatures between -17° C and -7° C (or -30 F and -12 F), depending on the location. Not surprisingly, much of the region is covered with permafrost, ground that remains frozen all year long.

The Arctic tundra also has the lowest precipitation of any of Canada's ecozones, yet is host to a surprising number of plants and animals, and represents a fascinating testament to nature's adaptability, as well as its cruel beauty.

The tundra begins north of the point at which frigid winter temperatures and low precipitation make it impossible for trees to thrive. This is the tree line; despite its name, however, the transition between taiga and tundra is not stark, and north of the line, small clumps of stunted jack pine or hardy spruce are found in sheltered sites.

The southern Arctic extends from northeastern Alaska, home to the Arctic National Wildlife Refuge, to Ungava Bay in northern Quebec, but 80 per cent of this area is located west of Hudson Bay. The northern Arctic incorporates the Arctic Islands, the District of Keewatin on Hudson Bay's northwest coast, and the northern tip of Quebec. Both regions

DENNIS FAST www.dennisfast.smugmug.com

experience long, dark, cold winters and short, cool summers with sunlight virtually 24 hours a day, as well as very little precipitation.

Shaped by glaciers, which retreated from the area just 8,500 years ago, the bedrock is exposed in some areas and elsewhere is layered with glacial drift, moraines or eskers that sometimes stretch for great distances and have been used for millennia by many creatures, including humans, as a natural highway system. In many places are glacial erratics, huge chunks of rock picked up by growing glaciers and dropped, anywhere from a few metres to a few thousand kilometres from their origins, as the glaciers melted.

To many, the rolling landscape of the southern arctic may seem barren and lifeless, but come spring it explodes into life. Though the growing season is brief, the long hours of daylight spur the growth of dwarf birch and willows. Ponds, lakes and wetlands dot the region, for permafrost prevents the soil from draining and in most places the land is moist or sodden during the warm season. Surface features such as mud boils and patterned ground are created by repeated freezing and thawing of the soil.

The northern arctic, which covers some 1.5 million square kilometres or 579,000 square miles, has fewer shrubs. Instead, it's dominated by lichens and herbs, such as purple saxifrage, mountain avens, and arctic poppies. More than 600 species of mosses and lichens are found in the northern arctic, but other plants are rare. The hardy species that do grow in the extreme north are usually found in sheltered valleys and along rivers and streams.

Most of the Northwest Territories, extending from west of Great Slave Lake to the Yukon border, primarily consists of undulating lowland plains, the legacy of the Western Interior Seaway, an ancient sea that bisected North America for millions of years. Over the last two million years, the limestone bedrock thus formed has been alternately scraped clean and reburied by glacial activity. In the east, this landscape merges into rocky hills and plateaus where granite bedrock predominates. Over much of the region, the soil is permanently frozen to depths of hundreds of metres, covered by a thin layer of soil that thaws during the summer months. To the north, particularly in the last two decades, there is open water in the Arctic Ocean, though offshore pack ice usually remains. Farther north still, the ocean is frozen year-round and the average winter temperature is below -30° C or -22° F.

Above the Arctic Circle, the winters are long and dark; at precisely 66° 32' north, the sun does not rise at all on at least one day. Farther north, the number of days of complete darkness increases. Snow can fall in any month of the year, and usually covers the ground from September to June. Yet during the summer months the temperature can soar; Yellowknife, on Great Slave Lake, can experience sweltering mid-summer temperatures.

COURTESY OF JUDITH SLEIN

Precipitation is scanty; the northern arctic receives an average of 20 centimetres or nearly eight inches annually, about what central Nevada gets.

This unforgiving climate and the tundra's limited vegetation combine to limit the number of species that make it their home, and those that do often have amazing adaptations to help them cope with the northern conditions. The dominant large mammals are muskoxen, barren-ground caribou, and polar bears, with a limited number of grizzly bears and moose found in the western mountains and along the tree line to the south. Ermine, arctic foxes, arctic ground squirrels and lemmings also thrive here; the last of these is a very important food source for larger carnivores and several birds of prey.

Bird life is abundant here. Many species of geese, ducks, loons, and swans arrive to nest on ponds, wetlands and coastal areas. They are accompanied by shorebirds such as black-bellied plovers, ruddy turnstones and red phalaropes. Other birds, including horned larks and snowy buntings are, remarkably, able to nest on the open tundra. The Arctic has no reptiles or amphibians, and while the northern region has relatively few insects, the southern Arctic sees huge swarms of insects hatch during each brief summer.

The tundra region is sparsely populated, with a total of less than 30,000 people, including those living east of Hudson Bay. About 80 per cent of these are Inuit, who live in the scattered communities spread across this vast area. Fishing, hunting and trapping still make up an important part of the economy and daily life for most people and locally caught meat and fish supplement most people's diets.

The Arctic is resource rich, possessing 59 per cent of Canada's estimated oil resources, 48 per cent of potential gas resources, and considerable reserves of valuable metals. However, as can be seen in the sidebar on The Arctic Transformation on page 18, harvesting these resources could create devastating effects on arctic vegetation, wildlife and culture, and by extension, major environmental and problems for the Northern Hemisphere as a whole.

Opposite: Yellow poppies that were photographed on Ellesmere Island. Below left; a Hudsonian godwit, and below right, a lovely snowy owl.

IIMAGES: DENNIS FAST www.dennisfast.smugmug.com

Polar Bear

Ursus maritimus • Threatened (US); Special Concern (Canada)

POLAR BEARS, the world's largest land carnivores, have no natural predators, apart from humans. Yet many believe they will be extirpated from much of their current range before the end of this century and some believe extinction of the species lies right around the corner. As a result, these magnificent predators have become emblems of Earth's deepening environmental crisis for many organizations.

Today, an estimated 20,000 to 25,000 great white bears are found on the sea ice and coastal areas of northern Canada, Denmark, Norway, the United States and Siberia. But that sea ice, on which polar bears depend for hunting seals – their main prey – as well as for mating and migrating, is rapidly retreating. In fact, in 2005, it was reported that a number of polar bears, despite being superb swimmers, had been found drowned off the north coast of Alaska. Clearly, as the ice floes on which they hunt shrink and drift farther apart, swimming greater distances across open seas far from shore leaves polar bears vulnerable to exhaustion, hypothermia, or being swamped by waves during storms.

At a meeting of the Polar Bear Specialist Group of the International Union for Conservation of Nature (IUCN) in July 2009, it was reported that the number of declining bear populations has increased to eight from the five listed in its 2005 report. Since nearly two-thirds of all polar bears live in Canada, the

Though superb swimmers, polar bears have been known to drown if the ice floes they seek have melted and disappeared.

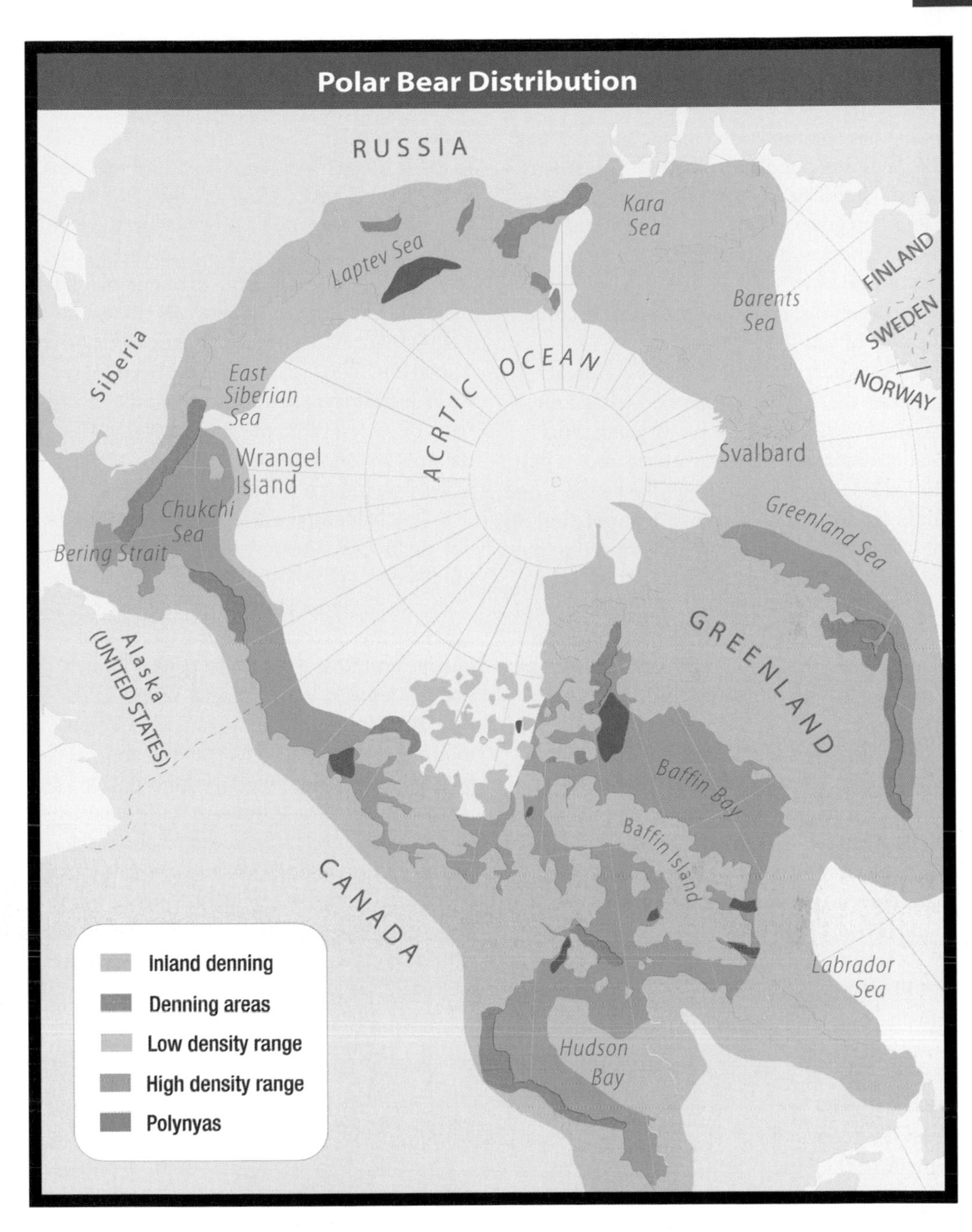

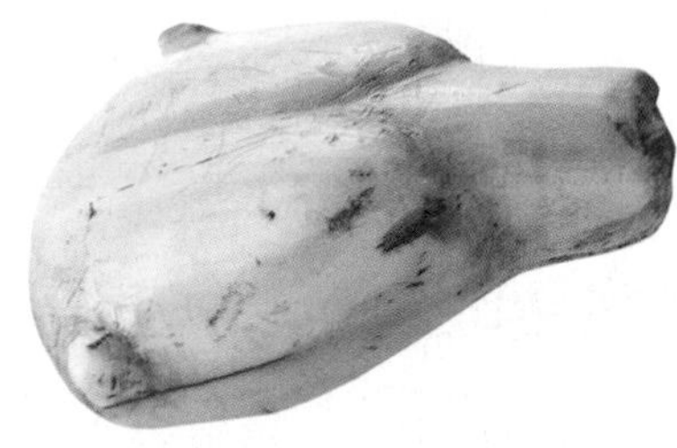

CANADIAN MUSEUM OF CIVILIZATION / CATALOGUE NO. X-B106/IMAGE NO.S90-3980

This beautiful ivory polar bear head was carved by an Inuit artist hundreds of years ago.

group has recommended that at least some of the declining populations be listed as endangered. Unlike its US neighbor, however, the Canadian government has decided that the bears remain a species of "special concern" – a determination based on the effect of harvest on populations, excluding the possible consequences of future loss of sea ice.

Polar bears are true creatures of the ice, a species that appears to have evolved in the Arctic during one of the many glaciations that have marked Earth's relatively recent past. Biologists theorize that sometime between one million and 200,000 years ago, a population of brown bears (*Ursus arctos*) was isolated in the far north by glaciers, and forced through a rapid series of evolutionary changes in order to survive. Even today, brown bears come in many colors, from blonde to dark brown and the same was undoubtedly true in the past. It's theorized that lighter colored bears, which would be less visible to seals and other prey, would be more successful hunters on the ice and therefore more likely to pass on their genes.

Fossil and DNA evidence indicate that polar bears were well established 100,000 years ago. These fossils, which are rare, include a jawbone, which was found in 2007 by a University of Iceland research team on Svalbard, an Arctic archipelago midway between Norway and the North Pole. The fossil was radiocarbon dated to about 130,000 BP, sometime during the last interglacial age. Since then, the global climate has been cooler than it is today much, but not all, of the time. A skull found in a Scottish cave has been dated to about 19,000 years ago – near the cold peak of the last glaciation. And a painting of two tear-drop-shaped bears on a wall in the Paleolithic cave of Ekain near the northern coast of Spain is believed to depict young polar bears that landed on the coast between 17,000 and 12,000 years ago. Later fossils, between about 11,000 and 8,000 years old, have been found from Scandinavia and Russia. And the inhabitants of a Mesolithic site about 8,500 years old on Siberia's Zhokov Island evidently relied heavily on polar bears for subsistence.

"Global temperatures have risen before during the polar bear's existence," writes author Rebecca Grambo in *Wapusk: White Bear of the North*. From about 1,150 to 750 years ago, "northern North America, Greenland, Europe and Russia enjoyed a period known as the 'Medieval Warm Period' or the 'Little Climatic Optimum'. Temperatures believed to range from about 1° to 3.5° C or about 2° to 7.7° F warmer than today reduced Arctic sea ice and allowed the Vikings to settle what really was a 'green' land. Little is known about the effects of the Medieval Warm Period on polar bears, although some evidence suggests that they moved north of their usual range. That would be expected if they were following the retreating sea ice."

That polar bears continue to survive is a demonstration of their remarkable ingenuity and adaptability, as well as their continuing evolution. Some biologists

DENNIS FAST www.dennisfast.smugmug.com

Relaxing in the snow, a big male waits for the ice to freeze on Hudson Bay.

BARBARA ENDRES

believe that it was less than 20,000 years ago that the polar bear's molar teeth became significantly different than those of brown bears or grizzlies. Their sharp, jagged molars, which are better suited to a carnivorous (rather than omnivorous) diet, are among a host of adaptations that aid a polar bear's survival in one of the world's most hostile environments. The most obvious of these specializations is the bear's white coat; a layer of clear, stiff guard hairs that tops a substantial, woolly layer of under hair, providing the bear with a layer of protection against the elements that is almost five centimetres or more than an inch thick. Coated with oil, the guard hairs are water repellant, and do not mat when wet. They also reflect light, while beneath its coat, the bear's black skin absorbs as much heat from the sun as possible in the far north. This is important, given that winter air temperatures in the Arctic average -34° C (or -29° F) and ocean temperatures drop below the freezing mark before the seawater actually freezes.

Polar bears' short tails and ears lie close to the body, reducing heat loss, and their comparatively large feet act as snowshoes on ice and snow or paddles in the water. Their claws are shorter and stronger than those of grizzlies, with long hair between the pads and the toes, for warmth, and their black footpads are covered in small, soft bumps, giving them extra traction on slippery surfaces.

That polar bears and brown bears are still genetically close enough to be able to produce viable offspring has been occasionally proven by captive bears, but it was not until 2006 that a wild cross-bred polar bear-grizzly, shot by an American hunter, was photographed. The widely distributed image of the "pizzly", as some dubbed it, clearly showed its dual ancestry (see page 38).

DENNIS FAST www.dennisfast.smugmug.com

Lounging bears could easily been mistaken for boulders, and in the illustration above, man-made stone sentinels are known as *innunguait.*

A warming Arctic has allowed grizzlies to expand their range north to the point where it overlaps that of polar bears; normally they live on the Arctic mainland and southern Victoria Island, but over the past decade, grizzlies have been sighted on Arctic islands farther north. There is also evidence that northern grizzlies have adapted to the conditions and begun hunting seals.

Having made his kill far out on the sea ice, Inuit hunter Peter Esau hooks a polar bear to his kayak to drag it to his waiting sled.

Inset: Sarah Kuptana and Shirley Esau clean a bear skin in Sachs Harbour.

R. KNIGHTS / NWT ARCHIVES / N-1993-002-0089

R. KNIGHTS / NWT ARCHIVES / N-1993-002-0131

However, grizzlies are not alone in responding with ingenuity to their changing surroundings. Polar bears on Hudson Bay's western shore – members of the world's most accessible population – are showing clear signs of stress due largely to climate change. Over the past decade, as ice in the bay melts earlier and freezes later, the bears have lost weight and cub survival has declined. The changing conditions have been particularly hard on pregnant females, for in late October or early November, when most of the hungry bears head out onto the newly formed ice to hunt following at least four months of relative fasting, pregnant females dig dens and go into semi-hibernation as they await the birth of their cubs. Following the delivery of one or more cubs in December or January, their mother must nurse them until they are big enough and hardy enough to follow her out onto the ice to hunt. This generally occurs in late February or early March, meaning that pregnant females must survive up to eight months without food – the longest known fast of any mammal.

Responding to these challenging circumstances, in the past several years, male polar bears have been seen taking matters into their own paws, as it were. Hudson Bay is home to large populations of beluga whales, and recently, young males have been seen working pods of these whales rather like cattlemen work herds of cattle. Gathering in groups of five or six, several of the bears take to the water of the bay and swim out into a pod of belugas, attempting to cut off a young whale and drive it toward shore. If they are successful, the remaining bears plunge in and together they drive the whale into shallow water where they can kill it. Then all join in a feast on the shore.

The bears on Hudson Bay have also begun to hunt harbor seals, in addition to their favorite prey, ringed seals, which they hunt from ice floes or ice shelves.

Arctic royalty

Male polar bears are truly imposing creatures. Adults usually weigh between 350 and 650 kilograms (or between 772 and 1,433 pounds) and are about 2.5 to three metres (or between eight and 10 feet) long. However, the largest polar bear ever recorded weighed 1002 kilos or 2,209 pounds and measured 3.7 metres or 12 feet long. Females are about half the size of males, but a mother protecting her cubs is nonetheless a formidable opponent.

QILD-1:819 / IMAGE NO. S90-2739

A polar bear has large, muscular hind limbs that are longer than its forelimbs, making its hind end stand higher than its shoulders. Because of their bulk and slightly lopsided gait, polar bears use more than double the energy of most other mammals to get about. Their

Above: This ancient ivory swimming bear almost perfectly echoes the real thing. Below: Young male polar bears learn hunting strategies at play.

DENNIS FAST www.dennisfast.smugmug.com

average walking speed is 5.5 kilometres or 3.4 miles per hour, but when disturbed, a polar bear will head for open water at a rolling gallop, reaching a top speed of about 40 kilometres or 25 miles per hour.

Polar bears have longer bodies and necks than other bear species and relatively small heads compared to their massive bodies. They also have an elongated muzzle with a slightly arched snout, unlike the concave profile of the closely related grizzly.

Polar bears spend much of their lives on offshore pack ice, often near areas with leads or polynyas, a Russian term for open areas of water surrounded by ice. These are the best places from which to hunt ringed seals as they surface for air. The bears also eat bearded, harp and hooded seals, and will occasionally capture a beluga whale or a young walrus. Given the opportunity, they will also scavenge carcasses of fully grown walruses, belugas, narwhals and bowhead whales. As the ice retreats in the summer, the bears travel great distances in pursuit of their prey. In the southern parts of their range, however, such as eastern Baffin Island and Hudson Bay, the pack ice disappears entirely by the end of the summer. In these areas the bears are forced to spend two to four months ashore, living off of stored body fat until the ice freezes again in the fall and they are able to resume hunting. Polar bears are able to slow their metabolism to conserve energy when food is scarce, unlike other bear species whose metabolism only slows in late fall. This is particularly advantageous in the Arctic, where the availability of the next meal can be uncertain at any time.

Not surprisingly, polar bears have large home ranges, though size varies from 50,000 square kilometres or 19,305 square miles to more than 350,000 square kilometres or 135,135 square miles, depending on the amount of food and number of mates and dens available. Polar bears are generally solitary animals, and are commonly seen in groups only when mating, when mothers are with cubs, or around a large food source such as a bowhead or grey whale carcass. Females reach sexual maturity at about four years of age, and males at about six; however competition for females is intense and males may not mate successfully until they are eight or 10. Mating takes place during the spring, and often involves several males following a female, sometimes fighting violently, until the largest, most dominant male drives the others away. Mating induces ovulation in female polar bears, and pairs will often stay together for several days, mating repeatedly in order to ensure impregnation. The gestation period is about eight months in total, often including delayed implantation, which ensures cubs are born at the appropriate time of year. During this time, the female needs to put on at least 200 kilos or more than 440 pounds of fat, to ensure that she can meet the demands of pregnancy, birth and nursing. If the female does not have enough fat reserves in the fall, she may not give birth or may only be able to raise on cub successfully.

Most polar bears don't hibernate, but pregnant females are the exception, entering maternity dens sometime in October and sleeping most of the time until the cubs are born. These dens are frequently dug in south-facing snowdrifts and are often on land, not far from the coast, although some are significantly further inland and some bears do den on sea ice. Pregnant females' body temperatures lower slightly, and their heart rates drop from a normal resting rate of 46 beats per minute to about 27 beats per minute once they enter their den for the winter.

DENNIS FAST www.dennisfast.smugmug.com

On a silent command, two young cubs scramble from their den to join their mother.

BARBARA ENDRES

Females usually give birth to two cubs, sometimes one, and rarely three, in their dens between November and January. Polar bear cubs are very small at birth, weighing about 454 to 680 grams or between one and 1.5 pounds and measuring about 30 centimetres or about a foot long. For the first month, the cubs are blind, have little hair, and spend nearly all of their time snuggled against their mother for warmth and nursing. They grow quickly on their mother's rich milk, however, and by two months they are walking inside the den, have teeth, and are covered in thick white fur. At this point, to acclimatize them, their mother begins to take them outside for brief outings. In late March or April, the family emerges from the den during the day; the cubs weigh between 10 and 15 kilos or 22 and 33 pounds by this time. The family spends a number of weeks near the den, sleeping inside at night, but eventually the mother's hunger forces the family out onto the sea ice to hunt. The cubs begin eating seal blubber when their mother catches her first seal of the year, but continue to nurse until they are between 18 and 30 months old. By the time they are 30 months old, their mother is ready to breed again and either she or a potential suitor will drive the cubs away.

The next few years are dangerous for the young bears, as they learn to hunt and defend themselves. If they survive their youth, they can live up to 30 years, but 18 is old for bears in the wild. The longest a polar bear is known to have lived in the Arctic is 32 years; in Winnipeg's Assiniboine Park Zoo, a bear died in 2008 at 41 years of age.

The polar bear's only natural enemies are humans, and for centuries, hunting has been the greatest cause of polar bear mortality. Polar bear hunts by Arctic peoples have taken place for at least 2,500 years; the bears were used for food, clothing, bedding and ceremonial purposes. Commercial polar bear hunting began in the 1500s, and by the 1700s polar bear hides were in high demand. During the 1950s and 1960s new hunting methods using snowmobiles, boats, and airplanes dramatically increased the number of bears being killed annually. International concern about the situation led to an agreement in 1973 that banned the use of aircraft or large motorized boats for polar bear hunts. Today, hunting is prohibited in Norway and Russia (though poaching is a problem in the sparsely populated Russian Arctic), and hunting is regulated by the governments of Canada, the United States and Greenland. In these countries indigenous Arctic people continue to harvest between two and three per cent of the total population for food, clothing, handicrafts and the economically lucrative sale of skins. However, Canada is the only country in the world that still allows hunting by non-native sport hunters under a quota system. Polar bears are also occasionally shot in defence of life.

Recent estimates place the global population of polar bears between 21,500 and 25,000 individuals, more than 60 per cent of which live in Canada. There has been a dramatic improvement in the population since regulations on hunting were put into place; in 1968 the worldwide population was estimated at only 10,000 bears. But the polar bear is now facing new, equally serious threats to its survival, especially the loss of sea ice habitat. Too, as Arctic oil exploration and drilling proliferate, especially offshore, the risk of spills, with potentially severe consequences, grows. Oil spills can not only decrease or contaminate food sources, but can also cause the bears to die from ingesting oil through grooming or from freezing because oil-covered fur loses its insulating properties. Toxic chemicals created worldwide are being transported to the Arctic on air and water currents, where they build up in the land, water, plants and animals. Because polar bears are at the top of the food chain, they ingest these chemicals from their prey, and the toxins bioaccumulate in the bears' bodies. Studies over the past three decades have shown that a polar bear's PCB level may reach three billion times the concentration of that found in sea water.

As research is conducted, evidence accrues, and predictions for the Earth's future climate become more complex, it appears that the greatest threat to the polar bear's continued survival may be global climate change. The results of increased greenhouse gases in the atmosphere are already being felt in the Arctic; the permanent sea ice pack has shrunk by 14 per cent over the past 20 years. Sea ice is melting earlier and forming later, giving polar bears less time each year to hunt for food. The effects of this decreased sea ice may already be showing up in polar bear populations: a National Geographic study published in 2004 showed that polar bears now weigh, on average, 15 per cent less than they did in the 1970s; the birth of triplets has become much less common; and only one in 20 cubs are weaned at 18 months, compared to half of cubs three decades ago.

POLAR BEAR / DESIGNED BY JOHN CROSBY (1952) / LIBRARY AND ARCHIVES CANADA / LAC 025 / CANADA POST CORPORATION {1953} / REPRODUCED WITH PERMISSION

Polar bears have adorned Canadian stamps on several occasions; this one was created in 1952.

Not surprisingly, perhaps, the impact of a warming climate is being felt most intensely in the southern portion of the polar bear's range. The U.S. Geological Survey and the Canadian Wildlife Service have published a study showing that the number of polar bears in Hudson Bay fell 22 per cent from 1,194 in 1987 to 935 in 2004. Experts are warning that at the current rate of climate warming, polar bears may be extirpated from southern Hudson Bay by 2050.

Some environmental groups have taken action, using the plight of the polar bear as a launching pad to try to force the American government to take action against global climate change. In February 2005, Greenpeace, the Center for Biological Diversity, and the Natural Resources Defense Council filed a petition with the U.S. Fish and Wildlife Service to list the polar bear under the Endangered Species Act. When nothing was done the groups launched legal action against the government and the court ordered the government to come to a decision. In response, following a status review of the species, in May 2008, the United States Department of the Interior added polar bears to the Endangered Species list, based on the threat posed by Arctic warming, providing them with concrete protection, including more intensive environmental reviews before oil and gas development can proceed in their habitat. More importantly, the ruling is raising awareness in the US about the reality of global climate change and the threat it poses.

Polar Bear Fast Facts

Ursus maritimus

WEIGHT & LENGTH

The world's largest land carnivores, males average between 350 and 650 kilograms (or between 772 and 1,433 pounds) and are about 2.5 to three metres (or between eight and 10 feet) long. The largest polar bear ever recorded weighed 1002 kilos or 2,209 pounds and measured 3.7 metres or 12 feet in length.

LIFESPAN

The oldest wild bear on record was 32 years old, and 25 is considered a good age for male bears. Females may live slightly longer. In captivity, one female lived to be 41 years of age.

HABITAT

Polar bears live throughout the arctic (see map), and over the past millennia have lived as far south as James Bay in Ontario and Quebec. Those southern populations are suffering now, as sea ice (from which they hunt) melts earlier and forms later. Bears adversely affected will either weaken and starve or move north. Polar bears travel great distances each day; even a mother with cubs can cover 30 kilometres or 18 miles a day.

MATING & BREEDING

Females reach sexual maturity at about age four, and males at about six; however competition for females is intense and males may not mate successfully until they are eight or 10. Mating takes place during the spring, but in a process called "delayed implantation," the fertilized eggs are put on hold until late August or early September. The mother's condition at the end of the summer determines how many cubs she will bear. If she is carrying enough fat, the usual two eggs will implant themselves; in an exceptional year, three cubs might be born, but if she is undernourished, just one or even no eggs will develop. The cubs, each the size of a pound of butter, are born in late December or January in a two-room cave of snow that their mother has built for the purpose. Cubs stay with their mothers for at least 18 months (in southern regions where food is easier to find) and up to three years in the far north; during this time, the youngsters must be taught everything they will need to know to survive. When the time comes to wean the cubs, she will abandon them or drive them away, causing a period of confusion before they eventually wander off to begin their own lives.

NAMES

These magnificent creatures are called *wapusk* ("white bear of the north") by the northern Cree of Hudson and James Bays and *nanuk* or *nanuq*, by the Inuit of the far north; this was anglicized as Nanook by filmmaker Robert Flaherty. The scientific name, *Ursus maritimus*, means "bear of the sea".

VIEWING POLAR BEARS

The most accessible site is Churchill, Manitoba, on Hudson Bay's western shore. A number of lodges and travel companies offer complete packages for viewing the bears as they gather near the bay between early October and late November, waiting for the sea ice to form. These packages, including one-day tours, offer direct flights from Manitoba's capital, Winnipeg, as well as accommodation, guides and daily transportation to the viewing areas in "tundra buggies," large buses on huge wheels that carry visitors safely above the level of even a big male stretching up on his hind legs.

In Alaska, polar bears are sometimes seen during the fall around Barrow, on the north tip of Alaska, and at Kaktovik, a remote village on Alaska's northeast coast. Though sightings are infrequent, they sometimes occur during the fall, when aboriginal Alaskans hunt bowhead whales and the carcasses draw hungry bears. Barrow has hotel accommodations and Kaktovik modest accommodations and a small general store and laundromat. Neither, however, has specialized vehicles for viewing polar bears, which have been known to attack humans and do damage in villages. Both villages are served by airlines from Fairbanks and Anchorage. For more information, see http://www.AlaskaCentres.gov

The Pizzly?

Or Grolar Bear?

CAN global warming change the course of love? Perhaps.

In 2006, on Banks Island, 2000 kilometres north of Edmonton, an American sports hunter shot a bear, believing it to be a polar bear. It wasn't. On closer examination, the animal exhibited the thick, creamy white fur and the long claws typical of the polar bear. But it also had the humped back, a slightly concave face and brown patches around the eyes, nose, back and one paw, typical of the grizzly bear. Subsequent DNA testing proved the creature was indeed a hybrid – its mother was a polar bear and its father a grizzly.

Hybrid bears have been produced successfully in zoos, but in the wild polar bears and grizzlies rarely cross paths, though there has traditionally been a slight overlap in their breeding seasons and habitats. But that may be changing. Warming Arctic air may be allowing grizzlies to expand their range farther north. They normally live on the western Arctic mainland and southern Victoria Island, but over the last decade have been sighted near the Beaufort Sea, Banks Island and Melville Island – traditional polar bear territory. There is also evidence that northern grizzlies have adapted to the conditions and begun hunting seals. Scientists are worried that crossbreeding will become more common as the grizzly's range expands – a possible threat to the polar bear's survival as a distinct species.

In the meantime, what to name the hybrid? "Grolar" and "pizzly" were first out of the gate. But "nanulak", an elision of the Inuit *nanuk* (polar bear) and *aklak* (grizzly or brown bear) is a more indigenous possibility.

A well-fed bear provides a striking contrast as it stro through a field of fireweed on an Arctic island.

DENNIS FAST www.dennisfast.smugmug.com

Caribou

Rangifer tarandus

THOUGH they eventually populated much of the Northern Hemisphere, from southern France (where piles of butchered bones more than a half-million years old have been found) and northern Spain (where caribou appear in magnificent 11,500-year-old cave paintings) to Scandinavia and Russia (where more recently they were semi-domesticated and are known as reindeer), biologists believe that caribou first appeared in North America. Studies of mammalian fossils in Eastern Beringia, the area of Alaska and Yukon that was largely ice-free during the glaciations of the past 1.8 million years, seem to show that these remarkable members of the deer family are at least as old as the Pleistocene era, or "Ice Age", as many know it. Perhaps it's not surprising, therefore, that they seem uniquely prepared for a life that's almost synonymous with winter. So highly adapted are they that they have been termed "chionophiles", or snow loving animals.

Ancient fossilized caribou bones, as well as those of voles, lemmings, ground squirrels, pikas, hares and weasels, have all been found at Fort Selkirk, on the Yukon River near its confluence with the Pelly River. The bones were buried in fine-grained wind-blown silt between a layer of volcanic ash that is at least 1.48 million years old (mya), and the lava bedrock below it, which dates to 1.83 mya.

A member of the deer family, caribou apparently originated in the mountains of the Americas sometime between 4.2 and 2.5 million years ago, during a period known as the Pliocene. As the global climate cooled nearly two million years ago, some caribou drifted south, ahead of the growing glaciers (see woodland caribou on page 88), ultimately reaching the eastern forests of the US and Atlantic Canada. Others, already well adapted to the cold, thrived on the unglaciated tundra. Over time, as the Earth cooled during a series of glaciations and warmed during interglacial periods, they spread west across Beringia to Asia and Europe, and east as far as the tundra of Greenland and Labrador, where the Mi'kmaq word *xalibu* or *qalipu* – "snow shoveller" – likely gave them the name by which we know them today. In the process, they evolved into non-migratory woodland and migratory barren-ground subspecies or ecotypes.

During periods of intense global cooling, caribou were pushed far to the south in many places around the globe. About 550,000 years ago, a band of Neanderthal trapped and killed a herd of about 40 caribou as they crossed the Verdouble River in southern France, and carried their butchered carcasses up to a large cave, now known as Arago Cave, above the river. After removing the hides and meat (and undoubtedly feasting on the latter), they left great piles of bones in the cave, where they were quickly covered by wind-blown sand and silt. Closer to home, almost 37,000 years ago, well before the height of the last glaciation, North American caribou wandered the woodlands as far south as the North Carolina-Virginia border, as evidenced by a cranial fragment found there.

But as the glacial ages came and went, caribou also continued to thrive in refugium – places that were free of the ice sheets even during the height of each glaciation – such as existed in northern Alaska and Yukon. As John Storer writes in his summary of discoveries for the Yukon government, at Thistle Creek, south of Dawson City near the Alaska border, caribou were found in sediments dated to

Brilliant autumn colors frame a magnificent barren-ground caribou bull.

DENNIS FAST www.dennisfast.smugmug.com

between 140,000 and 125,000 BP, a period of global warmth prior to the Early Wisconsin glaciation, along with the bones of horses and many smaller animals.

Caribou also evolved to be able to survive on North America's westerly and northerly islands, environments even less hospitable than the tundra of the northern mainland. The small, now extinct Dawson caribou (*Rangifer tarandus dawsoni*), which inhabited Haida Gwaii until 1908, was an example of the dwarfing process that typically accompanies long-term adaptation to island life. Once believed to have been introduced to the islands about 12,000 years ago, near the end of the last glaciation, the discovery of a heavily mineralized antler, which washed from river gravels in northeastern Haida Gwaii and was radiocarbon dated to 40,000 BP, has led biologists to reconsider; some now believe that small numbers of caribou may have inhabited the islands since well before the last glaciation.

Canada's endangered Peary caribou (see page 45) are an even more remarkable example of the species' ability to adapt, for they live in one of the world's most extreme environments. And at least some woodland caribou that live on the eastern slopes of the Rockies, were revealed in early 2009 to be a remarkable genetic melding of migratory barren-ground or tundra caribou and their non-migratory woodland cousins.

Caribou breed in the fall, and about 90 per cent of females aged three or over calve during a very brief period seven-and-a-half months later. Barren-ground cows calve in large herds

Caribou Fast Facts

Rangifer tarandus

HEIGHT & WEIGHT

Subspecies or ecotypes vary in size, but barren-ground, woodland and mountain caribou are bigger than mule deer, but smaller than elk; most bulls weight about 200 kilograms or 420 pounds, and cows about two-thirds as much.

DISTRIBUTION

Their body hair changes color with the seasons, but with the exception of Peary caribou of the far north, which are nearly white during the winter months, it is medium to dark brown in the summer and fall, with white around the rump and, in bulls, under the throat. The head and neck have a mix of gray, white and brown hair, but during the winter the darker guard hairs break off and the animals appear lighter in color. Unlike other species of the deer family, almost all caribou cows have antlers, which are generally not shed until summer, allowing them to defend their young. The antlers of bulls, particularly those in their prime, can reach two metres or six feet in width and are shed in the late fall or winter, when the mating season is over. Caribou have a keen sense of smell and specialized nostrils that warm frigid incoming air during the winter and help to condense and capture the moisture before it's expelled. Their coats are made for winter, with a dense woolly undercoat and a long-haired overcoat consisting of hollow, air-filled hairs.

LIFESPAN

Caribou can live up to 15 years, but their average lifespan is just five or six.

MATING & BREEDING

Mating takes place between late September and early November, when males battle for access to females. Dominant males collect as many females as possible; a barren-ground caribou bull's harem may number 25 or 30 females. During this rut, males almost completely stop eating and lose much of their reserve weight. Calves are born the following May or June, and though they continue to nurse until fall, within a month, the calves are able to graze with the rest of the herd.

NAMES

The word caribou comes from the Mi'kmaq *xalibu* or *qalipu*, meaning "one who paws", referring to its habit of pawing through the snow – known as cratering – to reach buried lichens, mosses, sedges or grasses below. In Europe and Asia, caribou are known as reindeer and are domesticated in many places.

VIEWING CARIBOU

A number of tour operators in northern Manitoba, as well as Nunavut, the Northwest Territories, Yukon and Alaska offer well-organized guided tours to view barren-ground caribou during the summer months. Peary caribou live so far north and are in such small numbers that they are rarely seen, except by inhabitants of the north, or biologists. Woodland caribou can be seen from the highways in several places, including Alberta's Hwy 11, which goes through the Rocky Mountain Forest Reserve in the west-central region of the province, linking Red Deer and the Icefields Parkway, and Hwy 40 farther north, which skirts the eastern edge of the foothills between Grande Prairie and Cadomin, Alberta. Woodland caribou can also be viewed in eastern Manitoba, along Hwys 314 and 304 east of Lake Winnipeg. The mountain ecotype are elusive and few in number, but can sometimes be seen along the seasonal road to the summit of Mount Revelstoke, in BC's national park by the same name, as well as in the Kootenay Valley west of Lake Kootenay.

Above: Free of insects at last, barren-ground caribou feed at the edge of Hudson Bay.

DENNIS FAST www.dennisfast.smugmug.com

DENNIS FAST www.dennisfast.smugmug.com

A barren-ground bull sports a new rack of antlers, covered with spring velvet.

between June 1st and 10th, usually in oceanside calving grounds such as Alaska's Arctic National Wildlife Refuge or the Tuktoyaktuk Peninsula in northern Yukon, where sea breezes help ensure that their calves will not be bothered by hordes of insects and where wolves, one of their three main predators (the others are bears and humans), are generally not found. When woodland caribou are about to give birth, they disperse into the forests, often calving in rugged terrain, near lakes or on islands to deter predators.

Caribou calves are almost always single births and are more physically precocious than other deer, able to stand and move about almost immediately. Caribou milk is also extraordinarily rich and speeds the growth of the calves. Nevertheless, though tundra caribou form huge protective herds as they seek rich pastures and their woodland cousins attempt to hide their young, wolves and bears take a high percentage of the young each year.

The long migration to the northern calving grounds combined with the birthing process take a large toll on barren-ground females and as soon as the young are able to travel, they head inland to find food. By mid-to-late June, however, clouds of insects emerge on the tundra, disrupting the animals' feeding, sometimes spurring the agitated herds into an exhausting race to escape the torment and or forcing them to seek relief by clustering in tight groups numbering thousands. When these strategies fail, the herds sometimes climb rocky ridges or invade snow banks where food, unfortunately, is generally scarce. Not until late summer, when nighttime frosts bring an end to the annual plague, are the animals free to feed almost non-stop, something that is crucial to their survival over the long winter ahead.

From the ancient Neanderthal of southern France to the modern Dene of Western Canada – some of whose lives were so intertwined with the barren-ground herds that they called themselves Etthen-eldeli-dene, "the people of the caribou" – these remarkable animals have been crucial to the survival of many societies. Even today, they are an inextricable part of the lives of the people of Nunavut, the Northwest Territories, Yukon and Alaska, as well as a symbol of Canada as a whole, one that decorates the nation's 25-cent piece.

Today, however, caribou of every description are under threat. Peary caribou cling to life on the Arctic Archipelago; within 200 years of European settlement, as a result of harvesting of old growth forests, woodland caribou were extirpated from Maine, Vermont, Michigan and Minnesota in the US, while in Canada, they were gone from Prince Edward Island by 1873 and from Nova Scotia and New Brunswick by the 1920s. A tiny remnant herd on the Gaspé Peninsula, the last of the southern Maritime populations, is listed as endangered.

Farther north and west, boreal populations of woodland caribou, which are found in Newfoundland, Labrador, northern Quebec and Ontario, as well as eastern and northern Manitoba, and northern Saskatchewan, Alberta and British Columbia are all listed as endangered or threatened.

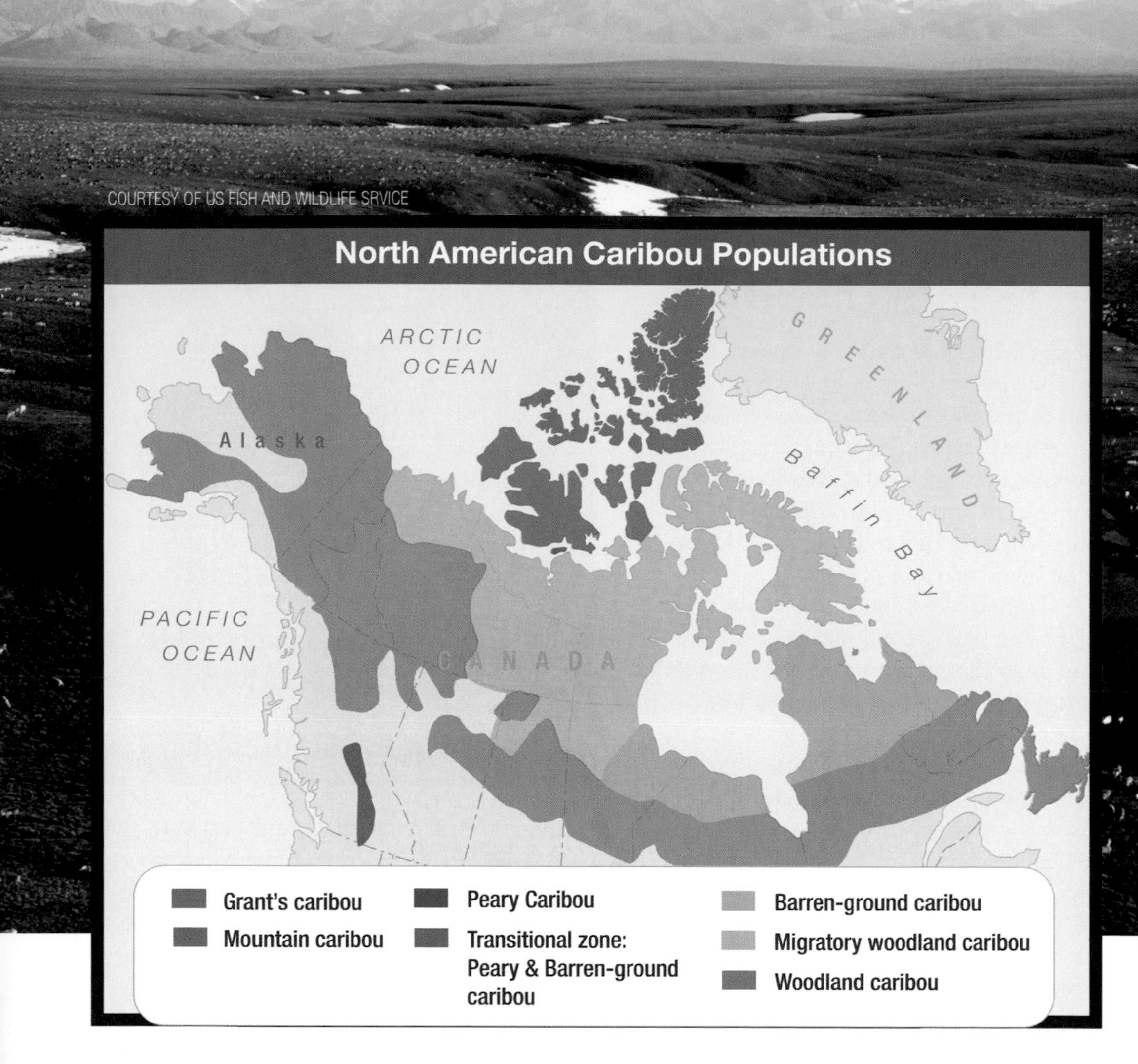

And the southern mountain populations, found mostly in British Columbia, are officially listed as threatened in Canada, though the Western Canada Wilderness Committee lists these small herds of "rainforest caribou", as endangered. Farther south, mountain caribou once ranged so widely that in Idaho a mountain, a national forest, a county and what was once one of the largest mining camps in the West – Caribou City – all bear their name. By the 1980s, however, these ancient animals had become the most endangered large mammal in the continental US, prompting a recovery effort that soon also involved neighboring Washington State (see page 200).

The northern mountain population, found in Yukon, Alaska and northern BC, is somewhat better off, but is listed as a special concern, as is the genetically distinct Dolphin-Union population, which inhabits Victoria Island and the neighbouring coast of Nunavut.

And while most barren-ground herds continue to number in the tens or hundreds of thousands, their calving grounds, particularly that of the Porcupine herd of Yukon and Alaska, have been under threat from oil exploration and development for at least two decades. Not surprisingly, Aboriginal cultures, with wildlife and environmental groups in both Canada and the US, have engaged in widespread efforts to halt the industrial development of these crucial regions.

The fifth subspecies, the Dawson's caribou, disappeared in the 1930s and has been declared extinct.

Peary Caribou

Rangifer tarandus pearyi • Endangered (Canada)

SMALL and delicate, with a coat the color of snow during the long Arctic winters, Peary caribou are often described with superlatives. They have been called the loveliest members of their species and they live in one of the Earth's most challenging environments – the far northern islands of Canada's High Arctic archipelago. They have survived where all but a handful of mammals could not, and offer a superb example of environmental evolution. They are also, alas, among Canada's most endangered mammals.

Not surprisingly, given their frigid homeland with its brief summer season, Peary caribou have never been as numerous as their larger barren-ground cousins. But historical evidence shows that they were once numerous throughout Canada's islands of the far north, and paleontological discoveries show that these delicate creatures inhabited the region both before and soon after the last glaciation. Whether they also found a glacial refuge during the intervening frigid millennia is not certain, though biologists have called them a "superb example of environmental evolution".

Whether that evolution took 120,000 years, as some have claimed, or 30,000 years, a date connected to the discovery of a fossilized antler, it's clear that they have survived intense periods of climate change. However, as Parks Canada biologist Micheline Manseau writes, "It shouldn't be inferred that they continuously occupied [the far north]. Their range contracted and expanded in response to the cooling or warming periods." Still, even in the past thousand years, they have survived a long period of global warming – the Medieval Warm Period, between 1150 and 650 BP – and an equally lengthy period of global cooling – the Little Ice Age, between 600 and 150 years ago. What is less well known is whether they can survive the impacts of climate change, complicated by human predation and modern weaponry.

Studies over the past 40 years have shown that Peary caribou are extremely susceptible to brief periods of alternate thawing and freezing, which can make it virtually impossible for herds to crater through icy crusted snow to reach food below. Such conditions in late fall, when the long winter lies ahead, or early spring, when the animals are weak from food deprivation, can cause a population that may have grown slowly over a period of decades to suddenly crash. And an annual series of such conditions can virtually wipe out Peary caribou in a particular area, and also dramatically affect herds of muskoxen.

DAVID R. GRAY / REPRODUCED BY PERMISSION FROM THE CANADIAN MUSEUM OF NATURE, OTTAWA, CANADA

This was the case on Bathurst Island in the south-central Queen Elizabeth Islands during the 1990s. Aerial searches over Bathurst in 1993 and again in 1998 indicated a cataclysmic decline over the six-year period, according to an article by Canadian Wildlife Service biologist Frank Miller, and his co-author Anne Gunn, who is with the Government of the Northwest Territories. In the December 2003 issue of *Arctic*, the authors write that in 1993, 2,400 caribou were sighted during 33 hours of low-level helicopter searches. Just six years later, in 1998, only 43 caribou were counted during 35 hours of flying time, a decline of 98 per cent. Moreover, the large number of carcasses found during the survey made it clear that the animals had largely perished in situ, rather than emigrating to other islands.

Likely because they generally go into the winter season in comparatively poor condition as a result of the rut, males in their prime died at a disproportionately high rate. Visual counts in 1993, for example, listed 691 cows and 310 bulls, along with calves, yearlings and juveniles of both sexes. In 1998, the survey found just 21 cows of breeding age and one lone bull. Starvation, brought on by severe snow and ice conditions in the winters of 1994-95, 1995-96 and 1996-97 had taken most of the rest; from the carcasses the team counted, Miller believes that less than 15 per cent of the herd had tried to flee to other islands.

Miller had also experienced the devastation in the summer of 1996 when, on touring the territory around his High Arctic base camp on Bathurst Island, he discovered nearly 300 carcasses of caribou and muskoxen. Most of the animals had died on land, but some of the muskoxen, in a desperate effort to find new sources of food, had moved, perhaps unwittingly, out onto the ice. Starving and weak, they'd tried to dig down for food and, discovering nothing but ice, seem to have given up. They froze in place, stranded in deep snow, as *Edmonton Journal* reporter Ed Struzik described them, "leaning against each other like statues that had been knocked over . . . by the wind." They were still there the following summer when Miller arrived.

As for the caribou, Miller was able to find just 91 in the area he surveyed; projecting that number over the south-central QEI region, he calculated that perhaps 500 caribou were left, a population decline of 85 per cent. Similar conditions the following winter apparently cut that number by another 90 per cent.

Good soil conditions and a rare wetland environment produce abundant vegetation, making Bathurst a major calving area. The island also boasts Polar Bear Pass NWA, a migratory route used by the magnificent white bears between March and November. There is also a long human history on the island, with Dorset and Thule habitation going back 4,000 years. In light of this long animal and human history, proposals have been made to set aside the northern half of the island as Tuktusiuqvialuk National Park.

Biologists studying Peary caribou have found small populations elsewhere in the Queen Elizabeth Islands and, through original research, have found possible explanations for the decline in numbers. Since the year 2000, a team led by Winnipeg-based Parks Canada ecologist Micheline Manseau has been studying past and present caribou populations, as well as the human history in the area. Focusing on Quttinirpaaq National Park, an area the size of Switzerland on northern Ellesmere Island, Manseau found only 10 caribou in 2000, and 35 animals, including one calf, the following year.

It was, Manseau told a reporter at the time, "a very empty landscape. There should be more caribou up here."

Searching the diaries of early European adventurers, she and her colleagues discovered that between 1898 and 1909 Robert E. Peary visited Ellesmere Island three times in his efforts to beat fellow American Frederick Cook

Two delicate females focus on visiting scientists at Alert Enclave on Ellsmere Island.

MICHELINE MANSEAU

to the North Pole. With the aid of Inuit hunters, who taught Peary's expeditions how to hunt caribou and live off the land, the expeditions eventually shot 260 caribou, along with 978 muskoxen, devastating both animal populations.

Many of the caribou pelts from Peary's expeditions, along with other body parts and the remains of other species, wound up in New York's American Museum of Natural History, where Manseau and her coworkers were able to take DNA samples. Based on a genetic assessment of mitochondrial DNA (mtDNA) – microscopic material from membranes outside a cell's nucleus, which is passed to future generations only by the mother – they compared the mtDNA diversity from historic specimens with contemporary samples from Northern Ellesmere Island. In a paper to be published in the *Journal of Mammology*, they write that Peary's harvest had resulted in the loss of one major mtDNA group and that historic and contemporary samples differed significantly. Combined mtDNA and nuclear DNA also indicated that contemporary samples from northern Ellesmere were most similar to those from central Ellesmere Island to the south, suggesting that following Peary's killing spree, animals moved north from central Ellesmere.

Moreover, the team found little genetic diversity among today's animals. In short, while the man for whom these lovely creatures are named may not have been responsible for their overall dramatic decline, he certainly affected Ellesmere's population. Not surprisingly, some of the Inuit of Ellesmere Island think the name by which the species is known ought to be changed. They suggest, following the lead of the region's other mammals, including the Arctic hare and Arctic fox, that they should be called the Arctic caribou.

In short, while Peary's expeditions significantly altered the caribou population on northern Ellesmere Island, it seems clear that the series of difficult winters during the 1990s had much more to do with what federal government surveys estimate is a 72 per cent decline in Peary caribou numbers since 1984. According to COSEWIC, overall numbers have slipped from just under 49,000 animals in 1961 to about 7,890 in 2001. In fact, the caribou are now so few in number that surveyors have often found them remarkably difficult to find, even when using joint ground and aerial surveys, satellite collars and remote sensing, as well as interviews with Inuit hunters and trappers organizations.

Inuit hunters, however, who have voluntarily limited their hunting in recent years to allow populations to recover, believe numbers of these beautiful, resilient creatures are once again on the increase.

"It's like a miracle, how Peary caribou multiply," Simon Idlout, an elder from Resolute Bay, said in late 2006. That's certainly what everyone hopes for.

The tundra is a riot of blossoms during Ellesmere Island's brief summer.

COURTESY OF JUDITH SLEIN

Peary Caribou Fast Facts

Rangifer tarandus pearyi

HEIGHT & WEIGHT

Covered from its short, rather square face to its fluffy tail with a dense, silky white coat during the nearly endless winter months, a Peary bull generally stands less than a metre or about three feet at the shoulder and weighs between 70 and 110 kilograms or about 150 and 240 pounds, about two-thirds the size of a woodland caribou bull, the largest member of the species. Females are two-thirds as large as males. Like other members of the species, the hair of Peary caribou is hollow, giving it excellent insulating properties; unlike others, it is almost pure white during the winter months, changing to slate gray in summer with white legs and underbellies. The velvet covering of emerging antlers is gray as well, compared to brown for barren ground caribou

MATING & BREEDING

Like other caribou, mating takes place between late September and early November, when males contest for access to females. During the rut, males almost completely stop eating and lose much of their fat reserve. Calves are born the following May or June, and though they continue to nurse until fall, the calves are able to graze with the rest of the herd within a month.

DIET

During the summer months, Peary caribou dine on flowers, mushrooms (when they're available), grasses, willows and sedges. During the winter, the wind over beach ridges keeps snow from collecting; there they paw or crater down to reach lichens and mosses, of which there are hundreds of species. The availability of winter forage is the main factor that limits populations.

BEHAVIOUR

Peary caribou live in small herds and while they do not normally migrate seasonally, they will move from island to island to find food if necessary.

DISTRIBUTION

Aside from a possible occurrence in Greenland, Peary caribou live only in Canada, on the islands of the Arctic Archipelago, with a small population on the Boothia Peninsula in Nunavut's far north.

HABITAT

During the summer months, they can often be found in river valleys and on upland plains, while in the winter, they often migrate to relatively snow-free hilltops and ridges.

THREATS

The Peary caribou is endangered, with a total population estimated to be less than 8,000; about 1,000 of that number live on Banks Island in the Western Arctic. Scientists speculate that changing climatic conditions have been to blame for a drop in numbers of almost three-quarters ; increasingly, a thick covering of ice on the snow has often prevented caribou from reaching the vegetation below, resulting in starvation. Wolf predation also affects the number of surviving calves; while hunting with modern weapons contributed to the decline over the past half-century. Low level flights of aircraft, ground vehicles and the construction of ground installations can all hamper crucial summer feeding or movement to better feeding areas.

VIEWING PEARY CARIBOU

A dozen or more communities and national parks offer opportunities for guided or self-guided arctic wilderness adventures. Among these is Nunavut's Bathurst Inlet Lodge, listed among *Travel & Leisure Magazine's* "World's 25 Top Ecolodges" in a region that has been dubbed the "Serengeti of North America". Once a trading post and mission, the lodge is now a community-partnered hub of commercial and cultural activities that serves eco-travellers drawn to the area's wildlife, including musk-oxen, peregrine falcon and geese.

Bowhead Whale

Balaena mysticetus • Endangered (US); At Risk (Canada)

IN May 2007, hunters off the coast of Alaska found the head of an explosive harpoon lodged in a bone between the neck and shoulder blade of a bowhead whale. Curators soon identified the arrow-shaped metal projectile as being of New England manufacture, and employed sometime in the late nineteenth century. Earlier, in the 1980s and 1990s, stone harpoons were found embedded inside several bowheads taken in the Aboriginal harvest, though the Inuit have not used stone harpoons for almost a century. In short, these bowheads had been hunted when Canada was a young confederation and had lived to tell the forensic tale. But they were not unusual. Through analysis of amino acids in the lenses of the whales' eyes, scientists have determined that while most bowheads live less than a century, others live much longer. One was estimated to be 211 years old when it died in the 1990s, which means it was swimming around before Napoleon even thought about conquering Europe.

But when France and America were having their revolutions, these Methuselahs of the Arctic waters, the longest-lived mammals on Earth, were in abundance. Bowheads had been hunted by indigenous people of the Arctic for a thousand years for energy-rich meat and skin with some underlying blubber, known as *muktuk,* and for baleen and bone, which were used to make tools and housing. Commercial exploitation of the species by Europeans began as long ago as 1611 in the eastern Arctic, and over the succeeding 200 years, the eastern Arctic stock was reduced from an estimated 30,000 to about 1,000. Nevertheless, before whaling reached its peak in the late 19th and early 20th centuries, an estimated 50,000 bowheads swam the northern polar region.

Huge and remarkably long-lived, bowhead whales are slowly recovering.

LIBRARY AND ARCHIVES CANADA / LIBRARY AND ARCHIVES CANADA / LAC 062

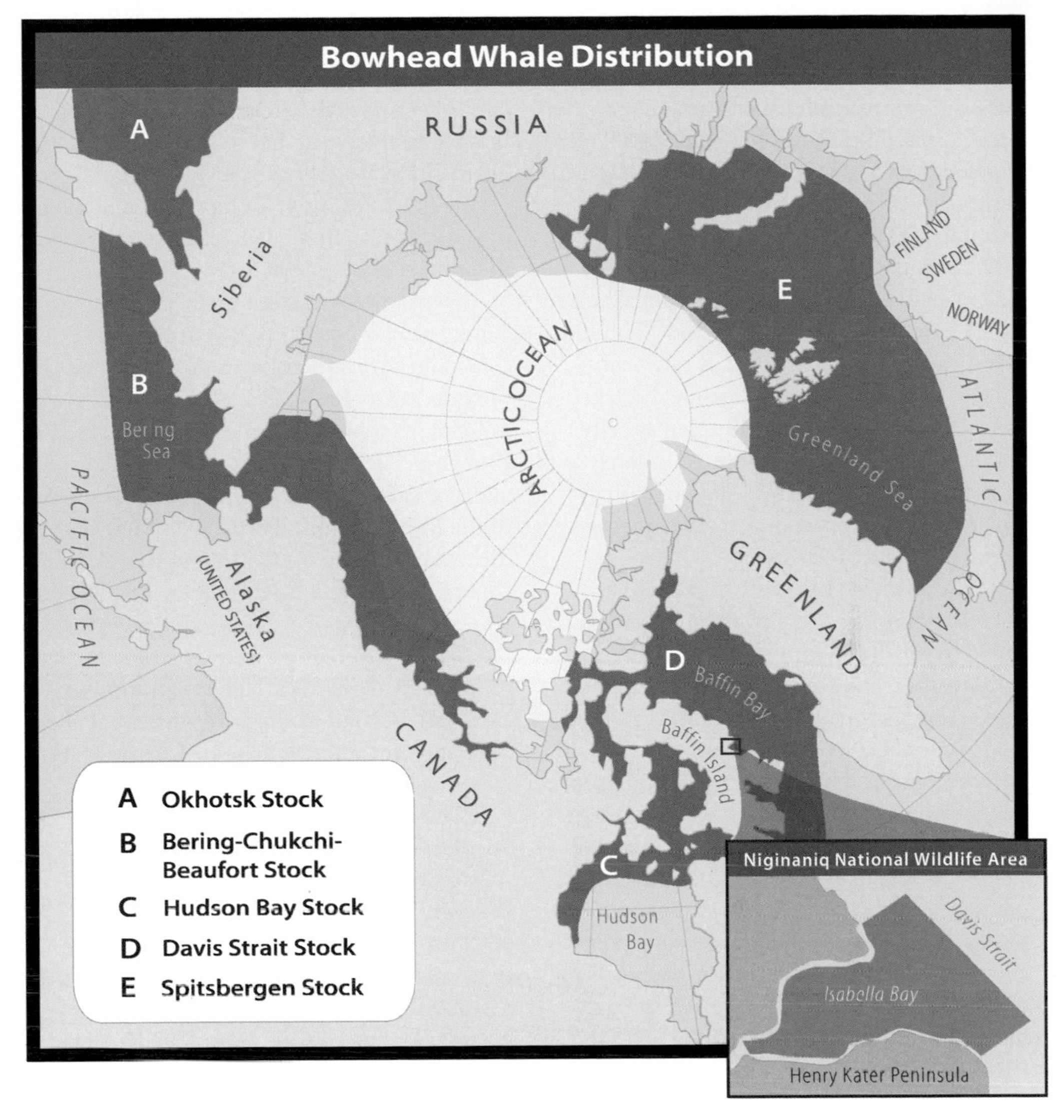

With great profits to be made from the sale of whale oil for lamps and the sturdy but flexible baleen or "whale bone" for corset stays, skirt hoops, buggy whips, umbrella ribs and many other consumer goods, hundreds of bowheads were killed each year. Finally, in the early 1900s, kerosene replaced animal oils for lighting and spring steel began to replace baleen. Nevertheless, thousands of bowheads were slaughtered during the height of the whaling industry, and once plentiful, they were hunted to near extinction.

Because of market changes, commercial whaling had virtually disappeared by 1910. In 1937, bowheads came under the protection of the International Agreement for the Regulation of Whaling, the first of several conventions designed to protect whale species from over hunting and establish international regulations. Full protection from commercial whaling was finally implemented in 1946. But today, there are only remnant populations in Canada's High Arctic (which have been listed as endangered by the Committee on the Status of Endangered Wildlife in Canada or COSEWIC) and in the waters off Russia's Kamchatka Peninsula. A once-bountiful North Atlantic population, which cruised the waters between Spitsbergen and Greenland, is believed to be extinct. Ninety per cent of the world's remaining bowheads, an estimated 8,000 animals, belong to the Western Arctic population, which migrates between the Bering and Beaufort Seas through the Chukchi Sea off Alaska's northwestern coast. Though this is the largest population of bowhead whales, and seems to be increasing, it is also the one most threatened by human activity and has been listed as endangered in the US. Moreover, research has indicated that only two young are produced each year for every hundred bowhead whales.

The bowhead has just two predators – orcas and humans. Orcas present no serious threat to bowhead populations, but human activity does. Hunting is no longer a major threat, though an indigenous hunt is allowed, based on quotas set and monitored by federal governments in cooperation with local people. But offshore oil exploration is already taking place in the Beaufort and Chukchi Seas, and may soon spread to other parts of the whales' habitat. The effects of these activities on bowheads are unknown, and are the subject of current research. Possible negative impacts include increased underwater noise that could interfere with the whales' ability to sense their surroundings and communicate effectively, causing them to abandon their habitat. Studies have shown that whales avoid loud sounds from vessels and hydrocarbon exploration, and move significant distances away from these human activities. Increased collisions with vessels are also a possibility. Though oil spills do not affect whales as much as they do other marine mammals and birds, because spilled oil accumulates at ice edges, where bowheads often feed, they are at risk of ingesting it. Also, synthetic toxins found in northern waters accumulate in the fat of arctic species, and because bowheads are so long-lived they may be at particular risk. The effects of these chemicals on whales are unknown, but they may impact reproduction or cause birth abnormalities. Nevertheless, it seems wise to err on the side of caution, for bowhead population dynamics mean that it will take many years for the impact of human activity to be fully felt.

Steps to protect these arctic giants are being taken, however. In August 2008, the Canadian government set aside a large and pristine bay on Baffin Island's northeast coast as Canada's newest National Wildlife Area. Known as Niginganiq in the Inuit language, and also called Isabella Bay, it is located south of the Inuit community of Clyde River and includes two deep offshore troughs rich in copepods, the main food for these magnificent whales. Serving as a crucial feeding area for the endangered Baffin Bay population during its late summer and early fall migration to Davis Strait, Niginganiq also features a shallow shelf at its entrance, which provides protection from predatory orcas.

The creation of the conservation area was the result of more than a quarter-century of work by the Clyde River Inuit and World Wildlife Fund Canada. In addition to its importance to the Baffin Bay bowhead population, currently estimated at 1,500 animals, the 336,000-hectare marine region is home to polar bears, ringed seals, Arctic char, halibut, narwhals, snow geese and king eiders.

Methuselas of the arctic waters, bowhead whales are also their giants. These massive whales with smooth, blue-black skin have white patches on the lower jaw, and some white markings on the belly and just in front of the fluke, or tail. This patch near the tail often grows and lightens as the whale ages, and the tails of very old bowheads may be all white. Adult females, which are slightly larger than males, grow up to 18.3 metres or 60 feet in length, and can weigh up to 70 tons. The head accounts for a third of a bowhead's total body length. Its upper jaw is arched like a bow,

A bowhead whale calf nurses from its mother, as another female keeps watch nearby.

L. DUECK, CANADIAN DEPARTMENT OF FISHERIES AND OCEANS

giving the whale its name. The bowhead's mouth alone grows to nearly five metres or 16 feet long, and its tongue can weigh up to 900 kilograms or close to a ton.

Bowheads are well adapted to life in the far north. They have no dorsal fin and are insulated from their icy surroundings by a layer of blubber up to half a metre or 1.5 feet thick. Their heads have a high bridge and are powerful enough to smash through thick sea ice to create breathing holes, when necessary. Bowheads have excellent vision, but may rely even more heavily on their superb hearing. They communicate with one another using low, moaning songs that travel long distances underwater. Though they do not echolocate like the odontocetes (toothed whales), they likely use their acute hearing to sense their surroundings and navigate through ice-choked waters.

Bowheads are mysticetes, or baleen whales, and are closely related to one of the world's rarest whales, the North Pacific right whale (see page 130). They were called Greenland right whales by European and American hunters, who found them the "right" whales to hunt, thanks to their great bulk and slow speed.

The right whale family evolved some 22 million years ago and features the longest, most elaborate baleen of any whales. These huge fringed plates, which hang from the upper jaw, allow the whales to capture up to 1,800 kilograms or nearly two tons of tiny planktonic organisms each day. They feed by swimming, at any depth, with their mouths open, straining the creatures they capture with their baleen. Interestingly, mysticetes' land-mammal ancestry is evident in their foetal development; in the womb they have traces of teeth, but these are reabsorbed before birth.

Bowheads live close to the edge of the Arctic icepack year round; they are the only baleen whales that do not migrate to warmer waters to calve. They do, however migrate seasonally. The whales of the western Arctic migrate 5,800 kilometres or 3,600 miles annually, spending the winters in the icy waters of the Bering Sea and summering in Canada's Beaufort Sea. Each spring in late March or April the whales follow the retreating pack ice through the Chukchi Sea and along the north coast of Alaska, arriving in Canadian waters from mid-May through June. When autumn comes, they migrate west along the continental shelf of the Beaufort Sea, cross the Chukchi Sea and then travel south along the Russian coast, passing through the Bering Strait by November. They are often accompanied by beluga whales, which follow them through leads they create in the ice and use the same breathing holes during this journey.

Bowheads swim slowly, usually between three and six kilometers an hour, about walking speed for a human, and often make long dives during their migrations. These dives generally last between six and 17 minutes, but bowheads can remain underwater for more than 30 minutes. While they usually live alone or in small groups of two or three individuals, they often gather in larger groups with animals of the same age and sex during migrations. Groups of young bowheads are the first to move north during the spring migration, followed by large males, and then females with calves. The fall migration follows the reverse order.

Mating usually takes place between late winter and early summer, but sexual activity has been observed at other times of the year as well. The bowhead's gestation period is somewhere between 13 and 14 months, and calves are born tail first, near the surface. Assisted by their mothers, they rise to the surface within 10 seconds of birth to begin breathing and can swim within a half-hour. They're generally born between April and early June, during the spring migration.

The calves are anywhere from 3.5 to 5.5 metres or 11 to 18 feet long at birth, weigh an average of nearly a ton, and are covered by a thick layer of blubber to protect them from the frigid water. They nurse for up to a year and grow quickly during this time; their growth rate slows considerably after weaning. Because of the long gestation and nursing period, most females only give birth to one calf every three or four years, beginning after they reach sexual maturity at between 10 and 15 years of age.

Bowheads grow, mature and age slowly; as noted earlier, they are considered to be the longest-lived mammals on Earth. Because their only natural predator, aside from humans, is the orca, many whales live to a very old age.

LIBRARY AND ARCHIVES CANADA / LAC 060

Bowhead Whale Fast Facts

Balaena mysticetus

LENGTH & WEIGHT

Larger than males, adult females can grow to more than 18 metres or 60 feet long and weigh up to 70 tonnes.

LIFESPAN

The average lifespan is believed to be 80 to 100 years, but if they can escape human predations, bowheads can live to be much older. One, which died in the 1990s, was believed to be 211 years old.

MATING & BREEDING

Mating probably occurs during late winter and early spring and the calves are born 13 or 14 months later, between April and early June. Bowheads calve at about three to four year intervals.

BIRTH STATISTICS

Bowheads weigh nearly a tonne or 2,000 pounds at birth and vary in length between 3.5 and 5.5 metres or 11 and 14 feet. Calves, which are shorter and heavier than other baleen whales, must begin swimming north with the herd almost immediately. Feeding on their mother's rich milk for a year, they grow quickly

FEEDING

Bowheads feed on plankton – small crustaceans such as copepods and amphipods – siphoning them from the water with their baleen, more than 300 overlapping gray or black plates of keratin, a material that is much like a fingernail. These huge plates, which measure up to 4.3 metres or 14 feet long and 30 centimetres or a foot wide, fray into fine hairs near the tongue. Swimming slowly with their mouths open, bowheads can trap and consume up to 1800 kilograms or nearly two tons of crustaceans daily.

DISTRIBUTION & MIGRATION

The main population of bowheads is found in the Bering, Chukchi and Beaufort Seas of the Western Arctic. Small populations, considered endangered, are found in Baffin Bay, Davis Strait and Hudson Bay of the Canadian Arctic and in Russia's Okhotsk Sea, between the mainland, the Kamchatka Peninsula and the Kuril Islands. They once populated the northern oceans between Scandinavia and Greenland, but experts believe that population was hunted to extirpation.

The only baleen whales that do not migrate to warmer waters to calve, bowheads do travel seasonally. The world's largest population moves from the Bering Sea coast through Bering Strait and the Chukchi Sea to the Beaufort Sea off Alaska's coast each summer to feed and return when winter threatens. They are often accompanied by beluga whales on their migrations.

POPULATION

Despite centuries of being hunted in the Atlantic and eastern Arctic, the bowhead population was estimated at more than 50,000 in the late 1800s. Today, more than 60 years after a moratorium on commercial whaling, the world population is estimated at 10,000 or less.

PREDATORS

Humans and orcas

NAMES

Bowheads are so called because of their huge upper jaw and mouth, which are arched like a bow. Yankee whalers, who hunted them in the North Atlantic, called them Greenland right whales, for they found them the "right" whales to hunt, thanks to their great bulk and slow speed.

VIEWING BOWHEADS

Travel companies offer expeditions to several places, including Auyuittuq National Park on Baffin Island's northeast coast, where whale watching of several species, including bowhead, blue, sperm and beluga whales, as well as narwhals, is offered.

Beluga Whale

Delphinapterus leucas • Endangered (Alaska's Cook Inlet population);
• Endangered (Canada's Southeast Baffin Island-Cumberland Sound population)

THE beluga whale's intelligence, sociability and expressive features have made it a darling of whale watchers and visitors to marine aquariums, but there are surprising gaps in our knowledge of this creature. Beluga pods live primarily in the arctic and sub-arctic waters of North America, Greenland and Eurasia, though an isolated and threatened population lives in the St. Lawrence River estuary near Tadoussac, Quebec, upstream of their winter home in the Gulf of St. Lawrence. Isolated populations also live in the Okhotsk Sea and in Alaska's Cook Inlet. Like the St. Lawrence belugas, these may be relics of the last glaciation.

Belugas are mid-sized toothed whales or odontocetes, a group that also includes killer whales, sperm whales, porpoises and dolphins. They belong to the Monodontidae family, along with their close relatives, the narwhals. They can grow to be five metres or 16 feet long, and males, which are larger than females, can weigh up to 1.4 tonnes or about 3,000 pounds. Their large, robust bodies are covered in a layer of blubber up to 12 centimetres or five inches thick, which accounts for nearly half their body weight. The blubber insulates them from the cold water in which they live, and provides a significant energy reserve. Their heads have a small "beak" below a bulbous "melon", and small black eyes located just behind the corners of their mouths, which are upturned in a characteristic smile.

Belugas are different from other odontocetes in several ways. Though the calves are gray-brown at birth, and adolescents are gray-blue, adults are pure white, which gives them their name, which is from the Russian *belukha*, derived from *belye*, meaning "white". They lack a dorsal fin, which is present in most whales, and instead have a dorsal ridge that runs from mid-back to the back of the tail flukes. This adaptation is thought to help to conserve heat by reducing the whale's surface area, and also serves as an "inverted keel", while allowing it to swim directly below ice sheets, making it easier to locate breathing holes. Male belugas have a marked upturn on the ends of their flippers.

Unlike most whales, seven vertebrae of the beluga's neck are not fused, allowing it to turn its head in all directions and increasing its agility. Also unlike other whales, the beluga moults the outer layer of its very thick skin annually. The skin of a beluga is 10 times the thickness of that of a dolphin, and 100 times thicker than the skin of most terrestrial mammals. In the early summer new skin cells grow rapidly while the whales shed their old skin, sometimes rolling on the rocky substrate of riverbeds to help speed the process.

The beluga's earliest known ancestor is the *Denebola brachycephala*, from the late Miocene period, about six million years ago. A single fossil was found on Mexico's Baja California peninsula, indicating that life was once lived in warmer waters. More recent fossils, as well as the existence of the isolated St. Lawrence estuary population, indicate that the beluga's range has likely varied with that of the ice pack, expanding during periods of global glaciation and contracting when the ice retreats.

Though well known to northern populations for millennia, belugas were first described for the benefit of biologists by Peter Simon Pallas, a German naturalist who spent much of his working life at the court of Russia's Catherine the Great.

Highly sociable, these white whales live in

Pure white and very sociable, beluga whales seem unbothered by the crowds of humans who travel to see them every year in Hudson Bay.

DENNIS FAST www.dennisfast.smugmug.com

pods with others of the same gender and age. Males often gather in pods of eight to 10 individuals, but can also be found in much larger groups. Females with calves usually travel in smaller groups, but pod membership is flexible, and individuals change pods from time to time. In the summer, belugas leave the frigid waters along the edge of the northern pack ice to congregate in warmer, shallower estuaries and adjacent waters. In some of these summering grounds, thousands of individuals can be found in a relatively small area. Many belugas return to the same summer habitat every year, and genetic studies have shown that whales in the same estuary are more closely related to one another than to the populations of different estuaries.

In the autumn, as ice forms in the shallow river and coastal waters, belugas begin a migration that can cover distances up to 3000 kilometres or about 1,600 miles. The Eastern Beaufort Sea population, for example, migrates vast distances offshore of Alaska and along the eastern coast of Russia to winter in the Bering Sea. As they are rather slow swimmers, generally travelling less than 10 kilometres or about six miles per hour, this migration can take more than a month, particularly when there are stops along the way. By contrast, the St. Lawrence River population spends the winter a relatively short distance away in the Gulf of St. Lawrence.

Belugas spend the winter in pack ice, which varies in concentration depending on the latitude of their winter range. They find gaps in the shifting ice, patches of open water, or pockets of trapped air under the ice in order to breathe. Occasionally, belugas become trapped under the ice and drown, or are surrounded by ice and vulnerable to hunters and polar bears. Recently, polar bears along the western shore of Hudson Bay have been seen using new techniques to hunt belugas from land during the summer months (see page 32).

At a global population of over 100,000, belugas are greater in number than many other cetaceans (whales, dolphins and porpoises), but are considerably fewer than they were historically before years of unregulated hunting. More than 40,000 live in the Beaufort Sea, with another 60,000 in Hudson Bay and 20,000 in the Canadian High Arctic. Their natural predators are polar bears, orcas and humans. Northern peoples have hunted beluga for centuries, eating their energy-rich skin and blubber, using their oil for fuel, and their bones and skin for tools and crafts. This level of use was

Beluga whale meat drying at Whitefish Whaling Station, also known as Nalruriaq, southwest of Tuktoyaktuk in the Mackenzie Delta, in 1955.

WILKINSON / NWT ARCHIVES / N-1979-051-0346S

sustainable, but in the eighteenth and nineteenth centuries commercial hunting by Europeans and Americans lead to a steep decline in the beluga population.

These whalers used the meat, blubber, and oil, and also sought out the belugas' fine melon oil as a lubricant for watches and machinery. Indigenous hunting continues in the Arctic today, where it is still an important part of many peoples' way of life. In some places, however, the declining populations can not sustain even the relatively small indigenous catches, which are estimated to be between 200 and 500 in Alaska (outside of Cook Inlet, see below) and about 1,000 in Canada annually. The Inuit are involved in discussions with both countries' governments to ensure that these numbers are low enough that they will not do any significant damage to whale populations.

Beluga whales are also considered vulnerable as a result of human-caused pollution, which in some areas, such as the St. Lawrence, is the most significant identifiable danger to their health. And global warming may also pose a threat. Because the impact of global warming is currently greatest at the Earth's poles, creatures like the beluga, which rely on pack ice and icy polar water for their habitat and food, may find themselves under environmental stress.

Several beluga populations are deemed particularly at risk. The Cook Inlet population, which lives in the large estuary that stretches from the Gulf of Alaska to Anchorage in south-central Alaska, is deemed in danger of extinction. One of five populations recognized in US waters, it was listed as endangered in October 2008. Despite a number of protective measures imposed over the past two decades, including a ban on all but Aboriginal hunting, numbers have plummeted, from an estimated 1,300 whales in the late 1980s to about 375 animals in 2008. Strong opposition to the listing came from former Alaska Governor Sarah Palin, but in the end, the George W. Bush administration finally bowed to the clearly demonstrated science and more than 180,000 letters of support to the National Marine Fisheries Service, the most the NMFS has ever received for a proposed action.

Since beluga females bear a single young only every three years on average, it will take time to determine whether the hunting ban will have any impact on population recovery. In the nine years prior to the ban, just five whales were taken under regulations on subsistence hunting. Clearly, there are other explanations for the declining numbers. James Balsiger of the National Oceanic and Atmospheric Administration (NOAA) Fisheries Service says recovery may be hindered by strandings, development along Cook Inlet, oil and gas exploration, industrial discharge or accidental spills, disease and predation by orcas. (The last of these is also a growing concern far to the east, where orcas are invading northern Hudson Bay, home to relatively healthy pop-ulations of belugas.)

Of Canada's eight populations, four – those off southeast Baffin Island, Ungava Bay, eastern Hudson Bay and the 650 whales living in and near the St. Lawrence estuary – are considered endangered or threatened by the Species at Risk Act (SARA) or the Committee on the Status of Endangered Wildlife in Canada (COSEWIC).

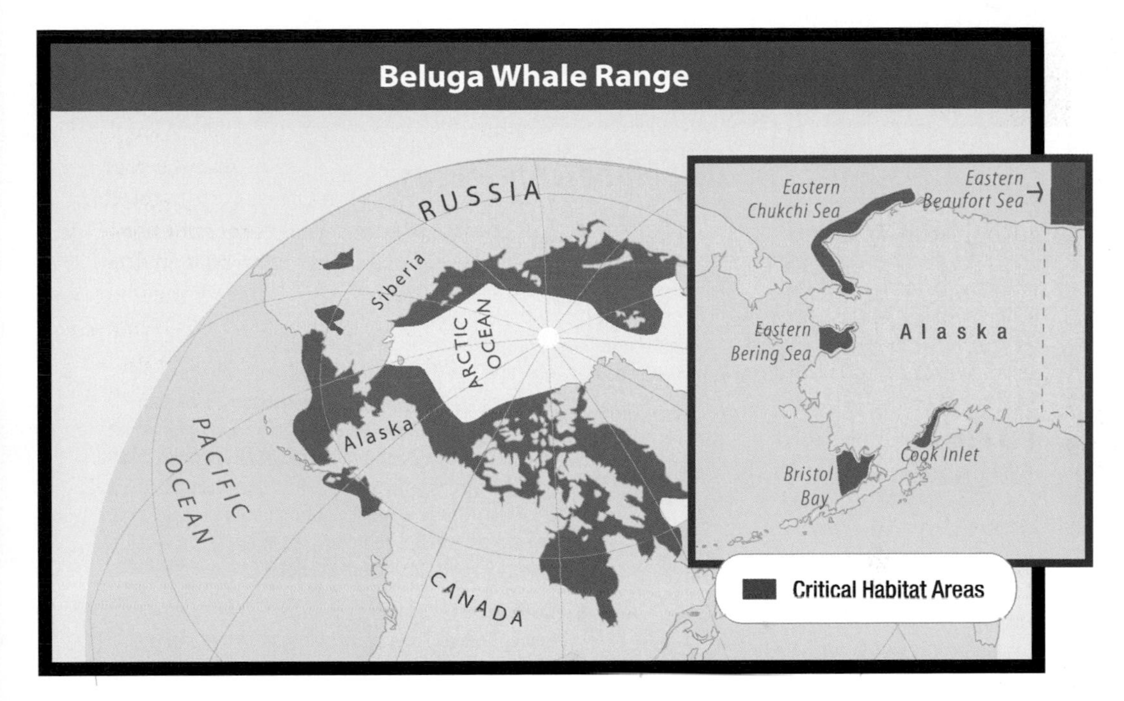

Chemical build-up in Arctic waters also threatens beluga populations and increased commercial shipping, shore development, oil and gas drilling and ice breaking can all disrupt beluga habitat. All have the potential to jeopardize these beautiful, accessible whales.

Though not strictly within the purview of this book, the St. Lawrence population is among Canada's most threatened. Numbering an estimated 10,000 animals in 1880, these animals were excessively hunted until 1950. Since then, bacteria, viruses, parasites and a very high rate of cancer have taken a toll on the animals. Autopsies conducted on 119 beluga carcasses stranded in the river estuary have shown that 27 per cent died of cancer, a rate higher than that found in humans and much higher than any other wild animals, with the exception of some species of fish.

These cancers, bacteria, viruses and parasites are all believed to have resulted from high levels of industrial contaminants from the large Canadian and US cities located upstream of the St. Lawrence. Despite protective measures in place for three decades, there has been no noticeable recovery in the population, which Fisheries and Oceans Canada puts at about 1,000. Nevertheless, the population was downlisted to threatened in 2004.

Beluga whales' bodies have a number of amazing adaptations that allow them to thrive underwater. They have acute vision both above and below water, and can see well in dim light. Belugas also have keen hearing, and like many odontocetes, they possess echolocation abilities that allow them to locate prey or other objects in dark, deep waters. The whales produce a rapid series of clicks, which are focused into a beam through their head "melon" and projected forwards. These sounds bounce off objects and return to the whales, where the vibrations are received primarily in a fat-filled cavity of the lower jaw bone, transmitted through the jaw to the ear, and interpreted to provide an image of what lies ahead. This ability may also help belugas to find areas of open water under the sea ice.

Beluga Whale Fast Facts

Delphinapterus leucas

LENGTH & WEIGHT

Males can grow to five metres or 16 feet long. Males, which are larger than females, can weigh up to 1.4 tonnes or about 3,000 pounds. Their large, robust bodies are covered in a layer of blubber up to 12 centimetres or five inches thick, which accounts for nearly half their body weight.

LIFESPAN

The average lifespan in the wild is between 35 and 50 years, though Fisheries and Oceans Canada reports that wild belugas have lived to be 75.

HABITAT & DISTRIBUTION

Belugas primarily live in the arctic and sub-arctic waters of North America, Greenland and Eurasia, though isolated populations live in Canada's St. Lawrence River estuary, Okhotsk Sea and James Bay. In the past, individuals were occasionally spotted off the British Columbia or even Washington State coasts, and a herd of 21 was seen in the late 1980s in Yakutat Bay, in southeastern Alaska. These were presumably from the once-numerous Cook Inlet stock. Occasionally, solitary belugas or small groups are reported along the coasts of Newfoundland, Nova Scotia and New Brunswick, and south along the US coast as far as Long Island, New York. These were likely strays from the St. Lawrence stock, although Newfoundland records may also involve strays from stocks along the Labrador coast or areas farther north.

MATING & BREEDING

Though they usually mate in March or April, belugas have been known to mate at other times of the year. A dominant male may mate with several females. Following a gestation period of 14.5 months, a single calf, measuring about 1.5 metres and weighing between 35 to 80 kilograms or 77 to 175 pounds, is born between May and July. Females with calves congregate in the warm, shallow waters of upper estuaries. Babies and yearlings stay close to their mothers, nursing for about one-and-a-half years. As a result, calves are produced on average every three years. Born gray, belugas reach their characteristic white coloring between 12 and 16 years of age. Females are sexually mature at about 10 years, while males mature at aout 17.

DIET

Belugas eat a wide variety of fish, as well as shrimp, crabs, squid and octopus. They often feed along the ocean bottom, diving up to several hundred metres or more than 1,000 feet deep.

NAMES

Belugas are also known as white whales and sea canaries, a reference to their wide range of vocalizations.

RECOVERY MEASURES

Despite the wealth of information about belugas and the threats to their survival, in Alaska and the St. Lawrence studies are ongoing to determine the best ways to promote an increase in these endangered and threatened populations. Several of Canada's Arctic populations are managed by limiting indigenous hunting.

Opposite: Feeding and travelling in pods with others of the same age and sex, belugas communicate with a complex range of sounds.

IMAGES BOTH PAGES: DENNIS FAST www.dennisfast.smugmug.com

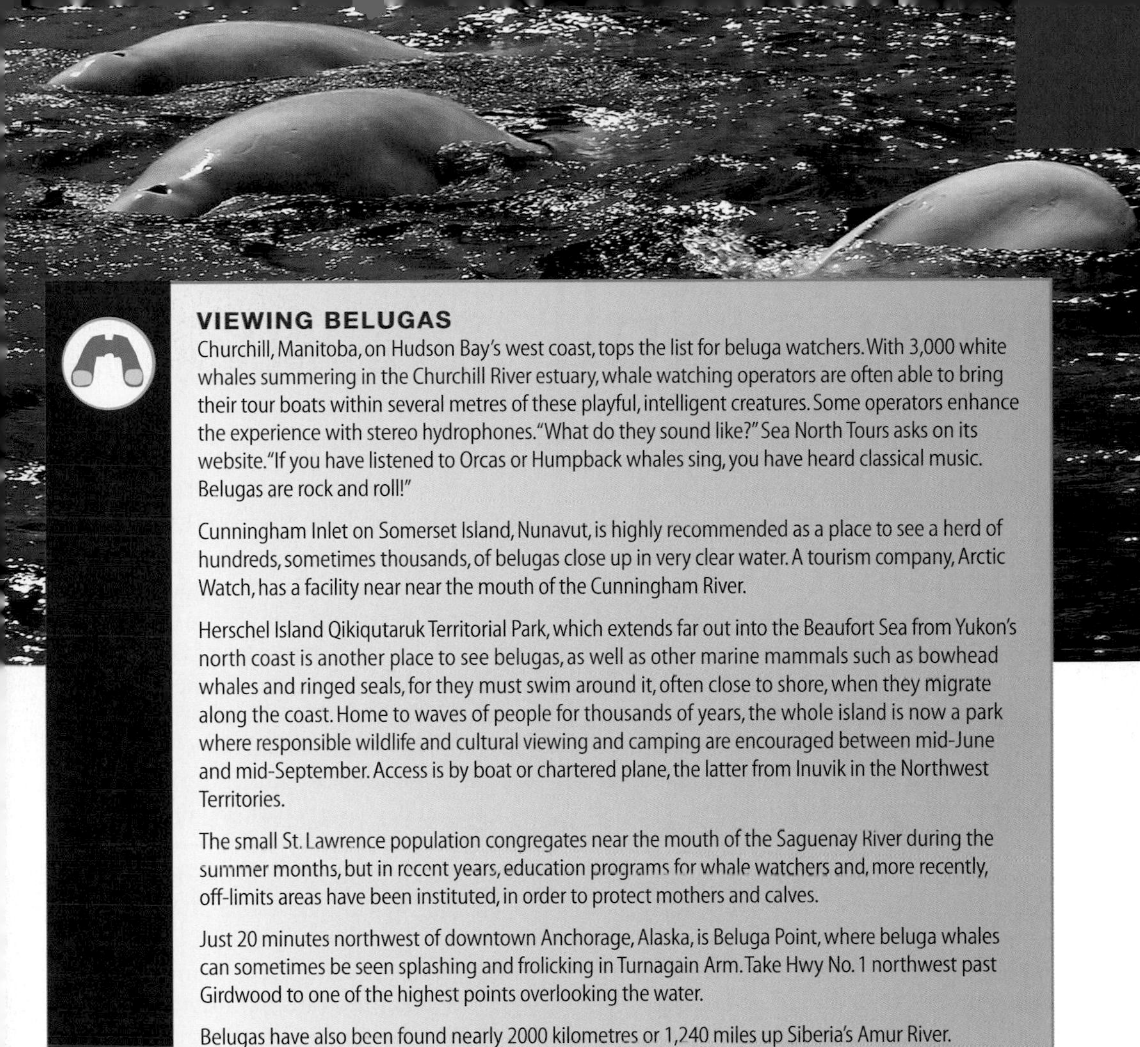

VIEWING BELUGAS

Churchill, Manitoba, on Hudson Bay's west coast, tops the list for beluga watchers. With 3,000 white whales summering in the Churchill River estuary, whale watching operators are often able to bring their tour boats within several metres of these playful, intelligent creatures. Some operators enhance the experience with stereo hydrophones. "What do they sound like?" Sea North Tours asks on its website. "If you have listened to Orcas or Humpback whales sing, you have heard classical music. Belugas are rock and roll!"

Cunningham Inlet on Somerset Island, Nunavut, is highly recommended as a place to see a herd of hundreds, sometimes thousands, of belugas close up in very clear water. A tourism company, Arctic Watch, has a facility near near the mouth of the Cunningham River.

Herschel Island Qikiqtaruk Territorial Park, which extends far out into the Beaufort Sea from Yukon's north coast is another place to see belugas, as well as other marine mammals such as bowhead whales and ringed seals, for they must swim around it, often close to shore, when they migrate along the coast. Home to waves of people for thousands of years, the whole island is now a park where responsible wildlife and cultural viewing and camping are encouraged between mid-June and mid-September. Access is by boat or chartered plane, the latter from Inuvik in the Northwest Territories.

The small St. Lawrence population congregates near the mouth of the Saguenay River during the summer months, but in recent years, education programs for whale watchers and, more recently, off-limits areas have been instituted, in order to protect mothers and calves.

Just 20 minutes northwest of downtown Anchorage, Alaska, is Beluga Point, where beluga whales can sometimes be seen splashing and frolicking in Turnagain Arm. Take Hwy No. 1 northwest past Girdwood to one of the highest points overlooking the water.

Belugas have also been found nearly 2000 kilometres or 1,240 miles up Siberia's Amur River.

The beluga is also able to dive deep under the water and go for long periods without breathing in order to find food. Belugas are opportunistic feeders and eat a wide variety of prey, which includes fish, such as Arctic cod and herring, shrimp, crabs, squid and octopus. Scientists do not know exactly what wild belugas eat during the winter months, or how much they eat on an average day; in captivity the whales eat from 10 to 15 kilograms (22 to 33 pounds) each day. Belugas often dive 400 to 800 metres (1,300 to 1,600 feet) under the water, and can stay there for up to 20 minutes.

To allow this, a beluga's body holds twice the blood that would be stored in a similarly-sized land mammal, and its blood cells can hold 10 times the amount of oxygen. Clearly, its muscles store and use oxygen efficiently. When it dives, blood is diverted to critical areas such as its heart and brain, and away from external tissues, which function even with reduced oxygen levels. To conserve energy, its heart rate slows from about 100 to between 12 and 20 beats per minute during a dive. When it finally surfaces for air, the beluga breaths through a blowhole, a modified nasal opening. A muscular, watertight flap covers the hole while the whale is under water, and the muscle is flexed as it nears the surface and begins to exhale.

Researchers have identified 16 distinct vocalizations including clicks, squeaks, whistles, squawks and a bell-like clang. These sounds are probably produced by moving air between nasal sacs near the blowhole, and may be altered by changing the shape of the melon. They are likely used to communicate among themselves; when danger approaches the level of communication increases.

Ross's Gull

Rhodostethia rosea • Threatened (Canada)

THOUGH his uncle, John Ross, is sometimes erroneously given credit for it, this small, elegant seabird was named for Lieutenant (and later Sir) James Clark Ross. The younger Ross, who served under Sir William Parry as expedition naturalist through the waters west of Baffin Island between 1821 and 1823, collected the first recorded specimen of these rare arctic gulls. Ross's specimen was collected during a land survey of Fury and Hecla Strait (which was itself named for the expedition's two ships) during the summer of 1823.

Ross's expedition returned home in the fall of 1823, where he learned that he had been promoted lieutenant. His efforts as a naturalist were further rewarded by his election as a fellow of the prestigious Linnean Society.

If Ross's gulls are found at all today – and they are among the rarest gulls in North America – they are still found not far from where Ross came upon them nearly 200 years ago. In fact, according to the public registry for Canada's Species at Risk Act (SARA), these unusual birds have been found nesting in just four places on the continent, including three sites in northern Nunavut: Prince Charles Island on the northeastern edge of Foxe Basin; the Cheyne Islands in Penny Strait and on an unnamed island in Penny Strait. The fourth site, in the Churchill Special Conservation Area near the town of Churchill, Manitoba, on Hudson Bay's west coast, was once the only known nesting site in North America. After having seen individual birds, but no nest, in 1978, a pair was found nesting at Akudlik Marsh, just four kilometres south of town, in 1980. Two years later, more than 10 birds were seen and these rare appearances brought flocks of birders to the site.

Even after other nests were found in Penny Strait, it was also the most accessible place to view and photograph these rare birds,

DENNIS FAST www.dennisfast.smugmug.com

This delicate gull is clearly marked as a breeding bird by its black neck ring.

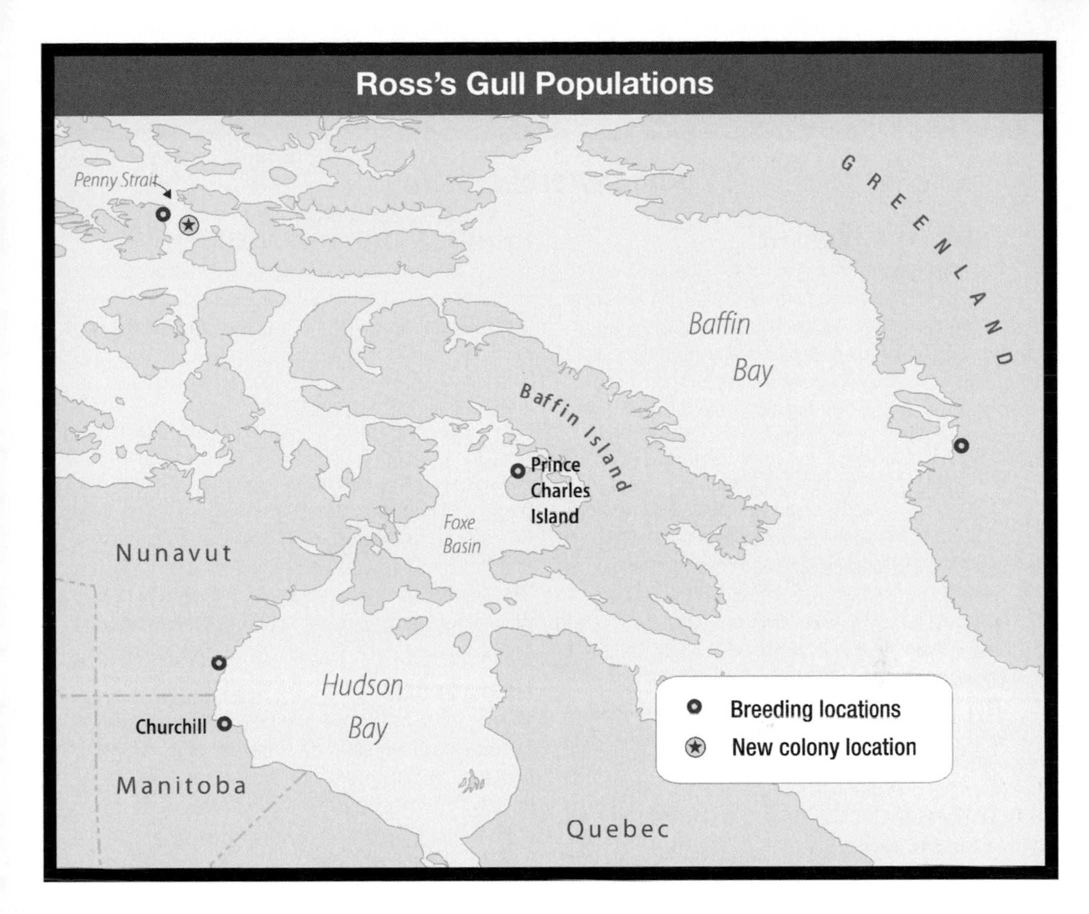

which may have been the problem. To protect the birds, which had abandoned Akudlik Marsh for a more remote site about 1999, a Special Conservation Area was established. Despite extraordinary care taken to protect the nesting birds from enthusiastic nature lovers, it seems that the gulls found the Churchill region too crowded for their liking. In 2009, it was reported that neither birds nor nests had been seen anywhere in the region for at least four years.

This is not atypical, however, for Ross's gulls are clearly elusive, both when nesting and feeding. For example, they have nested sporadically on the Cheyne Islands, sometimes in groups of up to seven pairs of gulls, since 1976. Yet they have also abandoned the site for several years, as they did between 2002 and 2005, before reappearing in 2006, when three pairs nested.

Nesting in Canada's High Arctic, they are usually found with arctic terns and Sabine's gulls, happily seeking anonymity in a crowd. This is also the case with non-breeding gulls, as a trio of scientists aboard the Swedish icebreaker *Oden* found during the summer of 1996. Small groups of non-breeding gulls were found in the waters north of Russia almost as far north as the North Pole, almost invariably in mixed company.

However, as indicated in Viewing Ross's gulls, below, these elusive birds are more likely to be seen at Point Barrow, Alaska. At this northernmost point on the Alaska coast, writes renowned scientist E.C. Pielou in *A Naturalist's Guide to the Arctic*, "Ross's gulls have been seen starting a fall migration in a northeasterly direction (!), heading for the polar pack."

Birding expert Bill Maynard, editor of *Winging It*, has witnessed thousands of Ross's gulls flying low just off Point Barrow in late September and early October. However, he has

Ross's Gull Fast Facts

Rhodostethia rosea

LENGTH & WEIGHT

Ranging between 28 and 34 centimetres, or 11 and 13.5 inches, Ross's gulls normally weigh just 170 grams or six ounces. However, juvenile birds captured in Siberia's Laptev Sea were found to weigh as little as 125 grams or less than 4.4 ounces and described as "very delicate and fragile" by Swedish ecologists who tagged them.

BREEDING & NESTING

They prefer to breed in regions of permafrost on small islets in shallow pools created by melting snow. Arriving in late May or early June, and nesting among other species, the pair constructs a nest of dry grass, sedge and moss, generally 40 or 50 metres (or nearly 150 feet) away from its neighbors. Nests are also sometimes simply scrapes in sand and gravel. Two or occasionally three dark, dull olive green eggs are laid – in Siberia, biologists have found that most nests have three eggs – and then incubated by both parents and vigorously defended with diving flights and loud, high-pitched calls. Pairs have been seen successfully defending their nest against jaegers.

Given reasonable weather, the chicks hatch in 21 or 22 days and are fed by both parents. Late frosts and/or snow can critically affect the eggs and heavy rainstorms have been known to drown small chicks.

RANGE & DISTRIBUTION

Though believed to breed in significant numbers in Siberia, as well as Norway's northern islands and Greenland, these are among the rarest breeding birds in North America. As indicated above, they are known to breed in three sites in Canada's High Arctic, and have nested in Churchill, Manitoba.

DIET

When nesting, the main diet is insects, but once the young are fledged, the family often feeds along mudflats, eating marine crustaceans, small fish and insects. Non-breeding birds and wintering birds apparently feed on small fish from ice floes. They also feed while swimming or dive from low flights or after hovering over the water.

BEHAVIOUR

Much is still to be learned about these delicate birds, but they are clearly easily discouraged from nesting by human or animal disturbance. The experience at Churchill demonstrated that they will not tolerate humans within 100 metres or 330 feet of a nest. In the High Arctic, even airboats or helicopters operated by tour companies have caused a nest to fail.

VIEWING ROSS'S GULLS

The best place to see gulls is Point Barrow on Alaska's northernmost coast. There, autumn migrants from Siberia have been witnessed going north to feed from ice floes in the nutrient loaded waters of the Beaufort Sea in late September. Later, in late October and early November, as winter begins in earnest in the far north and ice floes form farther south, the flocks return, apparently to feed in the Chukchi Sea, just north of the Bering Strait.

also seen large flocks returning several weeks later, heading southwest, "to points unknown", he writes.

This unusual migratory behavior can perhaps be explained by the bird's predilection for the polar ice of the far north as well as its circumpolar distribution. According to Canada's Species at Risk Act, there are an estimated 50,000 Ross's gulls worldwide; most breed in Siberia, with additional locations on Spitsbergen Island in Svalbard, Norway or islands off Greenland, or spend the summer feeding from drift ice.

This small dove-like gull – described by the Audubon Society as "a small elegant seabird" – is the only member of its clade, or group. Breeding adults are white, with pale gray wings and back, bright orange or red

During the breeding season, Ross's gulls are wary and easily disturbed; below, a lovely French painting showing the bird's seasonal rose coloring.

DENNIS FAST www.dennisfast.smugmug.com

legs and a black collar that encircles the neck. Non-breeding and immature gulls lack the black neck ring and instead have a black spot or smudge over the ear. During their first year, young gulls also have a black bar across their wings, which makes an 'M' during flight. Alone among all the members of the gull tribe Larini, Ross's gulls have a wedge-shaped tail. During the summer months, the birds often sport a rose-colored breast, the result of feeding on shrimp and other crustaceans in the ocean shallows, much as a wader does.

Despite centuries of study, relatively little was known about the evolution of the many species of gulls until the advent of mitochondrial DNA studies. The result has been fairly dramatic changes in classifying the world's 50 species of gulls. Where they had long been classified largely according to plumage characteristics – white-heads, masks or dark hoods, for example – mtDNA studies conducted by a trio of French geneticists and published in the *Journal of Evolutionary Biology* in 2000, showed that at least some of the classifications for the 32 species they studied were inappropriate. While all share webbed feet, short legs, long lives and an evolutionary history that goes back an estimated six million years, at least some appear to have evolved similar physical traits without sharing a recent common ancestor.

On the other hand, two of the Ross's gull's arctic neighbors – the ivory gull and Sabine's gull – demonstrated just the opposite, developing completely different plumages despite their northern breeding patterns.

Unfortunately, DNA studies were not completed on the Ross's gull, which has long been classified alone, but is nevertheless believed to be closely related to the little gull *(Larus minutus)*. A subsequent article in *Molecular Phylogenetics and Evolution* went somewhat further in reclassifying gulls, reaffirming the close relationship between Ross's and little gulls.

There are more unknowns about Ross's gulls, and particularly about the wanderers among them. In mid-November 2006, one was not only sighted, but photographed at the southern end of the Salton Sea, deep in the interior of southern California, just north of the Mexican border.

Ross's gulls have also been seen in Newfoundland and Nova Scotia in Canada, as well as in the American Midwest and New England.

COURTESY OF WWW.OISEAUX.NET

Eskimo Curlew

Numenius borealis • Critically Endangered (Canada and US)

PRIOR to 1860, the autumn skies of North America's northern and eastern coasts darkened each year with migrating Eskimo curlews. Millions of the pigeon-sized shorebirds embarked each fall on one of the world's longest known avian migrations. Travelling 14,000 kilometres or nearly 8,500 miles each way, they flew from the treeless Arctic uplands of Alaska and Canada's Northwest Territories, across the continent to the east coast and south across the Caribbean and along South America's Atlantic coast to the wet pampas of southern Brazil and Argentina.

Their great migration was a cause for wonder. John James Audubon described it this way when he visited Labrador on the north Atlantic coast in 1833: "During a thick fog," he wrote in his journal for July 29th, "the Eskimaux Curlews made their first appearance … They evidently came from the north and arrived in such dense flocks as to remind me of the Passenger Pigeons … The birds at length came, flock after flock … and they continued to arrive … for several days." The incredible bounty inspired Audubon's party to go hunting and he found them "extremely fat and juicy, especially the young birds …"

On being told about the numbers of migrating birds by local fishermen, the famed naturalist's first reaction had been disbelief. "The accounts given of these curlews," he wrote, "border on the miraculous."

The spring migration was equally remarkable, for the curlews flew west from coastal Argentina to the Pacific; following the coast north to Peru and Ecuador, they crossed Central America and the Gulf of Mexico to Texas and then moved north en masse over the centre of the continent to their Arctic breeding grounds.

But between 1870 and 1890, uncontrolled market hunting on the east coasts of both the US and Canada each fall and on the Great Plains each spring, combined with hunting during the winter on the South American pampas, devastated the huge flocks. Plump after a summer of feasting on blueberries, crowberries and insects on the tundra, or a winter of feeding on insects and grasshoppers on the pampas, these "doughbirds", as they were often called for the ease with which they were converted to cash, were much sought after. Particularly after the annihilation of the enormous flocks of passenger pigeons, they quickly fell victim to the period's unregulated commercial hunting. Small groups of hunters for the Hudson's Bay Company, for example, sometimes killed more than 2,000 birds a day to stock the store at Cartwright on the Labrador coast, and the same kind of carnage occurred in New England when flocks were forced to land to avoid storms at sea. Most were sent to market to be sold, though in the early years, between 1860 and 1875, too many birds were often killed to be collected and were left to rot.

Mottled brown on their wings and back, and buff-colored below, these members of the sandpiper family were not only tasty, but devastatingly easy prey. Their dense flocks offered undemanding targets, and they were remarkably

JOHN A. CROSBY / CANADIAN MUSEUM OF NATURE

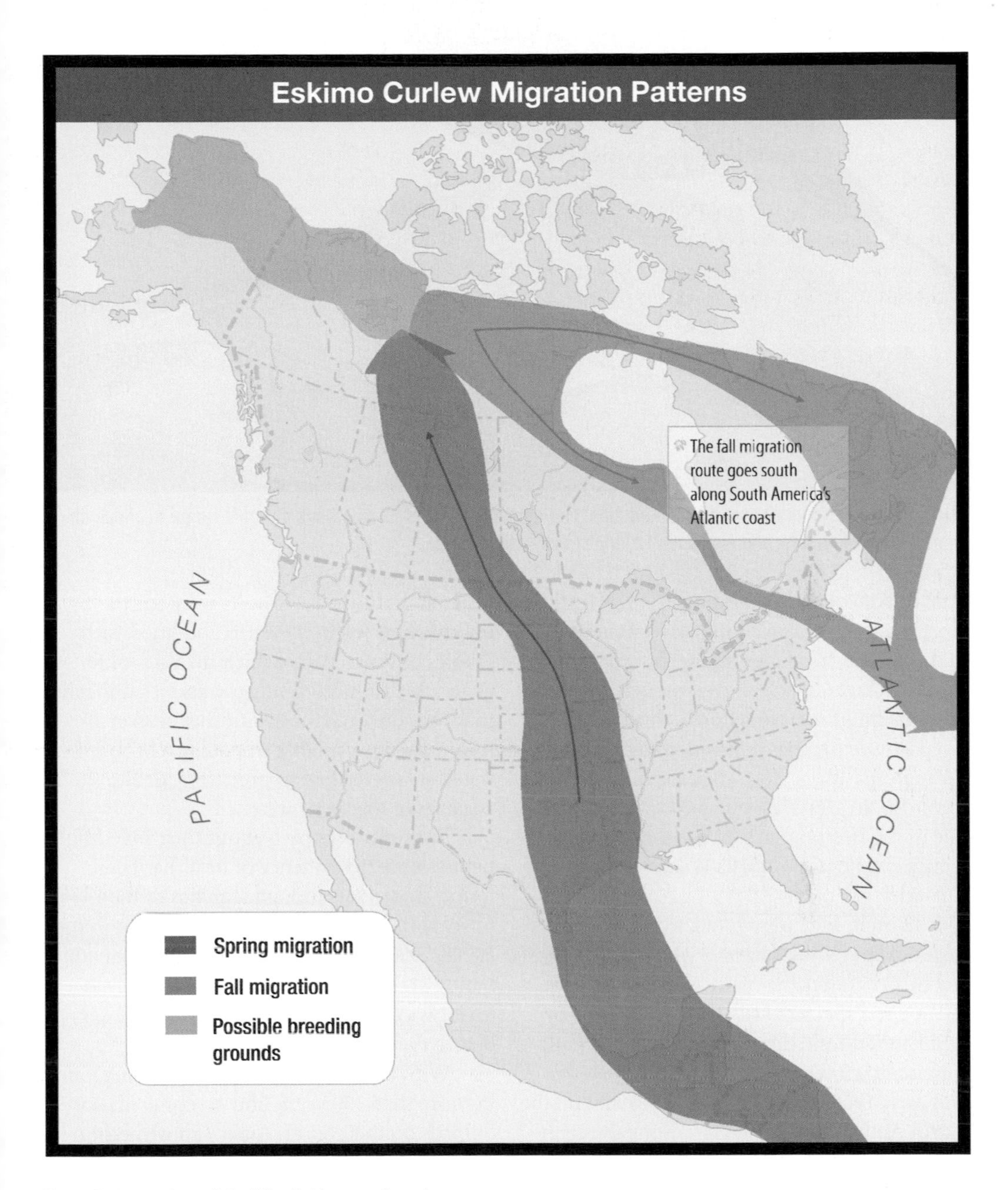

Few photographs exist of the Eskimo curlew; however, John Crosby's lovely painting, opposite, is almost photographic in its clarity.

unafraid of humans. Audubon, for example, witnessed young boys easily killing them with sticks on the Labrador coast; the birds were sold for six cents apiece.

When flying in migratory flocks, they also had a habit of circling back to within gun range when members of their flock were shot. As a result, according to COSEWIC, in the 1860s and early 1870s hundreds of thousands of Eskimo curlews were killed every year. Then, in 1875, the numbers suddenly plummeted. There was

JOHN JAMES AUDUBON / LIBRARY AND ARCHIVES CANADA / ACC. NO. 1970-188-1300 / W.SH. CLOVERDALE COLLECTION OF CANADIANA

Eskimo curlews, painted here by J.J. Audubon, had a deadly habit of turning back into rifle range to check on injured flock members, and would land to try to assist their mates.

little initial concern in Atlantic Canada or the Northeastern US, for it was simply believed that the birds had taken a different migration route. In Argentina, however, though numbers of wintering birds were down, those that arrived on the pampas continued to be killed through the 1880s. After that, the birds seemed to vanish. In 1894, just one bird was found for sale in the Boston markets, according to David Lank's excellent *Audubon's Wilderness Palette: The Birds of Canada.*

Though they were protected in North America under the Migratory Bird Convention Act of 1917, after the main killing spree any chance at a species revival was hampered by rapid agricultural development in their South American wintering habitat in the late 1800s and early 1900s, as well as at staging sites on the North American plains. As traditional staging sites disappeared, flocks were forced into the remaining areas, making them easier to kill. And the disappearance of important food sources during spring migration, such as the Rocky Mountain grasshopper, which became extinct in the early 1900s, made recovery more unlikely.

The breeding habits of the Eskimo curlew also hampered their resurgence. Like many other shorebirds, their nests were simple hollows scraped in the earth and lined with grasses or moss. Pairs, which mated for life, generally produced four eggs, and it's unlikely that the short northern summer season allowed a second chance at raising a brood if the first succumbed to predators, or the vagaries of the arctic weather.

Little else is known about their breeding habits, since no evidence of nests has been found since 1866, though searches of their historical breeding ranges were carried out in the 1970s, 1980s and 1990s. Environment Canada biologists believe that females were likely at least two, or perhaps even three years of age before they mated.

With no evidence of nests or nesting pairs in more than 150 years, and no confirmation of birds on their South American wintering grounds since 1939, it's hard to imagine that Eskimo curlews are still with us. Nevertheless, though the last specimen was obtained in the 1960s, between 1945 and 1985 there were no fewer than 80 possible sightings in North America. More recently, a group of three birds in southwestern Manitoba and a single bird in Saskatchewan, all believed to be Eskimo curlews and all seen along the birds' traditional

Eskimo Curlew Fast Facts

Numenius borealis

LENGTH
The Eskimo curlew is small – about 28 centimetres or 11 inches in length – with a slender, slightly down-curving bill and little or no eye stripe. Most of its relatives are larger; the similarly colored long-billed curlew, for instance, is 48 cm or 19 inches long, while the whimbrel, with which it is often confused, is 35 cm or 14 inches long and has bold stripes on its crown.

DESCRIPTION
Its back and wings are a mottled brown and its crown a solid brown color. Its breast and throat are buff colored, while its wing undersides are a rusty yellow to rich cinnamon. Its call is a soft, fluttering *tr-tr-tr.*

LIFESPAN
Not known.

HABITAT
According to Environment Canada, Eskimo curlews once used a variety of habitats during migration, as the map on page 67 shows. On their southward migration, they headed east across the Canadian Arctic and then south in two streams, travelling down the west side of Hudson Bay, across southern Ontario and the northeastern US to the Atlantic, or east across Labrador and Newfoundland and then down the Atlantic coast. During their northward spring migration in April and May, curlews were found in tallgrass and eastern mixed grass prairies, often near water or in areas disturbed by recent burns (or, in areas farther west, by bison, a clear indication of how long it's been since large numbers of curlews were seen during migration).

MATING & BREEDING
Nests were built on the upland tundra near the Arctic coast, and in grassy meadows farther inland. Created in mid- to late June, they were shallow scrapes in the earth, lined with grasses and moss, in which a clutch of four eggs was usually laid. The eggs hatched between early and mid-July and the down-covered young were able to move about almost immediately.

DISTRIBUTION
Experts believe the total world population is not more than 100 birds, if indeed Eskimo curlews still exist.

NAMES
In the mid-19th century, they were known widely as "doughbirds", for the ease that hunters could convert them to cash. In South America, where they once wintered, they were known as *Chorlito Esquimal, Chorlo polar, Zarapito boreal, Zarapito Esquimal* and *Zarapito polar.*

spring migration route, were reported in 1996. In 2006, a Wisconsin ornithologist spotted what he believed was an Eskimo curlew at Peggy's Cove, Nova Scotia. So, though many experts believe they might easily be confused with other shorebirds, including whimbrels, or other curlew species, hope lives on.

As a result, the most recent assessment, in 2007 under Environment Canada's Species at Risk Act (or SARA), continues to list this elusive (and possibly extinct) bird as "Endangered", while the 2008 Red List from the International Union for the Conservation of Nature lists it as "Critically Endangered".

In the faint hope that a last, tiny population still exists, the IUCN has suggested that historical breeding and wintering sites continue to be assessed; that the heath tundra along the Labrador coast be investigated on a regular basis during August and September as well as the beaches of Galveston Island during March; that credible sightings be carefully investigated; that mixed grass and tall grass habitat be encouraged on the Great Plains and regular burnings of these grasslands employed (for more information, see page 269).

Rock, water and coniferous forest are the primary components of the boreal forest wherever it appears, whether in eastern Manitoba, as above, or in Siberia.

PETER ST. JOHN

The Boreal Forest

Perched atop a bare branch, this great gray owl (one of the largest in the owl family) appears quite large with its dense, fluffy plumage. In fact, though it has a has a body length of about 72 centimetres or 28 inches, it weighs, on average, just 1300 grams, or 45.5 ounces.

DENNIS FAST www.dennisfast.smugmug.com

The Boreal Forest

Named for Boreas, the Greek god of the north wind, the boreal forest is draped, as *The Atlas of Canada* put it, "like a green scarf across the shoulders of North America."

The continent's northern forest is part of a global ring of fecundity that is under threat almost everywhere. Lying south of the treeless tundra of the far north, this vast swath of conifers and deciduous trees, runs from Alaska to Newfoundland. Often 1,000 kilometres (or 600 miles) wide, it is estimated to include 1.9 billion trees. On the other side of the globe, it once swept across Scandinavia (where just five per cent remains) and, known as the taiga, occupies much of Russia. In that huge nation it is increasingly fragmented, often by timber smugglers who send the wood to China, to be processed into goods that are shipped to the United States, the world's largest consumer of finished wood items, as well as – as indicated below – unfinished timber and paper products.

Today, about one-third of the world's land mass is forested. While that may seem vast, it is about half of what was there at the beginning of the Industrial Revolution. And the boreal region, punctuated by bogs, fens and marshes, and laced by magnificent rivers, contains one-quarter of the world's remaining intact forests. Covering an area more than 12 times the size of California, it easily rivals the world's rainforests. And like them, it is a rich pool of biodiversity, the world's largest storehouse of fresh water and the lungs of Planet Earth. According to the Boreal Forest Conservation Framework, a group of organizations and industries interested in boreal forest research and land use planning, locked in its "rich peat lands, mosses, soils and trees, the Boreal [forest] stores more carbon than any other terrestrial ecosystem, helping to regulate the Earth's climate." (As an indication of the growing awareness of the importance of the boreal forest ecosystem, the members of the framework not only include environmental groups such as the Canadian Parks and Wilderness Society, Ducks Unlimited Canada and World Wildlife Fund Canada, but also major forestry and energy companies such as Alberta-Pacific Forest Industries Inc., Domtar Inc. and Tembec Inc.)

Populated with an array of hardy plants from white spruce, which tower above the uplands, to lichens (including reindeer moss, an important winter food for caribou) that carpet the forest floor, it provides crucial habitat for large carnivores, including grizzly bears (see page 205), as well as the world's largest herds of woodland caribou (see page 88). And its mosaic of trees, rivers, lakes and wetlands creates nesting and breeding grounds for more than 300 species of birds; for many, including the beautiful, endangered whooping crane (see page 92), it is their only wild nesting place.

Though differences in climate and topography certainly create differences in vegetation across this green halo, as lead author Charles Krebs writes in *Ecosystem Dynamics of the Boreal Forest: The Kluane Project*, this is an eco-system that is "remarkably uniform in overall appearance, dominated by coniferous trees, and an Alaskan feels quite at home in northern Manitoba or eastern Quebec."

As immense as it may seem, North America's boreal forest – as well as forests everywhere around the world – is being constantly pared away by human development. In northern Alberta, for instance, are tar sands that hold great reserves of oil, a commodity for which demand never seems to cease. To get to that oil, the surface must be mined, and to get to the surface, the forest must be cleared. Complicating that destruction of a major storehouse of terrestrial carbon, as climatologist James Hansen wrote in the *Ottawa Citizen* in February 2009, "… producing oil from tar sands emits two to three times the global warming pollution of conventional oil." Alberta's tar sands range over 140,000 square kilometres (54,000 square miles) and current Alberta legislation does not require oil companies to restore the land to its original condition. Potentially, a portion of the boreal forest the size of the state of Florida may be lost to

The Athabasca River near Fort McMurray: Canada's tar sands were once locked beneath northern Alberta's forests, above; today, much of the once-pristine landscape seems an industrial wasteland.

IMAGES COURTESY OF DAVID DODGE / THE PEMBINA INSTITUTE / www.oilsandswatch.org

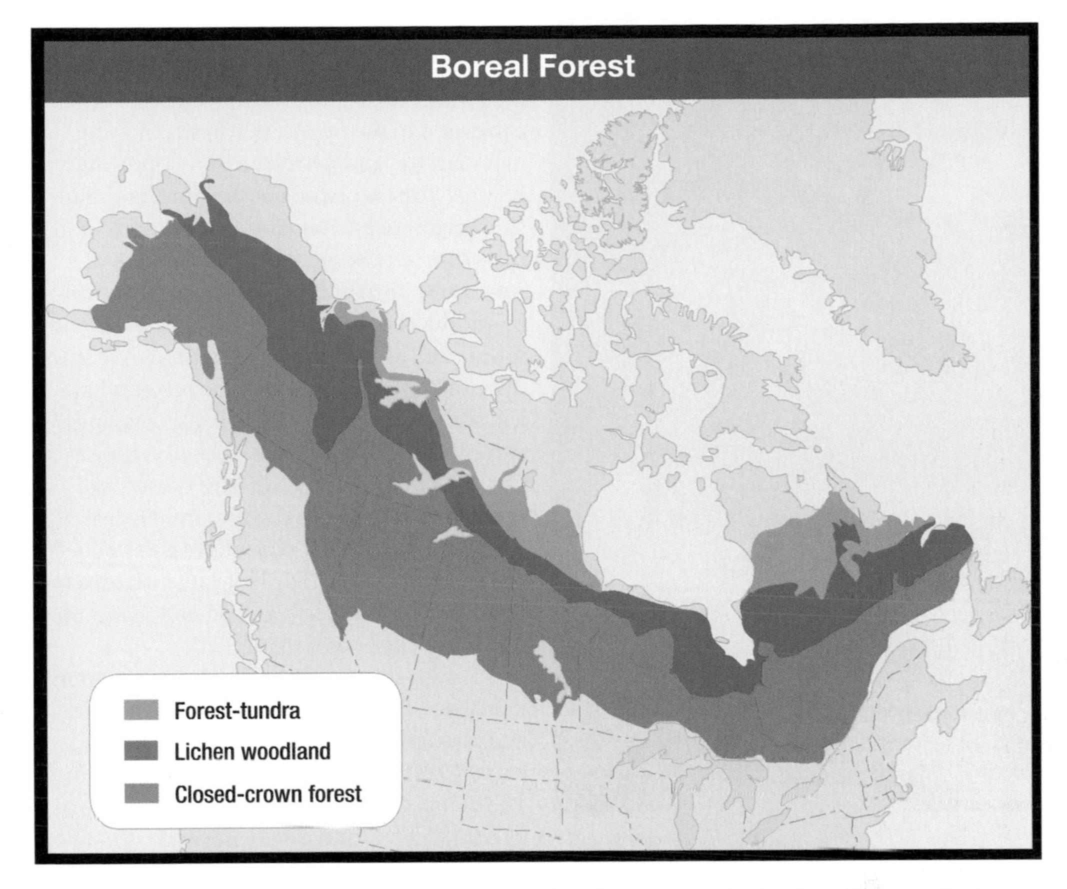

industrial production, which in the process creates excessive climate pollution.

But that is just one example. The Nature Conservancy estimates the rate of destruction of boreal forest at one per cent per year, similar to that of tropical rainforests. And globally, since 1990, the planet has lost a forested area nearly "twice the size of France", according to an October 2008 article in *The New Yorker*.

With just 12 per cent of North America's boreal forest currently protected, many believe this accelerating deforestation is unlikely to change before it's too late. In 2001, Klaus Topper, executive director of the United Nations Environment Program, wrote, "Short of a miraculous transformation in the attitude of people and governments, the Earth's remaining closed canopy of forests and their associated biodiversity are destined to disappear in the coming decades."

Topper's attitudinal transformation would take major shifts in our lifestyles and our values, but there is a growing clamor for change, not only in Canada, but globally. In 2004, the World Conservation Congress called on Canada to more stringently protect its boreal forests. Three years later, that call was echoed by 1,500 highly respected scientists from more than 50 countries, and in 2009 – following a poll that found that 81 per cent of Canadians were concerned about wilderness forests and 74 per cent said wilderness areas should be protected even when it meant job and investment losses – environmental and birding groups delivered 60,000 signatures to Canada's Parliament. The signatories

Laced with rivers and streams, the boreal forests contain more than 75 per cent of the world's unfrozen fresh water.

demanded that fully half of the remaining forest be protected, and that industries be required to abide by sustainable development practices in the rest.

Though biologists believe North America's northern forests may have taken thousands of years to reach their historic extent, they have nevertheless supported hundreds of cultures for millennia. Lakes and rivers provided natural transportation routes and trees created the materials for sleek canoes; bountiful fisheries, and plants and animals provided food and clothing, in short, the forests offered all the necessities of life, as well as abundant goods to trade. About 80 per cent of Canada's First Nations continue to live in this region, and their cultures and lifestyles have been closely tied to the environment for thousands of years.

At its southern edge, in a region that biologists call the mixed or transitional forest, some land is being cleared for agriculture, but the impact of farming is relatively small when compared to the impact of forestry, mining, hydroelectric and petroleum developments. Seismic lines, access roads, pipelines and mine sites fragment habitat and pose threats to animals that require large, unbroken areas of wilderness or migration corridors. Chemical by-products of industry can poison water and soil, as well as the animals and the people that depend on them. Hydroelectric projects have led to mercury contamination; mine tailings and smelter emissions have caused acidification, and logging activities have caused sedimentation and stream disruptions. Hydroelectric dams and reservoirs have also flooded vast tracts of land, destroying large sections of forest and affecting fish and other aquatic animals whose habitat is disrupted.

Forestry is one of Canada's most important industries, employing nearly 400,000 people, and about half of the nation's annual wood harvest comes from the boreal forest region. The global demand for forest products – two-by-fours, newsprint, catalogues and even toilet paper are still being made from virgin forests – has exploded in the last century, and technology has made harvesting trees much more efficient. And though the recession that began in the fall of 2008 slowed consumption somewhat, the United States is nevertheless a huge consumer of boreal forest products. Just as the bulk of China's finished goods are bought by Americans, a staggering 80 per cent of all Canadian forest products find their way to the US.

Many Canadian communities are dependent on the northern forestry industry for their livelihood, either directly or indirectly. Fire suppression, insect control, clear-cutting and single-species tree farming are widespread practices that may increase profits temporarily, but in the long term threaten to reduce both plant and animal diversity, making the forest more vulnerable to fire, disease or pests.

Though climate change is a natural part of the Earth's evolution, and both global warming and global cooling have occurred repeatedly over the past 10,000 years, the difference today is the enormous and still growing population our world is expected to carry. Today, even a slight increase in global temperatures – just 2.5°C or 4.5 °F above pre-industrial levels – could make disturbances from fire and insects more frequent and severe, releasing vast quantities of carbon currently stored in forest ecosystems, and accelerating, rather than slowing, the effects of climate change. The loss of forest ecosystems would also damage the quantity and quality of available water, the most crucial necessity of life for all species.

As a small compensation for these damaging effects, global warming would also allow the boreal forest to gradually spread northward, as it has in the past. A report issued at a United Nations forestry forum in April 2009 indicated that this silver lining would likely be experienced by northern regions in Canada and Alaska, as well as Sweden, Finland and Russia. Perhaps as early as 2070, during the lifetime of today's children, spruce forests could invade the northern shores of Hudson Bay, much of Quebec's Ungava Peninsula and even Baffin Island.

If what is now arctic tundra warms, creatures particularly sensitive to climate, such as fish requiring very cold water, could decline or disappear. A changing climate will also likely lead to a decrease in biodiversity, as extremely specialized species and those that do not move or adapt easily are replaced by those that are more aggressive.

The impact of climate change, whether naturally occurring or caused by human activity, is already being felt in Canadian forests. The mountain pine beetle has infested enormous tracts of forests in British Columbia, devastating swaths of lodgepole pine and leaving vast tracts of dead and dying rust-red forests in its wake. Allan Carroll, a research scientist at the Pacific Forestry Centre in

DENNIS FAST www.dennisfast.smugmug.com

Whooping cranes are among the most recognized of the many species that depend on the boreal forest.

MAIN IMAGE: MARK A. WILSON, http://commons.wikimedia.org; INSET: US DEPARTMENT OF AGRICULTURE

Easy to spot as the dying trees turn rusty red, great swaths of BC's pine forests, such as this tract south of Field, have been devastated by the mountain pine beetle, shown enlarged below.

Victoria, links the spread of the beetle to warmer winters; average winter temperatures have risen by more than four degrees in the last century. Many scientists are now worried that the beetle, which in the past was killed by winter cold in the mountains, may be able to survive today's comparatively mild temperatures; if so, it is likely to cross the Rocky Mountains and begin to infest the northern forests. In Alberta, forestry officials are already setting fires and felling thousands of trees, hoping that these actions will keep the beetle at bay.

Ninety-four per cent of Canada's northern forests are designated Crown land, most of it owned by the various provincial governments, and management of these forests is the responsibility of the provinces, which negotiate agreements with industry. Prodded by the public, recent court rulings that recognize constitutionally protected Aboriginal and treaty rights, including the right to protected forest ecosystems, and declining numbers of many species, provincial governments are finally taking steps to preserve the boreal forest. After nearly two decades of public protests, for example, in 2008 the Manitoba government announced it would no longer allow logging in provincial parks.

Thoughout the ecosystem, governments are being forced to encourage sound logging and reforestation practices, monitor harvests, and protect significant tracts of forest in national and provincial parks, ecological reserves, wildlife sanctuaries and conservation areas. Timber harvesting in areas vulnerable to erosion and those alongside roads, lakes and waterways is also increasingly prevented (though in BC, at least, harvesting the devastated pine forests has provided at least short-term compensation for logging companies).

Canada's boreal forest accounts for more than half of Canada's landmass and represents a quarter of the remaining large intact forests in the world. Thirty per cent of the region is covered by wetlands – marshes, fens, bogs, lakes, and river systems – which fill a number of vital ecological roles. They filter vast quantities of water each day, store carbon, help rebuild soils and restore nutrients, and mitigate flooding and drought. The forests themselves also store vast amounts of carbon and produce oxygen, playing a vital role in maintaining the Earth's climate.

Winters in the northern forests are long, cold and dry, while summers are short, warm and moist. The plant life that makes up these forests reflects these difficult growing conditions. Most of the approximately 20 tree species are conifers such as spruce, fir, pine and tamarack. The thick, waxy coating on their evergreen needles protects them from harsh winter winds, and their conical shape reduces the build up of snow so branches are less likely to break under the weight.

The make-up of the forests changes from south to north as temperatures drop. The southern portion sits on the ancient bedrock of the Precambrian Shield, forming an ecozone known as the Boreal Shield. It touches parts of six provinces, covers 1.8 million square kilometres and makes up almost 20 per cent of Canada's land mass. The rock below the forests was formed more than a billion years ago, during the Precambrian era. Many of the valuable minerals found in the Shield were created during the late Precambrian, as the rock, subjected to extreme heat and pressure from colliding continents, warped.

Much more recently, during the Wisconsin glaciation between 22,000 and 10,000 years ago, huge sheets of ice repeatedly advanced and retreated across these rocks, scraping and gouging them, blanketing them with gravel and sand, and creating the depressions that have filled to become lakes, ponds and wetlands. In places where the glacial scouring was particularly intense, the landscape is dotted with rocky outcroppings that support a wide array of lichens and shrubs. As the ice melted, soil slowly formed and forest cover gradually regenerated over the land, but it was not until about 5,000 years ago that the boreal forest took on its present character. In this southern band of forest, the trees grow taller and closer together, forming closed canopies that shelter abundant mosses, herbs and shrubs. Broadleaf deciduous trees such as trembling aspen, balsam poplar and birch join white and black spruce, fir and pine.

Farther north, as average temperatures drop, the forest canopy opens, the number of deciduous trees, which are confined mainly to the waterways, declines and other species are gradually replaced by stands of spruce and jack pine. Larch grows in low marshy areas and black spruce can be found in and around peat bogs. The northern edge of the boreal forest, where it meets the frozen, treeless arctic, is known as the taiga, after its Russian equivalent.

Rock soars from a northern lake in this dramatic boreal forest cliffscape.

JERRY KAUTZ

Here, snow and ice can be found for eight months of the year. Shrubs, such as dwarf birch, Labrador tea and willow, become more common and trees are smaller and slower growing than to the south. Yellow, green and gray lichens cover the ground and hang from trees, forming an important part of the woodland caribou's winter diet. These woodlands, wetlands and meadows mingle with, and eventually give way to, the open arctic tundra.

Unlike British Columbia's coastal and inland temperate rainforests, or the coastal redwoods of California, where trees more than 500 years old can often be found, conifers in the climax boreal forest are rarely older than 250 or 300 years of age, though spruce clones, usually found at the far northern margin of the boreal forest, can be much older (see sidebar

DENNIS FAST www.dennisfast.smugmug.com

on page 82). Regeneration is usually sparked by fire; during hot, dry summers, fires are common and often burn over large areas. Burned areas are pioneered by mosses, the aptly named fireweed, and shrubs, and quickly recolonized by aspen and birch, which can reproduce from stump sprouts, as well as from seeds. White spruce seedlings thrive under a canopy of birch and poplar, until they eventually tower over the aspen, creating a shady environment the latter finds intolerable. As spruce mature at between 80 and 120 years of age, a thick carpet of needles develops below the conifers. The acid in the needles allows few other plants to grow and a stand of spruce can persist for hundreds of years. Individual trees are rarely more than 300 years of age (though a stand of climax spruce surveyed near Inuvik in the Mackenzie Delta just south of the Arctic Ocean included one tree 400 years of age and another more than 300). Mature height is about 30 metres or over 100 feet tall.

The northern forests support many different species of birds, mammals, insects and fish, and some hardy varieties of reptiles and amphibians. Because large areas of forest remain largely undisturbed, animals in this environment have fared much better than those of more developed regions, and only a few are considered endangered. These include the woodland caribou, wood bison and whooping cranes. The forests are home to more than 300 species of birds each year; nearly half the birds in North America can be found here at some point. Many small songbirds and waterfowl migrate north each year to breed, while other species, including woodpeckers, finches, nuthatches, chickadees, owls, grouse and ravens, live in the northern forests throughout the year. Many mammals, both large and small, call Canada's forests home, and some have become emblematic of the nation.

Wolves, grizzly and black bears roam the forest, as do elk, moose and Canada lynx. Trade in beaver pelts was largely responsible for opening the country to Europeans, and other smaller mammals such as snowshoe hare, lemmings and voles are abundant and have adapted to fill differing niches in their woodland environment. About 130 species of fish are found in the cold, clear waters of Canada's northern forests, from small minnows and deep-dwelling stickleback to popular game species such as walleye, northern pike, lake trout and yellow perch.

In all, about two thirds of Canada's species of plants, animals and micro-organisms live in the northern forests. Preserving this habitat is critical to supporting biodiversity in Canada and the world.

Opposite: Feathered from head to toe, a snowy owl swoops over a wintery field. Its coloring, superb eyesight and soundless flight allow it to spot its prey from afar and take it unawares.

DENNIS FAST www.dennisfast.smugmug.com

Spruce Clones

IN the northern boreal forest, and particularly in the region known as the forest-tundra or taiga, black spruce are by far the most common trees. Like their white spruce cousins, most trees do not live longer than 200 years. Remarkably however, in the far north, where frigid winter temperatures and harsh winds last for months, black spruce far older than their taller southern relatives have been found. In Churchill, on the west coast of Hudson Bay, clumps of low, spreading spruce dating back to the Medieval Warm Period between 850 and 1350 A.D. have been found. And in northern Quebec, a similar rambling clump of stunted black spruce was found to be 1,750 years old.

These spruce achieve their great longevity through layering. Prevented from significant vertical growth by the fierce winters, they instead sprawl along the ground; staying below the average depth of the winter snow, they renew themselves by sending roots down from their sprawling branches. The result is vegetative propagation, which essentially creates clones of the original tree, or mother unit, as researchers term it. The result appears to be a clump of stunted black spruce, but in fact is a genetically identical, generational iteration of the same plant, known as a krummholz tree island.

Biologists were drawn to the krummholz islands on Hudson Bay because in many places they were found growing where black spruce are no longer able to thrive. After studying them, researchers discovered that each of these islands of trees were in fact hundreds of years old.

Very few have living centres that stretch back almost 2,000 years, but a significant number, Marie-José Lamberge, Serge Payette and Nadia Pitre discovered in their study in northern Quebec, were 1,000 or more years old. In short, they began life during another period of significant global warming, an era when the green pastures of southern Greenland drew Norse settlers, a time when grapes grew in Scotland and a period when the northern tree line marched hundreds of kilometres north.

DENNIS FAST www.dennisfast.smugmug.com

Wood Bison

Bison bison athabascae • Endangered (US); Threatened (Canada)

DARKER and heavier than their plains bison cousins, and never as numerous, the wood bison of North America's boreal forest shared the fate of their southern relations following the arrival of Europeans. As fur traders, adventurers and miners moved into the northern forests, wood bison numbers dropped precipitously. Though pre-contact numbers are not known, it's been estimated that in 1860 there were perhaps 168,000 wood bison in what are now Alaska, Yukon and Northwest Territories, as well as northern Alberta and British Columbia and northwestern Saskatchewan.

Over the next 40 years, both wood bison and plains bison were hunted almost to extinction. Efforts to stop the slaughter began in 1877, with an attempt to prohibit hunting; this, however, was soon repealed. Only in 1893, when wood bison numbers dropped below 300 was the Buffalo Protection Act enacted. Still the poaching continued, and in 1897, the North West Mounted Police were given authority to enforce the act. Nevertheless, by 1900, these magnificent creatures – the largest land mammals in North America – were gone from the wetlands and forests along Alaska's mighty Yukon River and from the boreal forests of Alberta and Saskatchewan. In 1906, the last wood bison in British Columbia was shot near Fort St. John.

At that time, less than 250 animals remained in a last wild herd south of Great Slave Lake in the Northwest Territories. To protect them and to enforce the hunting ban, buffalo rangers were appointed in 1911; by 1922, when Wood Buffalo Park was carved out of northeastern Alberta and the southern Northwest Territories, the wood bison population had recovered to about 1,500 animals.

However, just as its revival seemed assured, the herd was hit by another blow. In the mid-1920s, plains bison – which were also enjoying a recovery – were shipped from parks in southern Alberta north to Wood Buffalo. Unfortunately, the introduced bison carried with them bovine tuberculosis and brucellosis, which they had contracted from domestic cattle. These diseases, along with interbreeding with the plains bison, meant that by 1940 wood bison were believed to be extinct.

Then, in 1959, an isolated herd of some 200 healthy wood bison was discovered in a remote area of the park. Given a second chance to preserve the subspecies, Canada moved quickly, capturing 42 animals; 24 were released in a large, fenced enclosure in Elk Island National Park, east of Edmonton, Alberta, and the other 18 into the Mackenzie Bison Sanctuary near Fort Providence northwest of Great Slave Lake. Over the next two decades, the Canadian government also relocated animals to the Nahanni National Park Reserve, near the Yukon-NWT border. All these sites are in the wood bison's historic range.

These tiny herds were the beginning of the long road back to a sustainable wood bison population. But if anything, the path to recognition as a separate subspecies was even longer.

With their heavy winter coats, wood bison are perfectly at home in the snow.

DENNIS FAST www.dennisfast.smugmug.com

DENNIS FAST www.dennisfast.smugmug.com

Wood bison were first identified as a distinct subspecies by biologist Samuel Rhoads in 1897. Rhoads decided that their larger size, darker coat and more sharply defined hump merited subspecies designation. Most scientists agreed, though some debated the classification vociferously. In 2001, wood bison, which are non-migratory, were determined to be "sufficiently distinctive to warrant conservation as an entity separate from plains bison", as authors George Feldhamer, Bruce Thompson and Joseph Chapman put it in *Wild Mammals of North America.*

In 2008, a study based on mitacondrial DNA that was published in the *Canadian Journal of Earth Sciences* determined that "all North American and Siberian bison shared a common maternal ancestor some 160,000 years ago". This agrees with the theory that ancestral bison, *Bison priscus,* entered North America across Beringia, the Bering Straits land bridge, between 160,000 and 120,000 years ago, during the Illinoian glaciation. After the ice melted, descendants of this migration moved south and gave rise to a giant bison, *Bison latifrons.* This remarkable species stood 2.14 metres or more than seven feet at the shoulder and had horns that spanned the same distance.

For tens of thousands of years, these magnificent bison – the largest North American species found – survived in an environment that included huge and vicious predators such as saber-tooth cats and North American cheetahs. Living in small groups and preferring open woodland environments, *Bison latifrons* gradually increased in number during the long Sangamonian interglacial period. At the height of their distribution, they could be found from California to Florida and from North Dakota, Wisconsin and New England south into Mexico. Then, between 30,000 and 20,000 years ago, they began to disappear. For many years, it was believed they were overtaken by a smaller (though still large by comparison with modern bison) species with smaller horns, which biologists know as *Bison antiquus,* the ancient bison. Now, however, scientists believe *B. antiquus,* fossils of which date to nearly 30,000 BP at California's La Brea Tar Pits, in fact evolved from *B. latifrons*, becoming smaller, more versatile creatures that were well adapted to temperate climates. Over the next 15,000 years, these bison would change again, and yet again, growing larger in the immediate aftermath of the glaciation, as the melting ice sheets opened new territory, and then diminishing in size as human hunters were added into the equation.

However, as the Wisconsin glaciation began, as the massive ice sheets grew and sea levels worldwide dropped, a new stock of *Bison priscus* was moving from Siberia through Beringia and eventually into the glacial refuges in northwestern Alaska. These were northern bison, with thick coats and a taste for tundra plants, animals seemingly impervious to the long, cold winters and brief summers. Ultimately, scientists now believe, over the 15,000 or 20,000 years of the Wisconsin glaciation, these two very differently adapted bison – *Bison antiquus* and the second influx of *Bison*

Opposite: Despite its docile expression, wood bison bulls are astoundingly strong and determined.

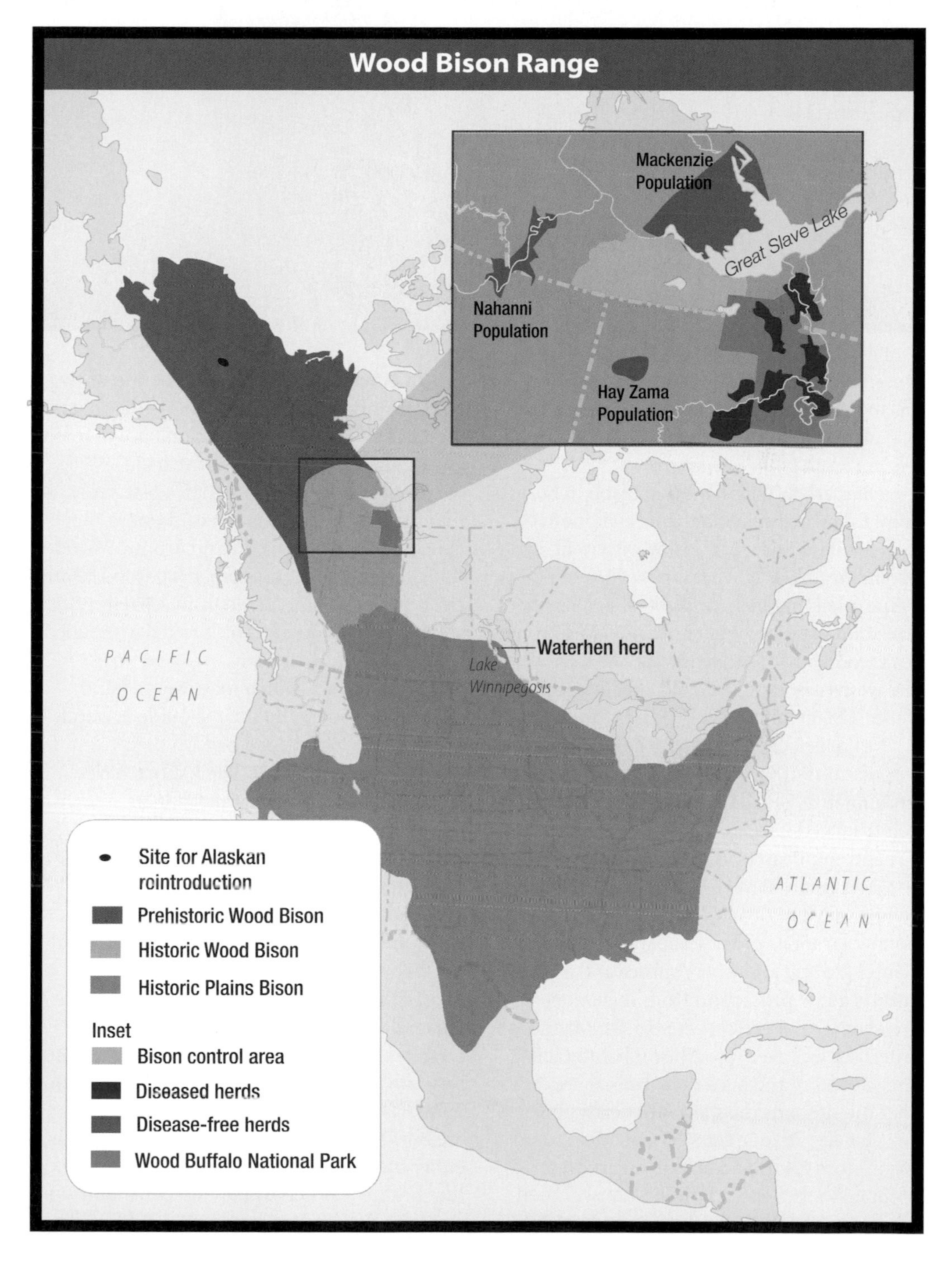

Dressed for all seasons, bison shed their winter coats for much lighter summer garb.

DENNIS FAST www.dennisfast.smugmug.com

priscus – respectively developed into the two bison subspecies that still live in North America: plains bison (*Bison bison bison*) and wood bison (*Bison bison athabascae*).

Remarkably, it seems that bison in Eurasia were following much the same evolutionary pattern. The European bison, or wisent, evolved in the forests of what is now Eastern Europe, while fossils in Siberia indicate that a northern variation of bison, very similar to North America's wood bison, evolved from *Bison priscus* on the western end of Beringia, only to disappear about 1,000 years ago.

In order to protect the relatively small populations of wood bison from disease, beginning in 1971, additional small herds were reintroduced into northeastern British Columbia and southeastern Yukon, both once home to wood bison herds, as well as the Interlake region of Manitoba, where wet meadows and willow savannas provide superb summer and winter grazing and where spruce and aspen forests allow protection from inclement weather. It's not certain whether the Manitoba Interlake was ever wood bison habitat, but the success of the herd there makes it clear that the site suits the species to perfection.

Finally, in June 2008, 53 Canadian wood bison from Elk Island National Park were shipped to the Alaska Wildlife Conservation Center just south of Anchorage, in Portage, Alaska. After an acclimatization period of two years, the bison will be released into the wild in the Alaskan interior, where bison have roamed for hundreds of thousands of years.

Today, there are more than 4,000 free-roaming wood bison, most of them in the Northwest Territories. In and around Wood Buffalo National Park, another 500 wood bison can be found. This is down from 2,500 in 1970, and all are considered to be infected with tuberculosis and/or brucellosis.

Dark brown, with a massive head and a large hump between their shoulders, wood bison have long, shaggy hair on their front quarters and front legs. Their short, black horns curve inwards on males, but are straight on females. Unlike most hoofed animals, which use their front feet to paw through snow when foraging, wood bison sweep their heads and necks from side to side to clear snow. In short, they are extremely well adapted to cold northern climates.

Listed as endangered until 1988, they are currently classified as threatened in Canada. Despite this, Canada's goal of establishing four separate, healthy herds, each with at least 400 animals has been reached and several herds are now being culled or considered for culling, either by Aboriginal populations or to ensure that healthy wood bison do not come into contact with the diseased herds of Wood Buffalo.

Wood Bison Fast Facts

Bison bison athabascae

HEIGHT & WEIGHT

Adult males measure between three and four metres or 10 and 13 feet in length, and are between 1.5 and 1.8 metres or five and six feet at the shoulder. They weigh between 350 and 1000 kilograms or between 800 and 2,000 pounds.

HABITAT & DIET

These magnificent animals prefer open boreal and aspen forest with large, wet meadows and slight depressions caused by ancient lakes. They eat primarily sedges and grasses, but will also eat leaves, usually of willow trees, and shrubs. They will also feed on lichens in the fall, but do not compete in a significant way with other grazing mammals.

BEHAVIOUR

Unlike their plains cousins, which historically roamed great distances seeking food as the seasons changed, wood bison move in more localized patterns, between open meadows and the surrounding forest. For most of the year, the main herds, comprised of adult females, subadults and calves, are separate from the mature bulls. Mature cows and bulls mix only during rutting season.

LIFESPAN

Wood bison can live to be 40.

MATING & BREEDING

The rut, or mating season, takes place each year in August and early September. Females begin breeding at two to four years of age and usually give birth to a single calf the following May, a gestation period of 270 to 300 days. Males usually do not mate until they are six years or older, for they must compete with the larger bulls for females.

THREATS

Wolves and bears (and of course humans) have historically been the main predators of wood bison and continue to prey on calves, as well as adults nearing the end of their lives. Free-ranging herds in some areas, particularly northern British Columbia, are also prone to collisions with motor vehicles as they cross highways. Herds in the southern Northwest Territories and the Hay-Zama region of northern Alberta are also threatened by contamination by diseased animals in Wood Buffalo National Park. To minimize this, Wood Buffalo is surrounded by a Bison Control Area; animals are shot if they enter the zone. Wood bison in northern BC are also threatened by interbreeding with plains bison, particularly in the Pink Mountain area south and west of the Hay-Zama and Nordquist herds.

VIEWING WOOD BISON

One of the most accessible places from which to view wood bison is along Highway 3 in the Northwest Territories, known as the Frontier Trail. This paved highway passes through the Mackenzie Bison Sanctuary, which is on the northwest side of Great Slave Lake. Travellers on the Frontier Trail from the Mackenzie River Ferry Crossing to Yellowknife, may catch sight of wood bison grazing by the roadside, or even crossing the highway.

Wood bison can also be seen along the Alaska Highway in northern British Columbia, approximately 80 kilometres or 50 miles north of Nordquist Flats. Remember to keep your distance, however – they can be dangerous if disturbed.

Members of a captive herd can be seen in the southern section of Elk Island National Park, just east of Edmonton. (To the north is a large herd of plains bison.) The two, very separate herds are bisected by the Yellowhead Highway (No. 16) and animals from one or another can often be seen from the road.

Both captive and free-ranging herds can also be found in Manitoba's Interlake, thanks to the efforts of the Skownan First Nation on the Waterhen River between Waterhen Lake and Lake Manitoba. The captive herd can be viewed with permission from the First Nation; the wild herd, which is centred around Chitek Lake in the Central Interlake, is deliberately located far from the region's only highway. The Chitek Lake region is also home to threatened woodland caribou, as well as elk, moose and white-tailed deer.

Woodland Caribou

Rangifer tarandus caribou • Endangered (US); Endangered & Threatened (Canada)

TALL and majestic, with its gray-brown coat, blaze of cream from throat to chest and enormous rack of dark amber antlers, a woodland caribou bull is one of nature's most magnificent autumnal sights. The largest and darkest of North America's three remaining caribou subspecies (see caribou introduction on page 40), woodland caribou are almost exclusively adapted to life in North America's old-growth boreal forests.

Because of their elusive nature, little is known about the historic abundance of caribou, but they were once found throughout North America's boreal forests, across Canada from Yukon to the Maritimes and Newfoundland and across the northern states from New England to Washington State.

The last 150 years have changed all that. Caribou were extirpated from Prince Edward Island by 1873 and from New Brunswick and Nova Scotia by the 1920s. In Maine, which has a town named Caribou, a few stragglers from the endangered Atlantic-Gaspé herd are rarely seen; in Wisconsin some consideration was given to reintroducing woodland caribou in the late 1980s, though elk were chosen instead; in Minnesota, caribou tracks (with hoof prints the size of salad plates) are occasionally found along the state's border with Canada, while in Montana, caribou – as well as grizzlies – are

Like his celebrated father, John Woodhouse Audubon painted from nature, and did remarkable work in an era before photography allowed intimate wildlife portraits.

JOHN WOODHOUSE AUDUBON / CARIBOU OR AMERICAN REINDEER, 1847 / LIBRARY AND ARCHIVES CANADA / PETER WINKWORTH COLLECTION OF CANADIANA

As they have in other places, Manitoba's woodland caribou, includng this herd in Atikaki Provincial Park east of Lake Winnipeg, have dwindled in recent years.

JERRY KAUTZ

sometimes spotted in the Yaak Valley in the Purcell Mountains in the state's northwestern-most corner.

However, in recent years, the only woodland caribou remaining in the lower 48 are the mountain caribou population (see page 200) that extend into Idaho and Washington. With fewer than 50 animals remaining in the South Selkirk herd and just 20 in the South Purcell herd, caribou are considered to be among the most endangered large mammals in the lower 48 states.

Woodland caribou are divided into five main populations: the northern mountain population in Yukon, Northwest Territories and northwestern British Columbia; the southern mountain population in British Columbia, Washington State and Idaho; the boreal population in northern forests across Canada; the Newfoundland population, and the small Atlantic-Gaspésie population in Quebec.

While each is intimately associated with North America's boreal forests, each is also adapted to its specific habitat and is unique in its habits. As indicated on page 200, some, such as the southern mountain caribou of the Selkirk Mountains, move up and down the alpine slopes to take advantage of food sources or to avoid areas of deep snow, while others, particularly the large herds of northern Quebec and Labrador, wander long distances between forest and tundra, and the woodland caribou of eastern Manitoba and north-central Saskatchewan move very little at all. As a result, each population is now treated individually; the relict herd on the Gaspé Peninsula south and east of the St. Lawrence River, for example, is listed as endangered; the southern mountain is classified as endangered in the US and threatened in Canada and the boreal population, which has declined by more than 50 per cent in the past half-century over much of its range, is classed as threatened.

The reasons for these declines virtually everywhere have largely to do with North America's diminishing old-growth forests, combined with the fragmentation of what remains by roads and blocks that have been logged. With the exception of the large, mobile eastern boreal populations, woodland caribou live in small groups and prefer extensive, undisturbed tracts of old-growth forest interspersed with marshes and bogs. Dense tree cover provides a degree of protection from predators such as wolves and bears, and mature forests contain large quantities of lichens, which grow on the ground and on mature trees and are crucial to caribou survival. With

Lichen come in dozens of colors and species, but all grow slowly and almost exclusively in old-growth forests.

DENNIS FAST www.dennisfast.smugmug.com

specialized bacteria in their stomachs and intestines that allow them to digest lichens, caribou are able to occupy a unique ecological niche.

During the winter months, caribou feed almost exclusively on these carbohydrate-rich lichens, digging craters in the snow to expose terrestrial lichen, along with other plants. In fact, the word "caribou" is believed to be derived from the Mi'kmiq word *xalibu,* meaning "one who paws".

When the snow deepens, caribou look for arboreal lichens such as old man's beard and witches hair, which grow exclusively on trees 80 years or older. In the spring and summer months, they graze on many kinds of fresh green vegetation, making up for the protein deficiency in their winter diet.

Though other factors have contributed to the decline of woodland caribou populations, the primary threat is the loss and fragmentation of these old-growth forests, bogs and marshlands, through forestry, agriculture and mining, as well as oil and gas development. Logging has reduced old-growth forests in size but perhaps worse, fragmented what remains, not only making it difficult for woodland caribou to find enough food to survive the winter, but creating small, polarized populations that are prone to extinction. As well, even when old-growth habitat is available, new growth in recently logged regions often attracts larger populations of moose, deer and elk, which bring with them an increase in predators.

Other deer species also compete with caribou for food during the spring and summer months and can carry meningeal brain worm, a parasite that is fatal to caribou.

Road construction, the building of pipelines and even opening old-growth forests to recreational use, particularly to snowmobiles, contributes to these problems, disrupting the feeding of these timid and elusive animals, and forcing them into open areas, where they are easier prey for wolves, bears and coyotes.

More roads and more development, which inevitably lead to more traffic, also create collisions with vehicles that kill or injure many caribou every year.

Climate change, biologists feel, is also likely to work against woodland caribou. Dale Seip, a biologist with the BC government explains, "We are seeing a northward expansion of deer and elk into caribou range and an associated expansion of coyotes that prey on caribou. Also, the major epidemic of mountain pine beetle in BC, which is now spreading into Alberta, is associated with a lack of cold winters over the last few decades. The beetle has killed hundreds of thousands of hectares of mature pine forest that was caribou winter habitat."

Though unregulated human hunting dramatically reduced woodland caribou populations in the past, today, they're protected throughout most of their range and harvest quotas have been set for Aboriginal hunters. Nevertheless, poaching remains a concern.

Many initiatives are underway at the community, provincial, state and national levels to try to protect the remaining woodland caribou. Most of these programs have research

Woodland Caribou Fast Facts

Rangifer tarandus caribou

HEIGHT & WEIGHT
Woodland caribou average 1.3 metres or four feet at the shoulder and weight between 110 and 180 kilograms or 250 and 400 pounds. Males are larger than females.

LIFESPAN
Caribou generally live between 10 and 15 years.

MATING & BREEDING
The mating season is generally early to mid-October and a single calf is usually born in early June. This represents the lowest reproduction rate of any member of the deer family. Where possible, cows will calve on islands in large lakes, to avoid wolves.

DESCRIPTION
The largest, darkest caribou, they have large concave hoofs that allow them to travel over deep snow. Unlike other members of the deer family, both males and females have antlers; the antlers of a prime bull are huge and complex, while females use their antlers mainly to sweep away snow in order to get at lichen below.

VIEWING WOODLAND CARIBOU

Because they are few in number over much of their range, and because most populations are not migratory, viewing woodland caribou is more difficult than planning tours to view their barren-ground cousins. Still, there are areas where canoe or kayak trippers and hikers can still hope to see these shy, elusive animals. These areas include Atikaki Provincial Wilderness Park in eastern Manitoba; in fact Atikaki, pronounced: ah-*tick*-ah-kih, means "country of the caribou". This rugged, beautiful park, which abounds with magnificent scenery and some of Canada's finest water routes, comprises about half of the Atikaki-Berens Caribou Range, which has a population of between 300 and 500 woodland caribou. Though there is no direct access by road, lodges, outfitters and air charter companies in Bissett, Riverton and Lac du Bonnet can provide air charters of an hour or less into the park's many lakes. An overland canoe route, complete with portages, begins at Wallace Lake Provincial Park, just off Provincial Road 304. Caribou often seek refuge from wolves on islands in Sasaginnigak Lake and can sometimes be seen swimming from one island to another.

In north-central Saskatchewan, Prince Albert National Park and Model Forest are located north of Prince Albert. The national park, which is nearly one-third water, is another place where canoe trippers might spot woodland caribou (some of which have been collared for research purposes), as well as one of the continent's few free-roaming herds of plains bison, timber wolves, black bears and Canada's second largest colony of pelicans.

For those who don't like canoe tripping, caribou can sometimes be seen from highways in several places, among them Alberta's Highway 11, which crosses the Rocky Mountain Forest Reserve west of Red Deer, and Highway 40, which traverses northeastern Alberta south from Grande Prairie.

and monitoring components, along with the plans for their recovery. Areas of critical habitat are being identified, and conservationists are having success in petitioning for the protection of public and private lands. However, unless Canadian and American governments act immediately to protect the mature, healthy forests upon which woodland caribou depend, and place the importance of these habitats above commercial development, these magnificent animals will continue to dwindle.

Whooping Crane

Grus americana • Endangered

FOR many – the disappearance of millions of plains bison aside – it was the whooping crane that marked the beginning of a real awareness of the negative impact humans were having on what had long been believed to be North America's unlimited bounty. Nearly five feet or 1.5 metres in height, elegant and distinctive, whooping cranes are the tallest and most easily identifiable birds in North America. Unlike their smaller sandhill crane relatives, however, they have never been particularly numerous.

It's not known how many whooping cranes populated the continent prior to the arrival of Europeans, but by 1850, when western settlement (with its accompanying draining of wetlands) began in earnest, an estimated 1,500 whooping cranes nested in shallow marshes across north and central North America and wintered as far south as Mexico. Over the next 80 years, as wetlands were drained, forests felled and grasslands converted to farm fields, the number of cranes steadily declined. Because of their small populations, they were not hunted to any significant extent. Nevertheless, records in the US show that 67 birds were shot in the 1890s, 59 in the 1900s and 55 between 1910 and 1920.

By 1929, Fred Bradshaw, director of what is now the Royal Saskatchewan Museum in Regina and a keen and knowledgeable birder, had become alarmed about the state of the whooping crane. Seven years earlier, he had come upon two nesting pairs in west-central Saskatchewan. Following a practice employed for centuries, he'd killed one of the females and removed the eggs for scientific study. Now both nesting sites had been abandoned and, aware of no other nesting areas, Bradshaw sent a questionnaire about the birds to 1,500 correspondents from the Arctic Circle to South America. Their responses over the following months led him to believe that the whooping crane was hovering on the verge of extinction.

North America, however, was mired in the Depression, and the centre of the continent was suffering the worst drought since the beginning of the settlement period. Even getting the word out was a huge challenge. When he was laid off for a month in 1931 due to financial constraints, Bradshaw set off for northern Saskatchewan to give a series of lectures illustrated with lantern slides. Unpaid, and funding his own travel and expenses, he began a long education process aimed at winning farmers and small communities over to the need for wetland restoration and conservation.

In the US, meanwhile, though a hunting ban had been instituted in 1916, by 1937 only 29 whoopers could

The rusty head feathers of this juvenile whooper will disappear as it matures.

DENNIS FAST www.dennisfast.smugmug.com

The majority of wild whooping cranes, includng this one, winter at the Aransas Reserve in southern Texas.

Inset: Whooping cranes have adorned many of Canada's postage stamps.

INSET: LIBRARY AND ARCHIVES CANADA / COPYRIGHT CANADA POST CORPORATION (1953) REPRODUCED WITH PERMISSION
MAIN IMAGE: COURTESY OF RYAN HAGERTY / US FISH AND WILDLIFE SERVICE

be found. Eleven of those wintered in Louisiana, with the remaining 18 at Aransas, Texas, on the Gulf of Mexico. To protect the tiny remnant in Texas, US President Franklin D. Roosevelt signed an executive order to purchase land for a whooping crane winter refuge.

Though, as indicated below, over the past quarter-century, efforts have been made to restore populations elsewhere, that refuge, which today encompasses 22,500 acres (9112 hectares) of rolling hills, marshes and salt flats, still provides a winter home for North America's only wild migratory population of whooping cranes. In addition to the original refuge, adjacent units total 115,000 acres (4650 hectares) and offer protection for American alligators, javelina, roseate spoonbills and armadillos.

Despite the efforts at education and conservation on both sides of the border, the decade-long drought of the 1930s, a hurricane in 1940 that devastated the tiny wintering population in Louisiana, and World War II continued to diminish the number of whooping cranes. In 1941, biologists believed there were only 15 or 16 of these majestic birds left.

Undeterred, the Audubon Society, along with American scientist Robert Porter Allen and Canadians Robert Smith, a biologist, and Fred Bard, who succeeded Bradshaw as director of the Royal Saskatchewan Museum, continued to work tirelessly to find and protect the tiny remnant population. Working with government departments, schools, naturalists, organizations such as Ducks Unlimited, and

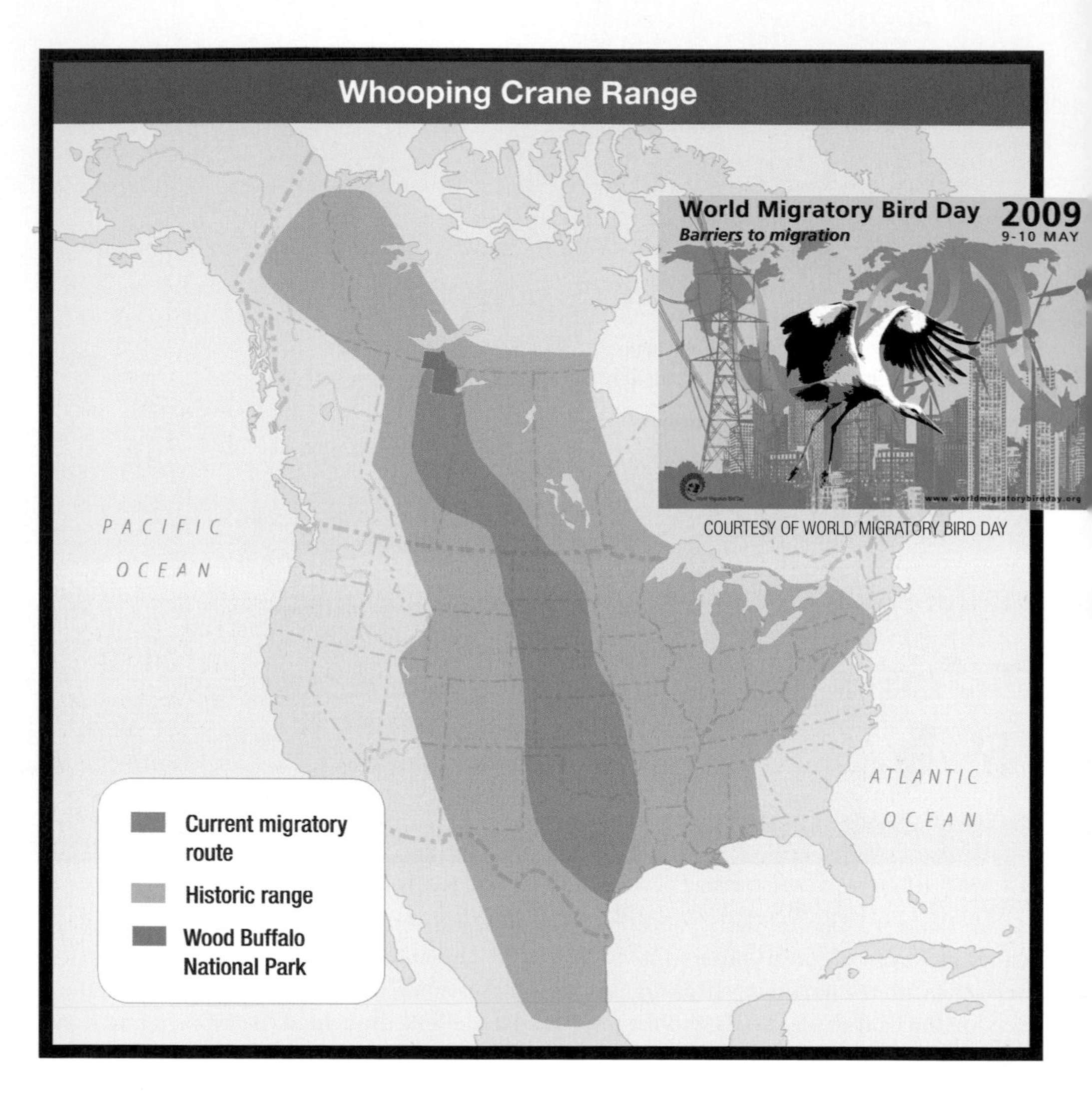

COURTESY OF WORLD MIGRATORY BIRD DAY

using radio and increasingly, television, they spread the word about whoopers during the post-war years. Given their subjects' magnificence and desperate straits, they found their audiences easily inspired.

Still, as Ed Struzik wrote in a Canwest News Service article in April 2009, "Smith was still determined to find out where the Texas birds were spending their summers in Canada. So for several years, he and Allen followed leads, flying over vast stretches of wilderness in the Yukon and Northwest Territories for signs of the big white birds and their nests. Smith got tantalizingly close in 1952, when he spotted two big white birds south of Yellowknife. Unfortunately, he didn't get close enough to confirm that they were whoopers."

In the end, it was a pair of fire fighters that discovered the nesting site in Wood Buffalo National Park, on the border between Alberta and the Northwest Territories. As George Wilson and pilot Don Landells returned from fighting a fire farther north, they spotted a family of whoopers near the Sass

River in late June 1954. The discovery was confirmed the same day by Canadian Wildlife Service zoologist Bill Fuller, who found other nesting pairs the following year.

Nevertheless, even for this poster child for endangered species, the climb from the edge of extinction has been slow. A decade after the discovery of the nesting site, the whooping crane population was just 42 birds. Over the next decades, though they varied from year to year, the numbers slowly climbed.

In an attempt to introduce a second migratory flock, biologists collected one egg from a number of productive nests in Wood Buffalo NP (whooping cranes lay two eggs in each clutch, but often only successfully raise one youngster) and introduced the pilfered eggs to the nests of sandhill cranes at Gray's Lake National Wildlife Refuge in southeastern Idaho. The sandhill surrogate parents successfully raised a number of adolescent whoopers and took their young to the Bosque del Apache National Wildlife Refuge on the Rio Grande River in New Mexico. However, instead of mating with other whoopers, these fostered cranes remained as solitary individuals and one formed a pair bond with a sandhill crane; they also suffered high mortality rates. The program was discontinued in 1989 and the last surviving member of this population was killed in 2002 when it struck a power line.

In 1993, following studies on sandhill cranes, which have both migratory and non-migratory populations, the recovery team tried again with 33 whooping cranes bred at the U.S. Geological Survey's Patuxent Wildlife Research Centre in Maryland. The plan was to develop a population of non-migratory whooping cranes, much like the flock that had existed in Louisiana until the 1940s. Installing the birds at Kissimmee Prairie in central Florida (where a large flock of non-migratory sandhills lives), it was soon discovered that captive breeding had not prepared the whoopers to avoid or defend themselves against predators; in the first few years, nearly 60 per cent of the birds were killed by bobcats. Also, since 1998, drought in Florida has reduced the cranes' nesting efforts. The campaign continued, however, and in June 2003, a chick, nicknamed Lucky, was fledged. Born in the wild of captive-raised parents that had been released into the wild, it was the first of a second generation of non-migratory whooping cranes. Since then, more birds have been released in Kissimmee Prairie and in 2006, the flock raised four chicks and another was fledged a year later. Despite these successes, the total number of birds declined from 37 in 2007 to 30, including 12 pairs, in September 2009.

In the fall of 2008, the Aransas Reserve estimated its wintering population at 270 birds, and the world total at 539, including a captive-bred population that migrates from Necedah NWR in central Wisconsin to the Chassahowitzka NWR on Florida's Gulf Coast with help from volunteers from Operation Migration.

However, no one associated with this long battle back from the edge of the abyss is taking anything for granted. Accidents, natural disasters, climate change and even apparently color-blind hunters continue to create problems for these majestic birds. And the winter of 2008-2009 was filled with such problems.

While biologists at the Aransas Refuge had hoped to see a record 300 returning birds in the fall of 2008, instead, 34 birds failed to return. Then another 21 birds – six adults and 15 chicks – died over the winter, likely as a result of a severe shortage of blue

Intricate dances are part of the annual mating rituals of whooping cranes.

DAVE MENKE / US FISH AND WILDLIFE SERVICE

DENNIS FAST www.dennisfast.smugmug.com

VIEWING WHOOPING CRANES

The most accessible place to view whoopers is at the Aransas NWR northeast of Corpus Christi between January and April each year. Accessed by Hwy 35 from the city, the refuge offers interpretive van tours on weekends. The two-hour tours begin at 10 a.m. and 1 p.m. Reservations are required. Between mid-October and March, a pair or family of whooping cranes can often be seen from the Observation Tower, feeding in the marsh during daylight hours. Ask at the front desk of the Visitor Center for the latest sightings. To view many whoopers, boat trips are available from Rockport, just south of the refuge. Contact the Rockport Chamber of Commerce at www.rockport-fulton.org for information.

During their spring migration, whooping cranes stage along Nebraska's Platte River, with thousands of sandhills. The U.S. Fish and Wildlife Service has designated the central reach of the Platte River from Lexington to Shelton, Nebraska, as critical habitat for whooping cranes. The birds are easily disturbed.

crabs, the whoopers' main food in their winter refuge, brought on by drought on the south Texas coast. Worse perhaps, researchers from the US Geological Survey discovered the presence of a wasting disease in the flock.

As the birds left for Canada in the spring of 2009, Tom Stehn, coordinator of the US whooping crane recovery program for the US Fish and Wildlife Service, was feeling more than a little discouraged. "In the 26 years I've been involved in this effort," he told journalist Ed Struzik, "this is easily the worst year I've seen." Almost 20 per cent of the flock had been lost over the previous six months and he predicted that up to 40 per cent of the females, weakened by such a food shortage at Aransas that the staff had to resort to feeding the birds corn – as Japanese farmers do with red-crowned or Japanese cranes in Hokkaido – might not lay eggs in Wood Buffalo NP.

Stehn's dire predictions were worse than the reality. In 2009, the nests were down by four and fledging success was about 25 per cent lower than normal.

Yet whooping cranes have a lineage that stretches back into the mists of time. The Gruiformes order to which they belong, which includes other large or medium-sized marsh-dwelling birds such as rails, grebes and bitterns, has the most complete fossil record of any avian order. And the discovery of fossils throughout the Southern Hemisphere suggests that their distant ancestors date back to the final breakup of Gondwana, the ancient southern supercontinent, about 120 million years ago.

Some of these ancient relations were even more imposing than the whooping crane; the *Diatryma steini*, for example, stood almost seven feet or more than two metres tall and had a massive head and bill. It was not until the demise of the dinosaurs, however, that these birds became abundant; the ancestors of modern cranes and rails appeared about 56 million years ago. Today, however, not just whooping cranes, but several of the other 14 crane species are listed as threatened or endangered.

Whooping Crane Fast Facts

Grus americana

DESCRIPTION

Snowy white with jet-black wingtips visible only during flight, whooping cranes have a crescent of black feathers with a patch of red skin on the head. Juveniles have rusty-brown plumage; their white adult feathers begin to appear just before their southern migration. By the time they migrate north again the following spring, the adolescents are nearly all white, like their parents.

LENGTH & WEIGHT

The tallest flying birds in North America, whooping cranes are almost five feet or 1.5 metres in height. Males weigh an average of 16 pounds or 7.3 kilos; females are slightly smaller. Cranes fly with their long necks and long dark bills pointed straight ahead and their equally long, thin legs trailing straight out behind.

WINGSPAN

The wingspan measures almost seven feet, or more than two metres and the wings, which are tipped with black primary feathers, are used much like a glider. Spiralling on warm updrafts to great heights and then gliding forward, cranes can fly great distances without tiring. Flights of 750 kilometres or nearly 500 miles over 10 hours are possible, though not the norm.

LIFESPAN

The average lifespan in the wild has been estimated to be about 24 years, though a 28-year-old banded female – one of 134 juveniles banded with brightly colored bands between 1977 and 1988 – was found dead on the edge of Muskiki Lake in Saskatchewan in October 2005. Given that she was known to have migrated from her birthplace in Wood Buffalo NP to her wintering site in Aransas NWR in Texas and back every year, biologists estimate she covered 225,000 kilometres or almost 140,000 miles during a lifetime that added 11 offspring to the flock. In captivity, whoopers can live to be over 35.

MATING & BREEDING

Whooping cranes mate for life, but will mate a second or even third time if their mates die. They also occasionally mate anew if unsuccessful in raising a chick. Their distinctive whooping call, which gives the species its name, is most often heard during the spring courting period, and pairs often perform duets, or unison calls. Particularly during the early morning hours, these calls can be heard for miles. Beginning before they leave their wintering grounds, the courtship calling is accompanied by dancing and wing flapping, with remarkable leaps in which the male sometimes leaps over the bowed figure of the female. The dancing intensifies as the long migration north approaches about mid-March.

Breeding pairs arrive at Wood Buffalo NP between the third week of April and mid-May and immediately choose a nesting area, generally a marsh with knee-deep water – so that chicks can swim from predators prior to learning to fly – surrounded by bulrushes. Nesting territories vary in size, depending on the density of other nests in the region. Once the site is chosen, construction begins. Usually a metre or three feet across, the nests are built of bulrushes, sedges and cattails, with edges that rise above the water.

Two eggs are usually laid, and the parents share the month-long incubation period. The eggs hatch in early June, however, perhaps because of food shortages or predation, usually only one chick survives. Reddish-brown when hatched., the chicks keep their parents busy feeding them a banquet of dragonfly larvae, minnows, leaches, seeds and berries, warming them and defending them from predators.

HABITAT

Whooping cranes nest and feed in marshes, shallow ponds and patches of shrubs and woodland. They winter in salt or brackish tidal marshes, on tidal flats and in freshwater upland ponds. During migration, they appear to stop at specific wetlands and harvested fields to feed, though in recent years, young cranes have summered in marshes in Manitoba and Saskatchewan, where they have not been seen for nearly a century.

DISTRIBUTION

They once nested from Ohio west and north to Great Slave Lake in the Northwest Territories, but today, the only wild migratory flock nests in Wood Buffalo NP. Due to its fragile nature, there is no public access to the nesting region. Wood Buffalo is also home to two other threatened species, the wood bison for which the park is named (found on page 83), and peregrine falcons.

NAMES

These elegant birds get their name from their distinctive whooping courtship and alarm calls.

The Sea to Sky Region

Dramatic and beautiful, North America's magnificent West Coast has become a magnet for people from across both Canada and the US, at the expense of many other species.

JACK MOST www.themostinphotography.com

A deep canyon along the Englishman River on Vancouver Island is typical of the coastal forests: lush, dense growth and fast-moving water.

JACK MOST www.themostinphotography.com

The Sea to Sky Region

The northwest coast of North America is a land of clichés: majestic, snow-covered mountains; lush green rainforests; clear rivers and streams rushing to the emerald Pacific.

This verdant, productive region of beaches, fjords, forests and mountains is unlike anywhere else in the world. Not surprisingly, it is increasingly known as the Sea to Sky Region. Tectonic and glacial activity formed this stunning landscape, which has more variation in a limited area than anywhere else in North America. The Pacific Ocean keeps the climate mild year round; mean annual temperatures range from 4.5° C (or 40° F) in the north to 10° C (50° F) in the south, with minimal seasonal variation. Moist air from the ocean brings high humidity and plenty of rainfall; combined with the long growing season, this allows plant life to thrive.

The climate and geography of the Pacific Northwest support several ecosystems with distinctive plants and wildlife. The conditions in this region are unique and the Coast Mountains effectively separate the area from the rest of North America. Many of the plant and animal species found on the Pacific coast are perfectly adapted to their habitats, and as such are endemic to the region; that is, they're not found anywhere else in the world. A few of these are the common rainforest fern; Pacific dogwood; Sitka spruce; mountain beaver and Vancouver Island marmot. Similarly, many species with ranges that cover vast areas of Canada and the United States cannot be found west of the Coast Mountains. For example, white spruce and trembling aspen, two extremely tolerant tree species, are found everywhere in Canada except the Pacific coast. The same can be said of many wide-ranging birds and mammals, notably the black-capped chickadee, meadow mouse and lynx.

The Sea to Sky Region stretches from southern Alaska to Northern California and has three main geological features: the offshore islands; the coastal trough, and the western slopes of the Coast Mountains. The backbone of the offshore islands is formed by the Insular Mountains, which were created by tectonic activity at the edge of the North American plate and run under, as well as above, the Pacific Ocean. Vancouver Island, at 400 kilometres or 280 miles long the largest island on the coast of the Americas, belongs to this mountainous backbone. Farther north, the mountains resurface as Haida Gwaii (also known as the Queen Charlotte Islands), the islands of the Alaskan

Taken atop Mount Douglas Park in Victoria, a cloud of pollution hangs over BC's Lower Mainland (Vancouver) and Bellingham in Washington State. High pressure days such as this one, mean stagnant air is not being refreshed by ocean breezes, causing a visible mass of pollution to hover over the land.

COURTESY OF COLIN WELCH

Panhandle and the St. Elias Mountains, which stretch north along the Gulf of Alaska.

These offshore islands are ecologically very important. Many provide nesting areas for seabirds rarely found elsewhere because of human populations. The Peregrine falcon, which has adapted to urban living in many places in North America, has found a wild habitat in Haida Gwaii where it is able to thrive. The islands are also important because they are one of the few areas in northern North America to have escaped total glaciation during the Pleistocene. The subspecies of many animals found on Vancouver Island and Haida Gwaii are so different from their mainland relatives that many biologists believe they have evolved in isolation for 100,000 years or more.

The southeast coast of Vancouver Island is also home to the Garry oak woodlands (see page 112), one of the most at-risk ecosystems in the Pacific Northwest. The island is the northern limit of the Garry oak ecosystem; taking advantage of a moderate climate with dry summers and rich soils, it runs in a thin strip south through Washington and Oregon, just east of the Coast Mountains.

East of the offshore islands is the coastal trough, known as the Inside Passage. This long depression between the islands and the mainland begins at Washington's Puget Sound in the south and runs north along the coast as the Strait of Georgia, Queen Charlotte Sound and Hecate Strait. It was formed 150 million years ago as continental plates collided. By contrast, Juan de Fuca Strait, which separates the island from Washington's Olympic Peninsula, was created only 14,000 years ago by a glacier flowing from BC's Fraser Valley.

The Puget Sound-Georgia Basin region has a huge watershed, extending from the Campbell River in northern BC to the Nisqually River in west-central Washington. The mild climate, along with the fresh water, silt and nutrients carried into the basin by the rivers make this an incredibly rich ecosystem, supporting approximately 3,000 species of plants and animals. This includes some 200 species of fish, more than 20 marine mammals, over 100 sea birds, and thousands of marine plants and invertebrates. Of these plants and animals, 63 are listed by one or more jurisdictions as species of concern.

This region and its watershed is also home to an estimated 10 million people, a figure that has risen dramatically over the past 20 years. This massive population poses serious environmental problems that are threatening the health of this ecosystem; some of the most serious are the release of toxic chemicals into the water, sewage pollution (until recently, neither Vancouver nor Victoria, BC's two largest cities, had even begun to plan for sewage treatment systems – they simply dumped their waste into the ocean), oil spills, destruction of habitat, mismanagement of fisheries and netcage salmon farming. The bioaccumulation of toxic chemicals and heavy metals, declining food supplies, and a loss of critical habitat are endangering many of the animal species that call Puget Sound and the Georgia Basin home.

The Coast Mountains are an interconnected series of ranges that run along the coast forming fiords, islands and mountains. The moist western side of the mountain range is included in the West Coast region, because it has a climate and vegetation similar to the rest of the coastal area, but the ridge of the mountain range is considered the region's easternmost boundary. This is because the eastern slopes of the Coast Mountains have a very different landscape than that of the Pacific Coast, with plant and animal life more similar to that in the interior region west of the Rockies.

A dense wilderness stretches along the British Columbia coast from Vancouver Island's north tip to the Alaskan border. This "Great Bear Rainforest" is the world's largest remaining undisturbed temperate rainforest. These forests of Douglas-fir, Sitka spruce, redcedar and western hemlock feature some of the oldest and largest trees on Earth, up to 300 feet tall and over 1,500 years old. With the coastal fjords and rivers, they are home to a rich diversity of wildlife: salmon, bald eagles, whales, and timber wolves, along with black, Kermode and grizzly bears, to name but a few.

For a time, as logging companies began to look to the northern rainforest as a source of income, this last large wilderness region seemed at great risk. However, sustained political pressure on provincial and federal governments to preserve BC's wild spaces led, in February 2006, to an historic land use agreement. Brokered by the BC government between environmental and First Nations groups and the logging industry, it protects 18,000 square kilometres (or almost 7,000 square miles) of intact temperate rainforest, limits forestry and supports ecotourism. In 2007, the Government of Canada pledged $30 million towards conservation and economic development initiatives for coastal First Nations, including sustainable fisheries, forestry and tourism. Finally, in March 2009, the Great Bear Rainforest officially became one of the most protected forested areas, a move that was widely applauded. In total, 2.1 million hectares (or five million acres) have been legally protected from logging; $120 million is available to First Nations communities to support sustainable economic alternatives to logging; and where it is allowed, "lighter touch" logging is legislated, meaning that it must follow ecosystem-based management (EBM), in which 50 per cent of the natural old-growth forest must be maintained.

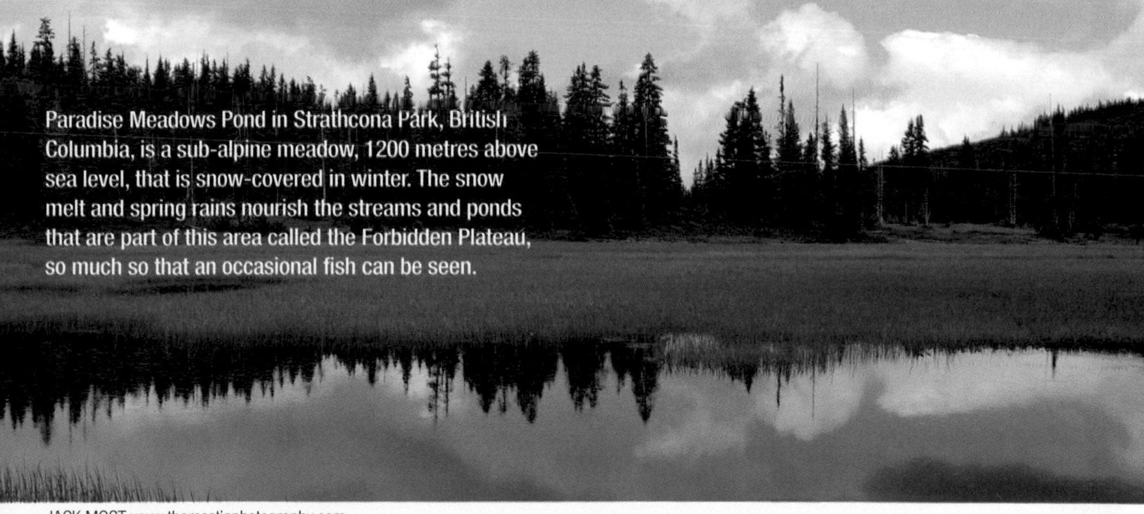

Paradise Meadows Pond in Strathcona Park, British Columbia, is a sub-alpine meadow, 1200 metres above sea level, that is snow-covered in winter. The snow melt and spring rains nourish the streams and ponds that are part of this area called the Forbidden Plateau, so much so that an occasional fish can be seen.

JACK MOST www.themostinphotography.com

Climax Douglas-fir

Pseudotsuga menziesii

THESE magnificent trees, which can grow to heights of more than 100 metres or 300 feet, are North America's second-tallest tree species, behind the coastal redwoods. And like redwoods, they can live for more than 1,000 years, making them among the continent's most impressive natural wonders. Unfortunately, Douglas-fir, particularly the coastal variety, is also among the world's best timber producers and though millions of seedlings are planted each year across North America (and even far beyond) more than a century of intense logging has left very few of these old-growth giants.

As recently as the late 1980s, motorists travelling off-season along scenic Highway 101 around Washington's Olympic Peninsula, would often find virtually the only vehicles on the road were logging trucks, and many of those were carrying not a load of logs, but just a portion of one, massive tree. North, on south-central Vancouver Island, is the magnificent and aptly named Cathedral Grove (see below) in MacMillan Provincial Park. The park is named for H.R. (Harvey Reginald) MacMillan, who donated the land to the province in 1944. While saving the spectacular grove was typical of MacMillan, who has often been called a visionary, it should be noted that MacMillan Bloedel, as his company was known at its height, was one of the largest logging companies in the world, controlling nearly a million acres of forested land in the 1950s. All the land around the park has been logged.

Of those that survived the onslaught, the tallest living Douglas-fir, at 328 feet or marginally over 100 metres in height, is the Doerner Fir, formerly known as the Brummit Fir, located in Coos County, in southwestern Oregon, which boasts the Douglas-fir as its state tree. The Queets Fir, located in Washington's Olympic National Park, may be the heftiest of the species, with a diameter at breast height (dbh) of 14.3 feet or 4.36 metres.

The oldest authenticated Douglas-fir in the US was slightly older than 1,400 years of age when it was harvested in the 1950s near Mount Vernon, Washington. Canada's oldest known Douglas-fir was one of a stand on Vancouver Island that was established after a fire about 635 AD. It blew down in a storm in 1985 and was afterward reported to have a ring-counted age of 1,350. Other trees are still alive in the same stand.

The oldest accurately-dated living Rocky Mountain Douglas-fir is a 1,275 year-old giant in New Mexico. This longevity is apparently an anomaly; growing on a relatively barren lava field has protected the ancient tree from fire, animals and humans.

Among living trees in Canada, some of the oldest and largest known are more than 1,300 years old, and can be found in the Stoltmann Wilderness, just west of Whistler, BC, (site of the 2010 Winter Olympics). And the Red Creek Fir, which towers over the surrounding forest near Port Renfrew, on Vancouver Island's west coast, is the largest member of the pine family in Canada.

Pine family? one might ask. And indeed, the Red Creek Fir, like all Douglas-firs, is not a fir at all, but rather, and despite its Latin name – *Pseudotsuga menziesii* or "false hemlock"– a species all its own, related to the larger Pinacea family. Also called red-fir, Oregon-pine, Douglas-spruce, and, in Spanish-speaking territories, *piño Oregon*, the Douglas-fir has cones that droop from the branches and fall whole to the ground when mature. By contrast, true firs have upright cones and scales with seeds that are shed from the cone, which remains, denuded, on the tree.

The fossil record of Pseudotsuga-like forms in North America goes back about 50 million years, to the early Tertiary period. The oldest known fossil cones, seeds and leaves were found in the Eocene flora of northwestern Nevada's mountainous Copper Basin. Remarkably, though the trees no longer grow where these ancient ancestors once did, the fossilized cones, seeds and needles are almost indistinguishable from those of modern Douglas-firs.

As time passed, and balmy Eocene temperatures slowly cooled, the range of the Douglas-fir expanded north, until about 20 million years ago, in the Miocene period, Douglas-firs could be found as far north as the Alaskan Panhandle.

Then, during the Quaternary – the past 1.8 million years that were marked by waves of intense global cooling – Douglas-fir moved south again, surviving each glaciation in small refugia far to the south, then marching north at speeds that have surprised paleontologists armed with theoretical models of migration. For example, during the last glaciation, Douglas-firs found refuge in a small area at the tip of the southern Rocky Mountains in what is now New Mexico. When the world warmed, they migrated out of their refugium northwest through Utah's Great Basin at an average rate of 286 metres per year, three times faster than scientists had predicted they would.

Over time, two varieties of the tree

Climax Douglas-firs dwarf a visitor to Cathedral Grove; inset: a vicious storm in 1997 took down many magnificent trees.

JACK MOST www.themostinphotography.com INSET: DENNIS FAST www.dennisfast.smugmug.com

Naming Douglas-firs

Douglas-firs were named for David Douglas, a Scottish botanist whose collecting trips were financed by the Horticultural Society of London. He arrived in British Columbia in 1825 at the age of 25. Travelling with a small terrier, living off the land, sleeping under his canoe or rolled in a blanket, he trekked thousands of kilometres through British Columbia. Following his adventures in BC, Douglas returned to England, bearing seeds from a Douglas-fir. They were planted at Scone Palace, as part of the collection of Kew Gardens. One grew and there it remains, one of the tallest trees on the grounds today.

Douglas died a violent and mysterious death in Hawaii in 1835. He was found dead in the bottom of a pit-trap created to capture wild bullocks. One of these huge creatures, also dead, was in the pit with him, but Douglas' wounds did not seem consistent with those he might have received from the bull. Some believed he had been murdered, but the cause of his death was never determined.

emerged, the truly enormous coast Douglas-fir and the tall, slender Rocky Mountain or blue Douglas-fir. Today, the natural range of the two occupies an inverted V in western North America, joining at about the 55 degree of latitude in central British Columbia. Coast Douglas-fir, which form the shorter western arm of the inverted V, are found from the western slopes of the Sierra Nevada in northwestern California north to central BC. Though mainly found from sea level to elevations of 5,500 feet (or 1677 metres), in the Sierra Nevada and Cascades, coast Douglas-fir can be found on mountain slopes 2,000 feet (600 metres) higher.

The Rocky Mountain Douglas-fir thrives from central BC south along the Rockies and into the mountains of central Mexico. Though they usually grow up to 9,500 feet (2900 metres), on Mount Graham in southern Arizona, the trees have been found growing at 10,700 feet or more than 3250 metres.

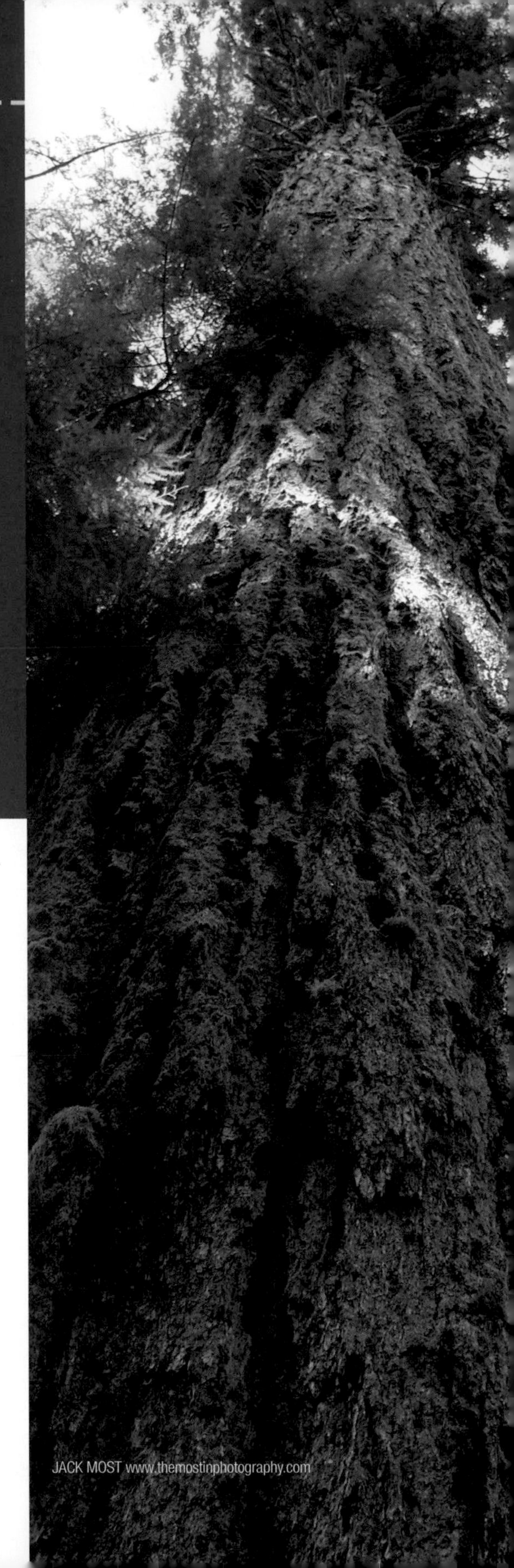
JACK MOST www.themostinphotography.com

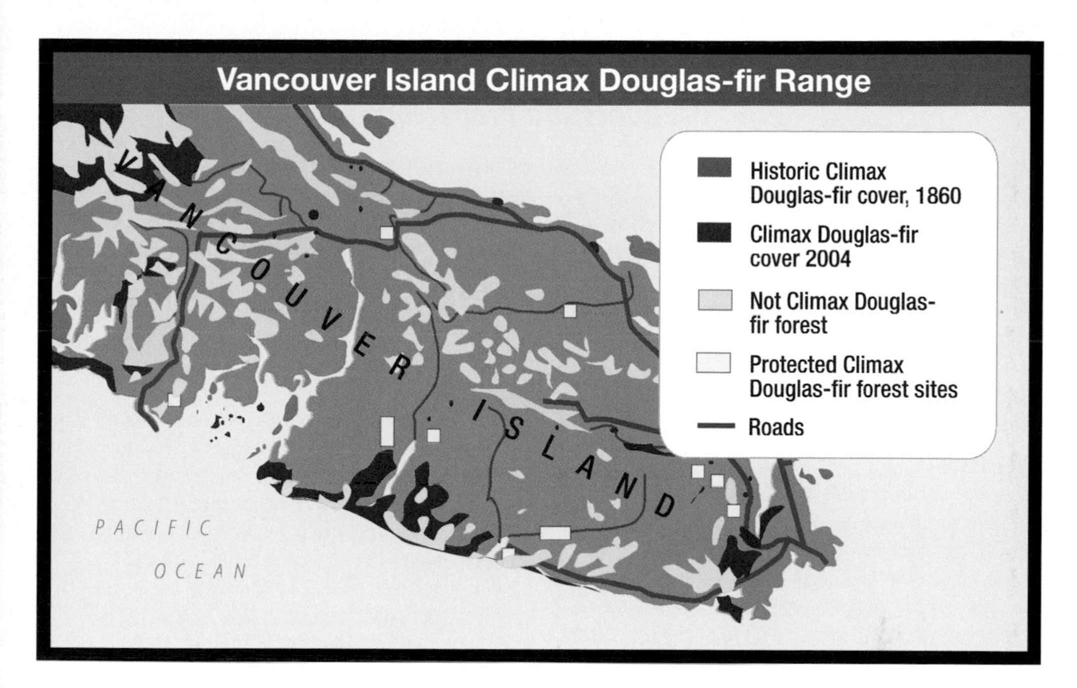

Their magnificent wood had an even greater reach. Though the native range of Douglas-fir never extended beyond western North America, the wood was appreciated, and used, as far away as Hawaii and Polynesia, where driftwood logs from thousands of kilometres away was prized for making bowls, paddles and even canoes. And today, nursery grown Douglas-fir are among the most popular Christmas trees.

A Douglas-fir can be easily identified by recalling the four Fs. Its needles are flat, with a single groove on the upper surface, fragrant and friendly – soft to the touch, unlike Colorado spruce, for example. And on mature trees the bark is deeply furrowed, with dark reddish-brown ridges.

In most places from Vancouver Island to northern California, coast Douglas-fir will eventually give way to western hemlock as the climax or dominant species. In a narrow, drier band on Vancouver Island's southeast coast, however, Douglas-fir is the climax species; and Cathedral Grove, in MacMillan Provincial Park on south-central Vancouver Island, is one of the most accessible stands of giant Douglas-fir on the continent (see page 108 for directions).

Native North Americans used the dense, durable wood of the Douglas-fir for a myriad of things, including bowls, spears, harpoon hafts and barbs, fish hooks and traps, snowshoes and burial caskets, as well as fuel for longhouse firepits. The boughs were used as floor coverings; the seeds were eaten and the sap was used for calking canoes. Pegs of pitchy heartwood were also used as torches that would burn even in the rain and the sap served as a medicinal salve for small cuts and sores.

As the "frosting" on this remarkable tree's contributions, each summer branches in sunny coastal locations exuded something that has been called "Douglas-fir sugar", which was much prized by people who lived without honey or cane sugar. Tests have shown that the frosty substance is 50 per cent trisaccharide sugar.

Insects, birds and mammals also rely on Douglas-firs for food and shelter, which may have given rise to the native Californian myth, which contends that each of the three-ended bracts on the cones represents a tail and two tiny legs of mice, who hide inside the cone's scales during forest fires; the tree is kind enough to be their enduring sanctuary.

Climax Douglas-Fir Fast Facts

Pseudotsuga menziesii

DESCRIPTION
One of the largest tree species in North America, reaching 100 metres or more than 300 feet or more, it has a stiffly erect crown and spreading or slightly drooping branches. Mature trees have very thick, ridged, rough bark and trunks that soar upward before branching out. The thick bark allows the trees to survive surface fires, but many groves of climax Douglas-fir are a legacy of ancient fires, for they grow quickly when fire creates an opening.

HEIGHT
The well-documented height of the tallest tree of any species, the Mineral Tree, from Mineral, Washington, was measured several times between 1911 and 1925 by University of Washington forester Richard McCardle. The height was 394 feet or 120 metres and the volume of wood was estimated to be 18,190 cubic feet or 515 cubic metres. New research suggests Douglas-fir could grow to a maximum height of between 430 and 476 feet (131 and 145 metres), at which point its water supply would fail.

HABITAT
The larger coast variety grows from the western slopes of the Sierra Nevada in northwestern California north to central British Columbia, generally from sea level to elevations of 5,500 feet, though in the Sierra Nevada and Cascades, they can be found much higher. The taller, slimmer mountain variety thrives from central BC south along the Rockies and into the mountains of central Mexico, growing from 1,800 to 9,500 feet, though in Arizona, they have been found growing at 10,700 feet.,

LIFESPAN
The oldest known Douglas-firs were more than 1,400 years old.

PRESERVATION
Since climax Douglas-firs go back between 800 and 1,400 years, recreating an old-growth forest is virtually impossible (though setting aside naturally reforested areas now would be a great help for our descendants of the distant future). What is needed is what growing numbers of organizations, native communities and individuals are doing now – insisting that logging be stopped in old-growth forests, as New Zealand did on its publicly-owned temperate rainforests in 2007.

NAMES
Douglas-fir, also called red-fir, Oregon-pine, Douglas-spruce, and, in Spanish, *piño Oregon*. Two varieties of the species are recognized: *P. menziesii* (Mirb.) *Franco var. menziesii,* called coast Douglas-fir, and *P. menziesii var. glauca* (Beissn.) Franco, called Rocky Mountain or blue Douglas-fir.

VIEWING DOUGLAS-FIRS
Cathedral Grove, in MacMillan Provincial Park, just past the western end of Cameron Lake on Highway 4 on west-central Vancouver Island offers a network of trails meandering through towering trees, some of which are more than 800 years old. This protected patch of climax forest also includes western hemlock, grand fir and western redcedar.

The Grove of the Patriarchs Trail in Mount Rainier National Park in Washington State, in the southeast corner of the park, just north of Ohanapecosh Visitor Center, is open from May to mid-October.

Coast Douglas-fir can be found (along with coast redwood) in Redwood National and State Parks in Northern California. A loop trail through Stout Memorial Grove in Jedediah Smith Redwood State Park – a 44-acre grove donated by Clara Stout to save it from being logged and as a memorial to her husband, lumber baron Frank D. Stout – not only reveals colossal redwoods, but also magnificent Douglas-fir, grand fir and western hemlock.

Visitors to Britain can also see a large Douglas-fir; in fact, the tallest tree in the United Kingdom is a coast Douglas-fir growing in Reelig Glen by Inverness. Known as Dughall Mor, it is 64 metres or 210 feet tall and was measured in 2005 by staff from the Royal Botanic Gardens, Kew, The Tree Register and The Forestry Commission.

Though snow weighs their uplifted branches down, walking through an old-growth forest of Douglas-firs is like being in a cathedral.

PETER ST. JOHN

For millennia, the trees' great friend was fire, for mature trees have thick bark that protects them from all but the most ferocious wildfires. And when fire did create openings in the dense coastal or mountain forests, Douglas-fir, as pioneer trees, took advantage of the change, growing more quickly in open sunlight than western hemlock or western redcedar. Forests of old-growth coast Douglas-fir provided sanctuary and bountiful harvests for the Douglas squirrel and served the primary habitat of the red tree vole and the spotted owl. The destruction of these forests in the past 125 years is the primary reason the spotted owl is today an endangered species (see page 217).

Despite millennia of use by native North Americans, vast tracts of climax forest greeted Europeans when they first arrived on the West Coast. With ever-improving logging equipment, they tackled the majestic Douglas-firs – among the world's best timber producers, yielding more timber per tree than another other North American species – as well as hemlock, western redcedar, Sitka spruce, sugar pine, western white pine, ponderosa pine, grand fir and, farther south, the enormous redwoods.

Few Douglas-firs were spared, even magnificent members of the species. In 1902, at Lynn Valley on Vancouver's north shore, a tree with a reported height of 126.5 metres or 415 feet and a diameter of 4.3 metres (14 feet three inches) was cut down. And in 1909, a Douglas-fir growing near Callam Bay in Washington State and named for US president William Howard Taft was logged, though not before it was immortalized with a photograph of local businessmen (including the owner of the logging company) sitting in the undercut. Written on the photograph is the following: "Considered the largest tree in the state." One can only wonder how large it would be today.

Almost too late, citizens of both Canada and the US have begun to realize the value of their ancient trees, as well as the lumber they produce. As the Vancouver-based Wilderness Committee explains, "In 2001, after years of passionate protest by concerned citizens from British Columbia, International Forest Products (Interfor) stopped logging in the core of the Stoltmann Wilderness, including the Upper Elaho River Valley and Sims Creek Valley."

Since then, the Squamish Nation has purchased Tree Farm License 38 from Interfor and declared the Elaho and Sims Valleys, as well as other areas within the wilderness region that are part of their traditional territory – Kwa Kwayexwelh-Aynexws, the "Wild Spirit Places" – off-limits to commercial logging. It was not only trees the Squamish people (and many others) were interested in saving; the Stoltmann Wilderness also encompasses the most southerly range of coastal grizzlies, as well as magnificent mountainous wilderness with clear, glacier-fed streams that are used by spawning salmon, and other species of fish.

Finally, perhaps realizing the embarrassing situation it might face as the world arrived for the 2010 Olympics at Whistler, in 2007 the BC government passed legislation fully protecting the Upper Elaho and Sims Creek Valleys.

The Living Forest

NOTHING distinguishes native North Americans from those who have arrived in the past 500 years more than their approach to nature. No matter where they lived, the continent's original inhabitants almost invariably viewed themselves as part of the natural world. By contrast, settlers from beyond North America's shores saw themselves as separate from nature and, perhaps more important, viewed nature's once breathtaking bounty as apparently endless, riches to be possessed and utilized without reserve or thought for the future. Today, the result is evident everywhere in seas that are empty of life, mountains denuded of trees, and overgrazed and infertile grasslands.

Despite centuries of determined pillaging, the continent's enormous size and magnificent diversity combined to protect pockets of nature's bounty almost everywhere. And thanks to its steadfast inhabitants, its challenging geography, an increasingly vocal environmental lobby and, in recent years, the reluctant support of its provincial government, many of these preserves are in British Columbia. From Gwaii Haanas and the Great Bear Rainforest to Clayoquot Sound and Cathedral Grove, they allow us glimpses of the magnificent world of the past.

Bark from cedars was taken in long strips that had been cut at the bottom. The tree was then left to heal.

AMANDA DOW

Many of us call these old-growth reserves "wilderness", assuming that forests with trees 500, 800 or 1,000 years or older are somehow wild, and therefore "untouched by human hands". As Jeffrey McNeely wrote in 2004, "The idea that nature exists in isolation from people has become part of the mythology of industrial society." Yet nothing could be further from the truth. In fact, as the Aboriginal inhabitants of many parts of the world have always known and as researchers are beginning to discover, old-growth forests have always been a crucial part of the lives of many cultures.

Northwest coastal peoples, for example, used more than 20 species of trees for almost every aspect of life. Most important was western redcedar, the "tree of life" for many coastal cultures. In a summary report for the David Suzuki Foundation, Arnoud Stryd and Vicki Feddema wrote that this magnificent tree, which can grow more than 100 metres tall and live more than a thousand years, provided bark for clothing, hats, mats, twine, blankets, diapers, towels and rope. It furnished planks for longhouses, logs for canoes,

DAWN HUCK / AFTER ILLUSTRATIONS BY MILLENNIA RESEARCH

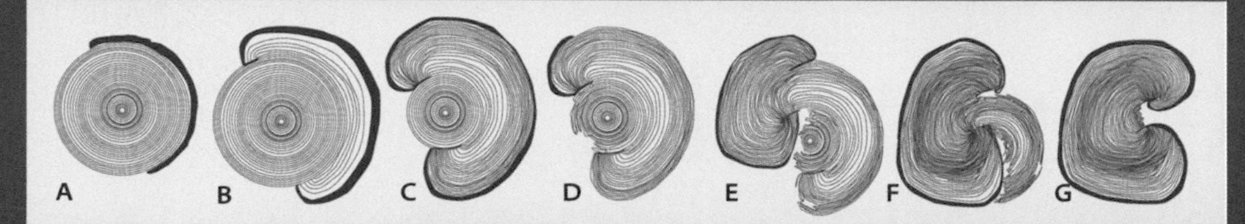

Beginning at left, these illustrations show how trees from which bark has been culled heal themselves. A) bark is stripped in 1799; B & C) scars form as the tree grows; D) bark is stripped again in 1817; E–G) first scar lobe grows into the original tree, while the second scar face decays.

branches for fish traps, roots for baskets and cradles, and wood for masks and totem poles, art that has come to symbolize North America's West Coast.

For at least the past 5,000 years, dozens of cultures extensively used western redcedar and many other species of trees. Yet when Europeans arrived, they perceived the forests to be "untouched".

These apparent contradictions can be quite simply explained. Coastal cultures harvested what they needed of a given tree and left it to heal itself. If, as in the case of the creation of one of the magnificent Haida sea canoes or a Nisga'a or Kwakwaka'wakw totem pole, an entire tree was needed, it was taken only after the appropriate rituals and prayers were performed. Clearcutting was not only impossible because of the available tools, it also was unthinkable. Destroy the tree of life and life itself would be destroyed.

Only recently have archaeologists begun to understand this cultural process and its place in reconstructing the past. Not surprisingly, perhaps, they have created their own label for the trees that have been thus marked by generations of judicious use – culturally modified trees, or CMTs. As Stryd and Feddema write, "For the Heiltsuk First Nation near Bella Bella [British Columbia], resource gathering sites (including CMT sites) are of spiritual significance because the resources were believed to be a gift from the Creator."

Such areas are also often an indication of long-term use, even where – as in Vancouver's Stanley Park or Cathedral Grove on Vancouvery Island – the original inhabitants no longer reside. Because trees were used in different ways, at different times, for different purposes over centuries, and because of their remarkable ability to heal themselves, detecting the trees thus used – the CMTs – is not always easy.

Nor are CMTs restricted to the forests of northwestern North America. Recent research by scientists in other parts of the world has revealed that living trees were also used in northern Scandinavia and southeastern Australia, among other places.

In his 2005 doctoral thesis at the Swedish University of Agricultural Sciences, Rikard Andersson recorded that in northern Sweden, the Sami, a reindeer herding society, used different tree species for a variety of reasons. Silver birches were scarred down to the cambium on one side of the trunk to produce – after 20 to 40 years – a thickening growth along the scar that made remarkably strong axe handles. Large Scots pines were sometimes carved near the base of the tree to produce handles for fastening reindeer cows during milking, or to attach a fence of some kind. During times of great scarcity, the bark of pine trees was also peeled so that the inner bark could be eaten. This practice apparently goes back at least 2,800 years, according to dating of subfossil logs.

In Australia, the Aborigines used bark from living red gum and gray box trees as roofing material, as well as to make canoes and to fashion shields and containers. Research into CMTs is ongoing in other parts of the world, including the boreal forests of Russia and the mountain regions of the southwestern U.S.

Visitors to almost any of the old-growth forest reserves that have been recently protected, to Cathedral Grove or even to Stanley Park may see trees with scars that record their long history in an integrated circle of life. It's an intensely spiritual experience that should make everyone think about today's disposable world.

Garry Oak *or* Oregon White Oak Woodlands

Quercus garryana

ONCE, the lower slopes of Beacon Hill on southern Vancouver Island were so bathed in blue camas blossoms in May that it must have seemed the sea had risen from the shore to immerse the land. This flood of flowers was followed, as the seasons progressed, by golden paintbrush, chocolate lilies, lupins, buttercups and violets. Flitting from blossom to blossom were clouds of butterflies – some 40 species of them. And framing this sea of colour, ancient Garry oaks grew in profusion, sheltering plants, mammals, birds, amphibians and insects, many of which are now rare or endangered.

Writing in 1843 about this parklike landscape, which was so different from the dark (and to European eyes, often forbidding) forests of much of the West Coast, Hudson's Bay Company Chief Factor James Douglas called it "a perfect 'Eden' in the midst of the dreary wilderness of the North". He further mused that it might have "dropped from the clouds into its present position". But this was not simply Nature's work; the magnificent May fields of blue camas, and their much sought-after bulbs, as well as the tall, spreading oaks, were the result of long-held agricultural practices – careful species selection, controlled burning, and the painstaking clearing of rocks and brush – on the part of British Columbia's Lekwammen or Songhees people.

Because the bulb of the deadly white camas is virtually indistinguishable from its edible cousin, eliminating the aptly-named death camas bulbs had to be done during flowering. Having culled the undesirable plants, the Lekwammen would return to their plots in the fall and, lifting small sections of soil, remove the larger bulbs and replace the earth about those that remained. The fields were then burned. The fires removed the dry grass and shrubs, acted as a fertilizer and promoted the growth of young grasses during the winter rainy season. The new growth, in turn, attracted deer and elk, whose droppings further fertilized the land.

This cycle produced huge harvests of blue camas bulbs. Steamed in large pits, often with salmon, the bulbs became soft, dark and sweet – comparable in taste, apparently, to a baked pear. The surplus was traded north to the Nuu-chah-nulth or other neighbours.

These practices – regular burning, aerating the soil and annual culling – not only dramatically increased the size of the camus bulbs, but greatly

HEATHER MOST

Left: A blue camas blossom promises bounty in the fall; above, mature oaks reach for the sky.

HEATHER MOST

enhanced the environment for Garry oaks, as well.

Farther south, where the trees are known as Oregon or white oaks, the woodlands were similarly assisted by native agricultural practices. The Cowlitz and Upper Chehalis of Washington's Puget Lowlands, and the Kalapuya of Oregon's Willamette Valley used regular controlled burns to keep competing shrubs and trees at bay. These fires little affected the oaks, with their thick bark and deep roots. Slow-growing, with small, sweet acorns, they can reach a height of 25 metres or nearly 80 feet. For millennia, the durable wood was used to make combs, sticks for harvesting plants, roots and clams, and clubs for sports or games, as well as for fuel. Acorns were roasted, steamed or pounded into a tasty mash.

Prior to European settlement, oak meadows or savannas could be found on low-elevation slopes and valley floors from the southeastern coast of Vancouver Island and British Columbia's Fraser Valley south to the Klamath Mountains of northern California. Everywhere, they served as a biological "hot spot", providing one of the most productive feeding, breeding and resting habitats on the West Coast, and supporting nearly 200 species of mammals, birds, reptiles and amphibians, as well as huge numbers of invertebrates. Species dependant on the rich environment included Columbian white-tailed deer, huge Roosevelt elk, black bears (as well as grizzlies, now extirpated from the region), several species of squirrels – including the western gray squirrel, which today is listed as threatened in Washington State, sensitive in Oregon and has been extirpated from parts of California – and amphibians and reptiles, including northern alligator lizards and rare sharp-tailed snakes.

Garry Oak Woodlands Fast Facts

Quercus garryana

DESCRIPTION

An attractive tree with heavy, gnarled branches and a round, spreading crown, the Garry oak has deeply lobed leaves that are bright green and glossy on the upper side, but pale with gold and rust hairs beneath. The gray bark of a mature tree is thick and scaly, with furrows and ridges. Growth is slow and trees may live 500 years or more.

HEIGHT

Mature oaks in prime conditions can grow to heights of more than 25 metres or nearly 80 feet.

HABITAT & DISTRIBUTION

White oaks grow best in open meadows with dry sunny summers and wet cool winters, but also grow in rocky upland habitats. Today, they can be found in a narrow band along the south eastern side of Vancouver Island, as well as in isolated areas on the BC mainland, as well as in a narrow band just east of the Pacific, from Puget Sound to northern California.

NAMES

The name commonly used in British Columbia – Garry oak – and the tree's Latin name, *Quercus garryana,* were bestowed by early botanist David Douglas, in honor of fur trader Nicholas Garry of the Hudson's Bay Company. In the US, the tree is known as white oak, Oregon white oak or post oak. When the trees grow over a larger area, the associated ecosystem has been variously called Garry oak woodlands or meadows, Oregon white oak woodlands or white oak savannas.

NOTES

Garry oak is the only native oak in British Columbia and Washington State, and the state tree in Oregon. Its woodland ecosystems are home to more than 100 species that have been designated at risk, including 23 that are threatened or endangered globally.

Above: At the feet of venerable Garry oaks, flower-studded meadows herald early spring.

IMAGES THIS PAGE: HEATHER MOST

Once also at home in white oak woodlands and now at risk are the western meadowlark, the state bird of Oregon, believed to be extirpated from Vancouver Island and the adjoining mainland BC coast; the streaked horned lark, believed to be extirpated from BC, rare in Washington and listed as critical in Oregon; the rare western subspecies of the vesper sparrow; and the Lewis's woodpecker – named after Meriwether Lewis of western exploration fame – which disappeared from southern Vancouver Island during the early 20th century and is declining in numbers in western BC, western Washington and southern California. Also at risk are the western rattlesnake and California mountain kingsnake.

Legend has it that carrying an acorn will help to preserve a youthful appearance.

HEATHER MOST

All these and many other species once thrived in the oak savannas. Studies in northwestern California have shown that "oak woodlands had significantly higher species richness [of grasses and forbs] than in grasslands, meadows or chaparral ecosystems." And in British Columbia, almost 700 plant species have been identified within the ecosystems associated with Garry or Oregon white oak. In other words, oak woodlands not only nurtured great animal diversity but plant species of many kinds as well. Little wonder the trees have been described in British Columbia as "our foundation native species".

The tree's tasty acorns are undoubtedly its most obvious blessing, but the tender leaves of spring provide browse for grazing animals, its summer growth gives shade for creatures of both land and water and the fallen leaves of autumn offer hiding places for small mammals, amphibians and reptiles, and even for fish, which shelter beneath them in streams that run past waterside stands of oaks.

Part of a hardwood forest that gradually covered much of the West Coast during the Hypsithermal, the period of global warming between 8,000 and 5,000 BP, Garry oak woodlands reached their greatest extent about 7,000 years ago, before gradually diminishing in area as the global climate cooled.

Coastal peoples greatly valued these open woodlands, with their flower-filled meadows, deer and elk grazing in the dappled shade, and birds nesting among the branches, and preserved them for millennia. And, as indicated earlier, some Europeans also found this ecosystem inviting. However, Charles Wilkes, travelling through the Oregon's Willamette Valley in 1841, described the landscape as "destitute of trees, except oaks" and unfortunately, it seems that Wilkes' view prevailed, for rather than protecting this bountiful environment, in less than two centuries, it has been all but destroyed by agriculture, settlement and urban encroachment. From an estimated area of nearly 1500 hectares or nearly 580 square miles in the Victoria area in 1800, Garry oak woodlands covered just twenty-one hectares or eight square miles 200 years later – a loss of 98.5 per cent.

In the US, oak woodlands once covered almost 1,600 square miles or more than 388,000 hectares in the interior Coast Range, and 400,000 acres or 162,000 hectares in the

Willamette Valley. Today, the Coast Range has less than four per cent of its historic oak woodlands and the Willamette Valley less than seven per cent. In Oregon's East Cascades and Klamath Mountains, however, habitat loss has been less severe, according to the Oregon Department of Fish and Wildlife, for fire suppression seems to have allowed oaks in some areas to expand into former grasslands.

Though white oak has recently been found to make superb wine casks, its beautifully-grained wood was difficult to season without warping and deemed of little value by foresters; trees were therefore rarely replaced when they were cut down. And settlers, with their great fear of fire, quickly abandoned the crucial annual burning of oak savannas. The result, according to the Oregon Oak Communities Working Group, is that where agriculture and urban sprawl have not destroyed the ecosystem, "valley woodlands once dominated by widely-spaced oaks are slowly [being transformed] into forests crowded with conifers and shade-tolerant trees." The oaks, which thrive on sunshine and space for their sweeping branches, find such conditions intolerable and slowly die.

Invasive foreign shrubs and plants, particularly Scotch broom, false broom, Armenian blackberry and English hawthorn, also threaten Garry oak woodlands, choking the land about their trunks. The trees are further imperilled by the spread of two introduced insect pests, the jumping gall wasp and the oak-leaf phylloxeran, which cause "scorching" of oaks, leaving them stripped of leaves and dead in appearance. In the past, this has often meant that trees were removed when still living.

Fortunately, in both Canada and the US, government agencies and civilian organizations are at last realizing the unique, species-rich ecosystem created by Garry or Oregon white oak woodlands and efforts are ongoing in both countries to rejuvenate existing woodlands and recreate new ones.

HEATHER MOST

VIEWING GARRY OAK or OREGON OAK WOODLANDS

In British Columbia, the north slope of Beacon Hill in Victoria, and the north end of the park together preserve Canada's only extensive relic of Garry oak woodlands, with several trees estimated to be more than 400 years old. Beacon Hill Park is south and slightly east of Victoria's Inner Harbour. Follow Douglas Street south of the Empress Hotel to Dallas Road (Kilometre 0 on the Trans-Canada Highway). The park is on your left; several roads lead into it from Douglas or Dallas. Garry oak woodlands can also be found on Hornby Island, in the Stralt of Georgia between Nanaimo and Courtenay on Vancouver Island's east side.

In Washington, State Hwy 142 runs along the Klickitat River, a tributary of the Columbia River, in the eastern Columbia Gorge. Klickitat County contains some of the most extensive oak woodlands and savannas in Washington State; many have relatively intact under-storey growth, dominated by native species. Western gray squirrels can be found here, as well as acorn woodpeckers and the largest breeding populations of Lewis's woodpeckers in the Pacific Northwest. The Columbia Land Trust is working to acquire oak woodland along the Klickitat River, both to improve the landscape and to enhance the river for spawning salmon. Northwest of Hwy 142, a 10-mile (or 16-kilometre) stretch of the river has been designated "wild and scenic".

In Oregon, Kingston Prairie is located on a plateau above the North Santiam River about 22 miles southeast of Salem in northeastern Oregon. This is one of the best examples of native prairie, with oak savanna, remaining in the Willamette Valley, where less than one-half of one per cent is left. From Salem follow Hwy 22 east 12 miles to Stayton. Go through the town, cross the North Santiam River and turn left onto Kingston-Jordan Drive for one mile. The road turns south and crosses a railway track. Take the next left onto Kingston-Lyons Drive and continue for 1.7 miles to a 90-degree right turn; a gravel road continues east. Park on a turnout. The preserve is both south of the gravel road and east of Kingston-Lyons and on the west side of Kingston-Lyons.

Basket Slough, in the west-central Willamette Valley about 10 miles west of Salem, not only provides winter habitat for dusky Canada geese (as it was established to do in 1965), but also has a white oak woodland, oak savanna and upland prairie for other rare birds. This is believed to be home to one of the largest concentrations of streaked horned larks on the West Coast, with a population of at least 10 breeding pairs and another 15 pairs on surrounding private lands.

Other preserves and restoration areas can be found online.

Oregon leads the way in these efforts, with major projects on both private and public land (see Viewing oak woodlands, above), while British Columbia's municipally managed Garry Oak Restoration Project is an ecological restoration and education program with nine sites on municipal parklands in Saanich, on southern Vancouver Island. Both the BC and Washington governments are also conducting research, which has produced hopeful signs that native parasitoid wasps may be used to check populations of invasive gall wasps.

The BC government offers a booklet entitled *Garry Oak Ecosystems,* while Washington State University, with state and federal assistance, has produced a series of booklets, including *Wildlife on White Oak Woodlands* and *Wildlife in Broadleaf Woodlands*, that are easily obtained online or by mail from the university. The former, by wildlife biologists Marnie Allbriten, of the Oregon Department of Fish and Wildlife, and Jim Bottorff, of the Washington Department of Natural Resources, identifies "the mighty oak tree"(including the Garry or white oak and its close cousins from the south, the black oak and occasional coast live oak) as "the premier wildlife tree in the region".

Kermode Bear

Ursus americanus kermodei • (At risk)

RARE and beautiful, the Kermode bear, or spirit bear, as it's often called, is legendary among the Aboriginal people of the Pacific Northwest. The Tsimshian people say that when Raven, the creator, melted the ice and snow and turned the world green, he decided that every tenth black bear would be white, as a reminder of when the Earth was pure and clean. The white bear, Moksgm'ol, was said to have special powers, and would use these abilities to help chosen individuals. Sadly, the land of the spirit bear is not as pure as it once was, and today the bear's future is largely dependent on human activities.

The Kermode is not a polar bear or an albino, but rather a subspecies of black bear produced when a variant of one recessive gene, when inherited from both parents, produces offspring with a white coat. As in the Tsimshian legend, about one in every ten Kermode bears is white. No one knows the total number, but on Princess Royal Island alone, scientists estimate that there could be 100 white bears. The entire population, both the black and white varieties, were named for naturalist Francis Kermode, who became curator of the British Columbia Provincial Museum (now the Royal BC Museum) in 1904.

The species' future depends on the preservation of large tracts of intact rainforest habitat. For the past two decades, environmental conservation groups have been lobbying for the creation and protection of preserves – areas of wilderness large enough to support the long-term survival of Kermode and grizzly bears, as well as wolves, Sitka deer and five species of salmon, "umbrella or focal species", as they are sometimes called. In the late 1980s, the Valhalla Wilderness Society, based in New Denver, BC, commissioned a report proposing that such species would require a 262,000-hectare (or 650,000-acre) conservancy area, arguing that this was the minimum territory these species required to remain viable.

In the years since the proposal was put forward, the British Columbia government and First Nations have, through intensive planning meetings and public input, come to a number of land use agreements. The most significant, announced in February 2006, provides protection for 34 per cent of the province's coast, including protection of 212,415 hectares (or 524,665 acres) of the Spirit Bear Conservancy Proposal, about 80 per cent of the area initially proposed. In addition to protecting the bears and other species, the creation of the conservancy demonstrated that public awareness of the threats that development poses to wildlife can sway legislators into action.

COURTESY OF WAYNE McCRORY

The spirit bear's history stretches back over several hundred thousand years. Black bears did not arrive in North America until the Pleistocene, the last 1.8 million years of alternating glacial and interglacial periods that many know as "the ice age". During the glacial

In 2006, the Spirit Bear was designated British Columbia's provincial mammal.

Because black bears can have white cubs, as shown at left, and vice versa, protecting only the white Kermode bears does little to sustain the subspecies.

PAUL SMITH

periods, black bears migrated across the Beringian land bridge from Asia, as did many other species. In modern day Canada and the United States they encountered another member of *Ursidae*, the bear family. The North American short-faced bear arrived on this continent sometime during the late Miocene, between 5.3 and 10 million years ago, and evolved separately from European and Asian bears. Standing 1.5 metres at the shoulder, with a short, broad muzzle and powerful jaws, short-faced bears were fierce carnivores, the largest in North America during the Wisconsin glaciation. They dominated the continent, ranging from Alaska to Mexico, and from the Pacific to Atlantic coasts. Despite their speed and power, short-faced bears died out during the Great Extinction at the end of the last glaciation, a relatively brief period that saw the demise of almost all of North America's indigenous large mammals. The bears' disappearance was probably due in part to the earlier extinction of many large prey animals – mammoths, mastodons, giant moose and others – that were major food sources. However, the demise of the ancient bears could also have been caused by increased competition for food and habitat with the growing numbers of black and brown or grizzly bears (see page 205).

The smaller black bear, in particular, flourished in the North American habitat. Well adapted to forested areas, it is an expert

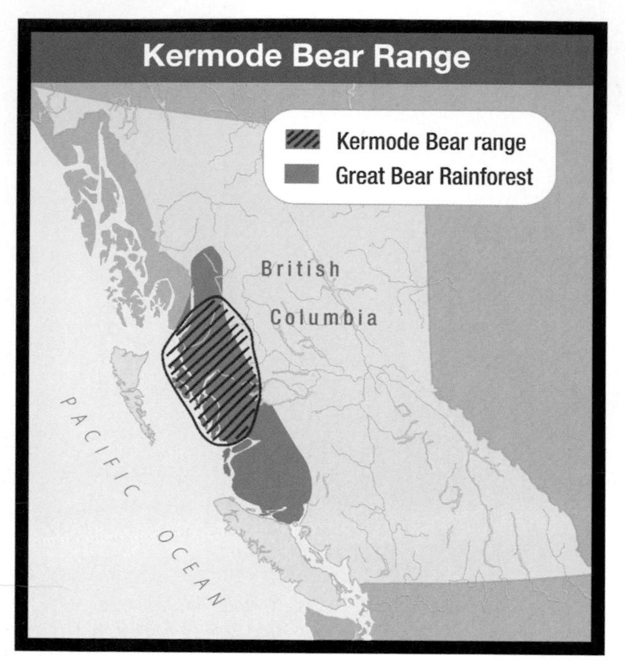

tree climber and swimmer, and can move very quickly over short distances. Although, like all members of Ursidae, black bears are considered carnivores, they are actually highly omnivorous, eating fruits, insects, plants and honey along with prey animals and carrion both small and large. Their opportunistic nature has allowed them to out-compete other bear species over the past 10,000 years, and is now both allowing them to thrive alongside human populations in many parts of the continent and, paradoxically, placing them in harm's way.

As the ice sheets waxed and waned, some of the black bears in western North America did not migrate south to escape the growing ice sheets, but rather survived on refugia – "islands" of land that escaped glaciation but were surrounded by ice or water. These refugia effectively isolated their populations and led to the creation of distinct subspecies. Of the 16 subspecies of black bears, eight are found in BC.

Biologists believe Kermode bears were likely isolated from other black bear populations about 300,000 years ago, and spent the Illinoisan and Wisconsin glaciations on the westernmost part of Princess Royal Island. The number of bears carrying the recessive gene for a white coat must have been unusually high among this small group of animals, because today one in ten black bears born on the island will have a white coat. However, blond or white black bears are very occasionally found elsewhere among North American black bears, including a population in northern Minnesota and another on Long Point which stretches out into Lake Winnipeg in Manitoba. Long Point also boasts bears with cinnamon, brown and even steel blue coats. No one has studied the genetics of white coat coloration elsewhere, to see whether it is similar to the Kermode of coastal BC.

Kermode bears have lived in close association with coastal First Nations for thousands of years and as a result, appear to have developed a tolerance for people, similar to that demonstrated by some animals of the Galapagos Islands.

Hunting the white bears has been illegal since 1967, but because black bears can carry the gene that produces white bear cubs, protecting only white bears does not protect the subspecies. Further, because Kermode bears depend on wild forests and clean rivers, with plenty of salmon, two far more serious threats to their survival are the plunging numbers of salmon, particularly sockeye and chum salmon (see page 223), and the loss of old-growth forests to British Columbia's logging industry.

In areas not designated as parks and conservation areas, the BC government and First Nations will allow some logging to continue, and much of this will be in old-growth forests. It would be foolish to deny that logging is an important industry, or that it plays a large role in British Columbia's economy. In 2009, the

Kermode Bear Fast Facts

Ursus americanus kermodei

DESCRIPTION

The Kermode bear is not a polar bear or an albino, but a subspecies of black bear which, after tens of thousands of years of evolution, has established a variant of one recessive gene that, when inherited from both parents, produces offspring with a white coat. About one in every ten Kermode bears is white, some with yellow or orange coloration on their backs; most of the rest of the population is black. Like other bears, the white-coated Kermodes have dark noses, footpads and eyes.

HEIGHT & WEIGHT

Fully grown bears weigh from 68 to 136 kilograms (or 150-300 pounds) and measure between 1.2 and 1.8 metres (or four to six feet) from nose to tail, standing less than a metre between 2.5 and three feet tall at the shoulder.

DISTRIBUTION

This distinctive subspecies is found only in British Columbia's coastal rainforests, from River's Inlet in the south to the Nass Valley in the north, and east up the Skeena River as far as Hazelton The Kermode bear population is largely concentrated, however, on Princess Royal, Roderick, Pooley and Gribbell Islands and along Douglas Channel.

MATING & BREEDING

Males have very large home ranges, up to 300 square kilometres or 115 square miles, which usually overlap with those of a number of females.

Female bears begin to reproduce at three or four years of age, mating in the spring, and giving birth to between one and four (usually two) cubs in the mid-winter, while the mother is in hibernation. About the size of a half-pound of butter, the tiny cubs are born blind and defenseless. They stay in the den until spring, nursing and growing, until their mother wakes and leads them outside for the first time. The cubs are weaned at about eight months, but remain with their mother for several years. In the wild, Kermode bears can live for more than 25 years. Elsewhere, the oldest wild black bear on record was shot in New York State in 1974; it was 41 years old.

DIET

Like all black bears, these bears are omnivorous, eating green plants, berries, fruit, insects and are opportunistic predators of larger mammals and carrion. From late summer through fall, salmon make up an important part of the bears' diet.

BEHAVIOUR

Except mothers with cubs, Kermode bears are normally solitary animals. They hibernate during the winter, often choosing to den in dry, protected cavities inside enormous old trees. When salmon are abundant, hibernating bears can live for up to seven months off the body fat they built up during the summer and fall.

NAMES

Kermode bears are named for naturalist Francis Kermode, but are often referred to as "spirit bears".

province announced a new type of logging, called ecosystem-based management. However, some scientists don't feel the guidelines go far enough, for they will allow up to 70 per cent of old-growth forests to be clearcut, including two other islands that are home to Kermode bears, Roderick and Pooley Islands.

In addition to their intrinsic and irreplaceable value, ancient forests are important for the very survival of Kermode bears, for they include massive hollowed-out trees where the bears den for protection from winter storms.

Currently, the new guidelines do not protect critical denning habitat for bears (including grizzlies); as a result, these crucial "winter apartments" will continue to be lost.

The diverse plant life of old-growth forests provides abundant berries and other sources of food. Trees also shade mountain streams, keeping them cool, and their roots

With a milky-pink nose and light blond coloring, this Kermode bear looks almost albino. Its coloring is typical of these bears, however, including its dark eyes.

DENNIS FAST www.dennisfast.smugmug.com

VIEWING KERMODE BEARS

Many tour operators are now offering spirit bear sightseeing cruises, often promoted as eco-tours, in the waters around Princess Royal Island and the inlets in the nearby mainland. Two groups dedicated to the preservation of the Kermode Bear and its rainforest home are the Valhalla Wilderness Society (www.vws.org) and the Spirit Bear Youth Coalition (www.spiritbearyouth.org).

help keep the water free of silt, providing suitable salmon habitat. These extensive root systems anchor the thin layer of soil to steep mountainsides; when the trees are removed, heavy rains and strong winds combine with the layer of bedrock just below the soil to lead to a much higher incidence of landslides, which can destroy critical bear feeding areas and salmon streams. Where the forest has been clear-cut, it is replanted with seedlings of a few tree species. These grow into dense but comparatively unvaried forests that don't provide the same habitat for bears, or many other species.

Ecotourism is currently being promoted as a key economic initiative in British Columbia. Many First Nations groups support it as a way to maintain their cultural integrity and protect species and wilderness, while providing opportunities for their people. An increasing number of charter boat operators are offering summer tours, many focused on bear viewing and whale watching. Huge Alaska cruise liners transport thousands of tourists along the coast almost daily, and Zodiacs and day boats raise the possibility of many of these passengers exploring the coastal inlets, bringing even more people into an area that was once virtually inaccessible, and disturbing the flora, fauna and sacred sites.

True ecotourism promotes biological

and cultural diversity, shares the economic benefits with local communities, increases social and environmental awareness, and minimizes its environmental impact. However, nature-themed conventional tourism is often marketed as ecologically-friendly, but is less focused on the integrity of communities and ecosystems than on making a healthy profit. If ecotour operators are responsible, and their numbers limited, ecotourism will be a way of allowing people to share in the beauty of the Pacific coast and its creatures; if not, this industry will contribute to the degradation of the world's last large area of intact temperate rainforest, and the disappearance of the spirit bear.

Other key Kermode bear habitats that require protection include Gribbell Island, where over 30 per cent of the bears exhibit the white coloring that makes them so distinctive. Mainland forest habitats must be protected as well; if they are degraded, the range of other subspecies of black bear found further inland could shift to overlap that of the Kermode bears; interbreeding with different subspecies would dilute the gene pool, making the white bears even less common.

IMAGES BY PAUL SMITH

Orca

Orcinus orca • Endangered (southern resident population, US & Canada)
Threatened (northern resident population, West Coast transient populations, Canada)

THOUGH sometimes called killer whales, orcas are in fact the largest member of the dolphin family. Among the most distinctive creatures of the sea, orcas are found in all the world's oceans, making them the second most widely distributed mammals on Earth, after humans. Though long seen as fearsome predators, they are also intelligent, social animals that are often friendly and curious towards humans both in the wild and in aquaria.

With their jet black bodies, white undersides and white oval-shaped facial patches behind and above their eyes, orcas are easily identified. Equally distinctive are their tall, triangular dorsal fins. They also have large paddle-like flippers and a dark gray area called a "saddle patch" behind the dorsal fin. The shape and markings of the dorsal fin and saddle patch are unique, and researchers use them to identify individual orcas.

In the northeastern Pacific, these magnificent animals have been revered for centuries by West Coast peoples, and feature predominantly in Aboriginal art and spirituality. More recently, they have become a symbol for biodiversity and the need for wildlife protection.

Yet West Coast orcas are declining in numbers for a variety of reasons. Three main populations reside year round off the Pacific coast: a northern and southern "resident" population and a "transient" population that differs from the resident populations in both genetics and behaviour.

Resident orcas live in tightly knit, multigenerational, matriarchal family groups; several families travel together as a larger unit known as a pod. Each pod of resident whales uses a different "dialect" to communicate with one another. This language of whistles, squeaks, whines and clicks is so distinctive that researchers can tell pods of orcas apart by the dialect they speak. They feed largely on fish, preferring salmon, and the southern resident population has a demonstrated preference for chinook or king salmon, the majestic salmon species that is itself endangered (see page 229).

The range of the resident communities overlaps somewhat, but researchers have never found the two communities in the same place at the same time. From June to September, the northern resident whales generally live in the waters between northern Vancouver Island and Alaska, while the southern resident whales can be found between southern Vancouver Island and Puget Sound.

Transient orcas can be found all along the Pacific coast year-round, generally close to shore. To the north, off Alaska's western and northern coasts, most transient populations are relatively healthy; however, one group of animals, in Prince William Sound off the southern coast of Alaska, was devastated by the 1989 *Exxon Valdez* oil spill. Numbering 22 animals at the time, it has not produced any calves since; just six orcas remained in the populations in 2008.

Transient populations are just that, with individual whales leaving their mothers and joining different pods for varying periods. Ranging widely, feeding on seals and other mammals, they live in small pods of between one and seven individuals and travel more widely than resident whales. They also dive for up to 15 minutes, while residents rarely dive for more than three or four minutes. Though quiet when hunting, transient whales are very vocal when attacking and killing prey. Possibly because of their looser social structure, all pods of transient orcas in the Pacific use the same dialect to communicate.

The number of orcas in all three West Coast populations is declining. At greatest risk – and indeed believed to be in significant danger of being extirpated – is the southern resident pod, which included just 83 orcas in 2008, down from 90 animals in 2004. In 2008, seven orcas vanished, including two breeding females and two calves. While there are many threats to orca survival, it seems the southern resident population is literally starving to death.

Studying their diet, University of British Columbia researchers Graeme Ellis and Dr. John Ford found that while 96 per cent of their prey is made up of various species of salmon, 72 per cent of that salmon is chinook, one of the least abundant (but largest) species in the northeast Pacific. So specific is their preference, Ellis and Ford found, that orcas will continue to hunt for chinook, which have a high fat content, even when other, smaller species are available.

Even worse, not only are chinook hard to find, it appears they are also highly polluted with persistent organic pollutants (POPs), toxins that are passed on to the orcas who feed on them. A study by Donna Cullon of Fisheries and Oceans Canada and published in January 2009 in the journal *Environmental Toxicology and Chemistry,* reveals that the southern resident population, which summers in the Puget Sound area of Washington and in the southern Strait of Georgia, have toxin levels four times higher than the northern resident population, which is found along the central and northern coasts of BC.

POPs include flame retardants, organochlorine pesticides and industrial chemicals, which do not break down in the environment. Some of these pollutants, such as polychlorinated biphenyls (PCBs), have been banned in

PAUL SMITH

Like most other toothed whale species, orcas like company; they hunt and travel together in pods of varying sizes.

While these orcas seem to be enjoying themselves at the edge of the arctic pack ice, whales can drown if caught beneath shifting ice floes.

COURTESY OF NOAA PHOTO LIBRARY / DEPARTMENT OF COMMERCE

Canada and the US for years, but persist in the environment and are still in use in other parts of the world. Worse, they are increasingly concentrated as they move up the food chain and like polar bears (see page 28), orcas are at the top of the food chain. This has different consequences for males and females. Males continue to accumulate organochlorines throughout their lives, while breeding females pass much of their accumulated POPs to their offspring, either during gestation or while nursing. Particularly at risk are first-born calves. The continued accumulation of POPs may contribute to their relatively short lives of males, compared to those of females. The average life expectancy for males is 29 years, while for females it is 50 years.

Research has also shown two other things that put the southern resident population particularly at risk. First, the pollutants tend to concentrate in the bodies of chinook salmon toward the end of their years in the ocean. And second, their long journey home to spawn uses much of their body fat, forcing orcas hunting close to their natal streams to consume more of them in order to compensate for the salmon's lack of fatty tissue. Appallingly, scientists recently declared that West Coast orcas are the most contaminated marine mammals in the world.

On the plus side, the US government has instituted much more stringent regulations for waste handling and cleanup efforts have led to marked improvement in water quality. In Canada, in Vancouver, two of the city's five sewage treatment plants treat waste only in a primary way, which removes only about half of suspended solids before dumping it into the Strait of Georgia. Victoria, the province's capital, dumps raw sewage into the ocean.

Also at risk and classified as threatened by the Government of Canada are both the northern resident and the northeast Pacific transient populations. The former had been growing slowly but steadily since 1981, when live-capture fisheries and commercial hunting were banned, but in recent years, the population has been in decline. From an estimated 216 orcas in 1997, it is now deemed to be about 200.

The transient population consists of about 220 animals. Though most individuals have been identified by color patterns and dorsal fin shape, counting these transient orcas is difficult; one went 14 years between appearances.

Other threats to orca survival include increased vessel traffic in the waters where they live, which not only raises the chance that they may be struck by a ship or a propeller, but also contributes to noise pollution. Other sources, such as marine industrial equipment, military sonar, and underwater construction, also contribute to this problem. Scientists do not yet know what kind of long-term effects noise pollution in orca habitat will have on the animals, but it has the potential to interfere with the whales' ability to communicate with one another and locate food.

Locating prey is increasingly difficult for orcas of both resident populations as salmon stocks decline. Whether orcas will be able to adapt to falling salmon numbers by changing their diet or residence patterns is difficult to tell. The marine mammals that are the staples of transient orcas' diets are still abundant in BC and Washington, but seal and sea lion populations are declining farther north in Alaska and the Aleutian Islands; this could have an adverse impact on orca numbers.

Even whale watching, which has been a staple of the tourism industry in many places has come under fire. Whale watching guidelines have been developed by Fisheries and Oceans Canada, and in July 2009, a protest was launched by residents of San Juan Island, Washington, against commercial whale watching boats that were following a pod of southern resident orcas. It was apparently a first in US, and perhaps North American, history.

Whales and dolphins belong to an order known as cetaceans, which are descended from hoofed mammals, resembling modern pigs or cows, more than 55 million years ago. They likely began their marine existence in shallow coastal waters, perhaps in an effort to escape land predators and take advantage of untapped food resources. Gradually adapting to their new environment, they ventured into deeper waters and began to dive for food. Modern marine mammals, which include the largest creatures on Earth, have adapted to be able to swim, dive, keep warm and find food in even the coldest, darkest reaches of the world's oceans. Many scientists believe that this evolution occurred 55 to 60 million years ago or more, during the Paleocene, following the extinction of the dinosaurs. Others feel that cetaceans' evolutionary history began in the more recent past, for the earliest fossils of toothed and baleen whales date from the Oligocene period, 27 to 30 million years ago.

Cetaceans have lived in the oceans longer than any other marine mammals. Because of this, whales and dolphins have developed more

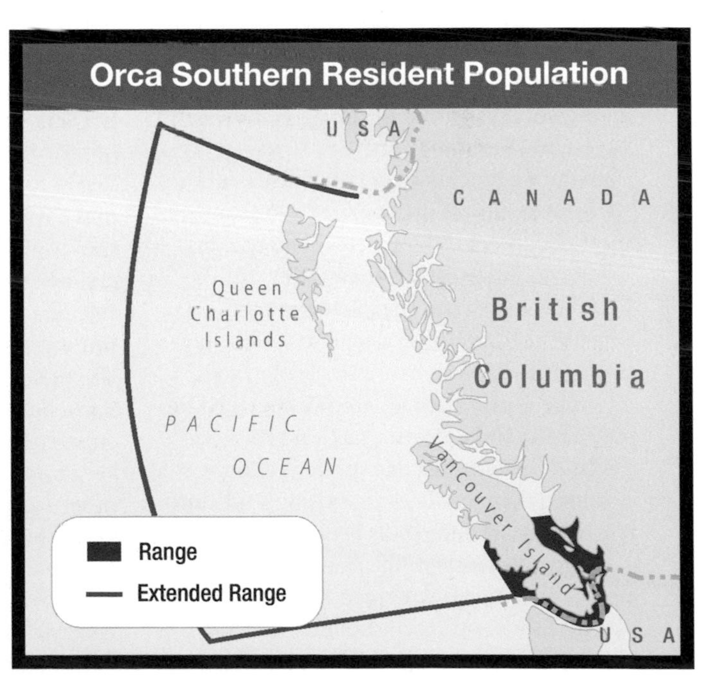

COURTESY OF NOAA / SWFSC / PRD

In the two resident West Coast populations, individuals can be identified by their distinctive markings and the shape of their dorsal fins.

pronounced physical adaptations to their ocean environment than sirenians, pinnipeds or sea otters. Like other marine mammals, cetaceans are warm-blooded, give birth to live young and nurse their offspring. But they also have sleek, streamlined bodies, rigid foreflippers and no hind limbs or dependence on land.

Cetaceans are divided into two groups: toothed whales (odontocetes) and baleen whales (mysticetes). Estimates vary, but the two likely diverged more than 30 million years ago. In addition to teeth, orcas, sperm whales, porpoises and dolphins have a single blowhole.

The *Orcinus* genus, which has existed for at least five million years, has long been believed to include just one species; recently, however, there has been increasing study into the significant differences between the wide-ranging omnivores of the transient populations, the fish-eating residential populations (such as those off North America's Pacific Coast) and, for that matter, the pygmy populations that live in the Arctic and Antarctic. Are they different subspecies, or are the distinctions merely cultural or geographic?

Genetic studies show that transient and resident orcas are as different in their genetics as they are in behaviour. The two groups likely began to diverge about two million years ago, and have not interbred for as many as tens of thousands of years. Northern and southern resident whale populations are more closely related to one another than this, but still have not interbred for hundreds of generations. Though they are classified as a single species, some believe they should be seen as distinct subspecies.

And as to whether or there have ever been other species belonging to this genus, large fossilized teeth between two and five million years old are thought to be those of *O. orca* or perhaps a related species, *O citoniensis,* an extinct whale with a higher tooth count and smaller size than today's orcas.

Orca Fast Facts

Orcinus orca

LENGTH & WEIGHT

Adult males are usually about eight metres or 25 feet long and weigh more than six tons. Females are smaller, about seven metres or 23 feet long and weighing four to five tons. The dorsal fin on a male is about 1.8 metres or about six feet high, while that of a female is half that.

DISTRIBUTION & HABITAT

Orcas are widely distributed, and are found in both cold and warm waters, but seem to prefer cooler waters and travel according to the availability of food.

DIET

North America's resident orcas feed on fish, primarily salmon and squid. They hunt in large pods, using echolocation to locate their prey and vocalizations to communicate with the other hunters. Transients feed on marine mammals such as seals, sea lions, dolphins, porpoises and small whales. They hunt in smaller packs and are generally quieter than resident orcas when travelling and hunting, using the element of surprise to their advantage when they attack their prey.

MATING & BREEDING

Orcas become sexually mature around 15 years of age, but males usually do not begin to reproduce until they reach 21. Females generally give birth to a single calf about every five years, after a gestation period of 16 to 17 months. Orcas are protective mothers, and the other resident pod members help them look after their young. Even so, nearly half of the calves born do not reach one year of age, greatly reducing the average life span of orcas. With no natural predators, if a calf survives infancy, a female could statistically live for between 50 and 80 years. However, the high mortality rates of orca calves, combined with the effects of POPs, cuts the average life expectancy to less than 30 years for males and about 50 for females on North America's West Coast. The life expectancy of captive whales is shorter than that of orcas in the wild.

NAMES

In Latin, the name *orca* means "barrel-like" or "in the shape of a cask", while *Orcinus* is probably derived from Orcus, a Roman god of the netherworld. Once referred to as "whale killers" by sailors, who witnessed their attacks on larger cetaceans – or *ballena asesina,* "assassin whale" by the Spanish – the name evolved into "killer whale". They are not known to kill humans, though occasionally they take moose or cows wading at river mouths. Orcas are also sometimes called "the wolves of the sea" because they hunt in well-organized packs.

VIEWING ORCAS

As numbers decline, watching orcas from boats is increasingly regulated and even frowned upon, though a number of companies in both British Columbia and Washington State offer whale watching tours. In the past, whale watchers (an estimated 400,000 annually in the two countries) were allowed to approach to 100 metres (Canada) or 100 yards (US) of orca pods. Proposed new regulations would nearly double that approach distance. The rules are designed to keep the surrounding waters quiet, for orcas use sonar to hunt and navigate. New rules also create a boat-free 800-metre or 850-yard zone on the west side of Washington's San Juan Island between May 1st and September 30th.

PAUL SMITH

North Pacific Right Whale

Eubalaena japonica • Endangered

"The Rarest Whale in the World"

NORTH Pacific right whales once cruised the waters of the Pacific Coast from Baja California to Alaska. Huge and robust, with enormous mouths and curtains of baleen like their northern bowhead whale cousins (see page 48), they can reach a length of 60 feet or 21 metres and weigh up to 70 tons. Like bowheads, they feed by swimming with their enormous mouths open, skimming copepods and other zooplankton from the water. However, unlike bowheads, which prefer arctic environments along the edge of the ice pack, northern right whales once populated warm coastal waters from California and northern Mexico to Alaska on the Pacific and from Florida to the Maritimes on the Atlantic. Though there are no known breeding or nursery areas on the Pacific coast, it's assumed that, like their Atlantic cousins, they raised their young in shallow, coastal bays.

Their great bulk, slow cruising speed and propensity for lingering on the surface made them the "right" whales to kill and virtually guaranteed their rapid demise once modern commercial whaling began. Beginning in the 1600s in the eastern Atlantic, tens of thousands were killed, not only for their oil, but for their baleen as well, which was used for corset stays, umbrella ribs and horsewhips.

When right whales were extirpated in the eastern Atlantic, whalers began chasing them

COURTESY OF JOHN DUNBAR / NOAA / OPR

Wart-like patches of roughened skin, or callosities, on the rostrum and above the eyes help researchers to identify individual right whales.

off the New England coast. In the Pacific, an estimated 40,000 right whales were killed during the 19th and early 20th centuries. The killing continued even after the Convention for the Regulation of Whaling took effect in 1935, for neither Russia nor Japan were signatories. Finally, in 1949, right whales were protected under an international convention, though the Russians purportedly continued to harpoon them illegally into the 1960s. By that time, however, many believed that they had been extirpated from US waters and in 1970, the US finally listed them as endangered off both coasts.

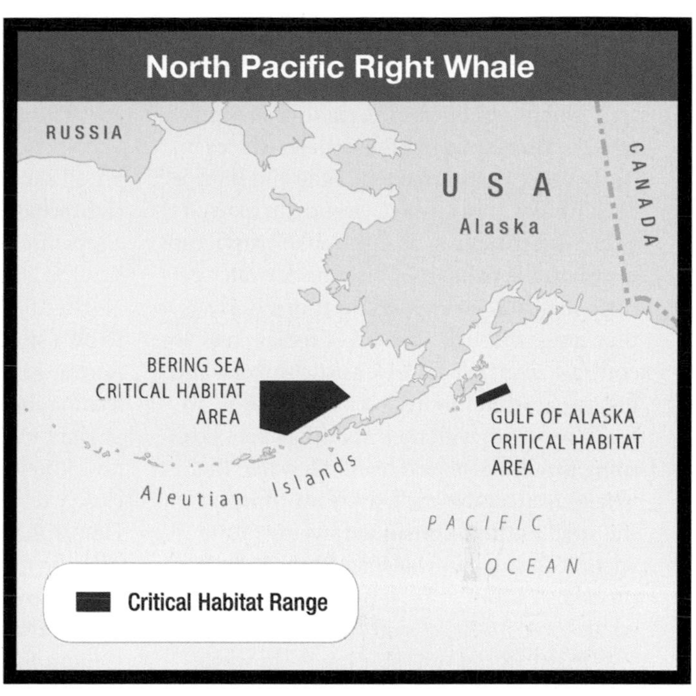

But a handful of North Pacific right whales clung to life. As The Centre for Biological Diversity notes, "In the 1980s, a sighting of a single individual was deemed worthy of publication in scientific journals. Beginning in 1996, scientists began to see a small congregation of right whales annually in the Bering Sea, and in 2005 they found more right whales in this area than in any of the previous five years."

These included at least a half-dozen females – critical for population growth – and possibly a calf or two. Only one mother and calf had been sighted prior to that, in 2002. Given that pregnancy is a year long, and that mothers nurse their calves for another year and then require a third year to regain their weight, the tiny North Pacific population cannot be expected to increase by more than one or two animals each year.

Following genetic testing, and sustained lobbying on the part of several organizations, the American government finally listed the North Pacific right whale as a separate, endangered species in April 2008, necessitating the designation of critical habitat in the Gulf of Alaska and the Bering Sea.

Despite these encouraging signs, experts estimate there are no more than 50 North Pacific right whales off the North American coast, and possibly an additional hundred off the coast of Siberia. And notwithstanding the protected "feeding box", as the Gulf of Alaska Critical Habitat has been dubbed, threats abound for the tiny population of whales. Foremost among them are ship strikes and net entanglements; reducing or eliminating both are among the main priorities of the American recovery plan for the species, which aims at down listing their status from endangered to threatened.

North Pacific right whales are substantially longer than fully-grown gray or humpback whales, and much more robust; humpbacks look positively svelte by comparison. Taxonomically, the right whale family, Balaenidae, consists of the right whales, bowhead whales of the Arctic, and the poorly known pygmy right whale of the Southern Hemisphere. Right whales are more distantly related to other, more familiar baleen whales such as humpback, gray and blue whales (see page 164).

Along with bowheads, right whales differ most noticeably from other baleen whales in the manner of their feeding. Blue and humpback whales have pleated, elastic throats and gulp food consisting of shrimplike crustaceans or fish in the manner of huge underwater pelicans, their throats expanding enormously as they lunge through a school of fish or krill. By contrast, right whales lack elastic throats, and instead have enormous, very tall mouths. To feed they swim with their mouths open, skimming tiny zooplankton from the water. Right whales feed almost exclusively on copepods, the smallest food consumed by any of the whales, so small the baleen of other whales probably cannot strain it out. Other baleen whales feed on larger krill and small fish.

Right whales resemble bowhead whales so much that whalers, and prior to the late 19th century even scientists thought they were the same species. However, bowheads occur only in the Northern Hemisphere, and have a distribution that tends to be considerably more polar than the right whales. Bowheads tend to be found very close to, or in the pack ice. Right whales, by contrast, prefer temperate waters; southern right whales can be found wintering in the waters off southern New Zealand, Australia and Argentina.

Despite the scientific confusion, modern observers can distinguish right whales from bowheads without much difficulty because of the wart-like patches of roughened skin on the rostrum and above the eyes of right whales. These white "callosities", which are usually covered by yellowish cyamids known as whale lice, give the whales the appearance of having large white patches on their heads. These patches are permanent enough to be used to individually identify right whales in numerous studies.

Bowheads not only lack callosities, but are larger and often have a large white patch on the lower part of their jaws. To distinguish the two species, the right whale has often been referred to as the "black right whale".

Though black right whales of the North Pacific, North Atlantic and southern oceans are all similar in appearance, the southern right whale, *E. australis,* has been considered a separate species for many years. However, prior to 2000, scientists considered both the North Atlantic and North Pacific right whales to be a single species – *Eubalaena glacialis.* Recent genetic studies have changed that, leading them to conclude that the populations of right whales living in the North Pacific – now known as *E. japonica* – are actually more closely related to right whales of the Southern Hemisphere than they are to their North Atlantic relatives, *E. glacialis.*

Many people are familiar with right whales thanks to Roger Payne's National Geographic Society television specials on the southern right whales. In late 2008, the society's very popular magazine, *National Geographic,* featured North Atlantic right whales, which annually run a gauntlet of ships, nets and lines as they migrate back and forth along North America's Atlantic coast between Florida and the Bay of Fundy. One of the busiest shipping and fishing corridors in the world, it's home to the tiny *E. glacialis* population, between 350 and 400 members of a species that once numbered 10,000 or more.

The southern right whale population was once considerably larger, numbering an estimated 70,000 animals. Though also devastated by whaling in the 20th century, the southern population is recovering at a rate of seven per cent annually and now numbers more than 7,500 animals. Southern right whales feed in the plankton-rich waters surrounding Antarctica and then winter in the temperate coastal waters off Argentina, southern Africa, New Zealand and Southern and Western Australia.

North Pacific Right Whale Fast Facts

Eubalaena japonica

LENGTH & WEIGHT
Males average about 50 feet or 15 metres long, while females, which are generally larger, can reach 60 feet or 18 metres, and weigh 70 tons or more. Large members of the species can weigh 100 tons, making them among the heaviest animals on Earth.

DESCRIPTION
Very dark blue or brown, stocky and rotund, right whales have long, fine curtains of baleen, no dorsal fin, broad tail flukes, and distinctive roughened patches of skin on their heads. Like fingerprints, these wart-like callosities are unique to each animal, allowing whales over a year old to be visually identified. Right whales also have two widely spaced blowholes, which give them a distinctive V-shaped blow that can reach up to 16 feet or five metres out of the water.

LIFESPAN
Like bowheads, right whales are long lived, but scientists believe an elderly right whale might be 100; bowheads more than 200 years of age have been found.

HABITAT
Historically, North Pacific right whales occupied a band of the Pacific that stretched north from Japan, across the Bering Strait and south along the North American coast to California. Until recently, they were believed to prefer coastlines and large bays between 20 and 60 degrees north latitude for breeding, much as their southern and Atlantic right whale cousins do. However, since they have rarely been spotted along the Pacific coast from California to British Columbia in either historical records or in recent years, marine scientists now believe that their breeding grounds may in fact be offshore, in the open ocean. Today, the tiny surviving population summers in the Bering Sea and the Sea of Okhotsk, but its present wintering habitat is unknown.

MATING & BREEDING
Right whales eight years and older mate between December and March and, following a year-long gestation period, a pregnant female will give birth to a 2,000-pound or one-tonne calf. During mating, the whales are flirtaceous and affectionate, nuzzling and stroking one another and rolling on the surface of the water, exposing flukes, bellies and backs. It's believed that right whales are polygamous. Females appear to breed only once every three to five years.

BEHAVIOUR
The wart-like callosities on the head of a right whale are often covered with a type of lice, and the whales are believed to sometimes swim upside down, scratching their huge heads on the ocean floor. Like most other whales, they are curious, intelligent and unaggressive.

RELATIVES
Right whales are closely related to bowheads, and more distantly related to blue, humpback and gray whales.

RECOVERY STRATEGIES
A sanctuary has been created in the southeastern Bering Sea off Alaska where right whales are known to feed during the summer months (see map on page 131). Canada's Department of Fisheries and Oceans undertakes annual surveys or, perhaps more correctly, scouting missions off the BC coast, but no North Pacific right whales have been seen off Canada's Pacific coast for more than 50 years. On the Atlantic coast, volunteers and scientists are part of a network of spotters that undertake daily visual inspections from beaches, cliffs, office towers and planes. Information gathered this way is phoned by a hotline to an early warning system, which passes it on to commercial and military ships, which can then alter their courses to avoid collisions.

NAMES
The name rings with irony, since thanks to their great bulk, slow speed and thick layer of blubber, along with a propensity to float after they die, whalers long considered this species the "right" whale to kill.

VIEWING RIGHT WHALES

The number of North Pacific right whales is so small that seeing them takes enormous time and expense; on the Atlantic coast, right whales can be spotted during the winter months off the coasts of Florida and Georgia. But perhaps the best place to spot right whales is in the Southern Hemisphere, where whales gather in significant numbers off Argentina's east coast, off the Auckland Islands and off Australia's southern and western coasts. In Australia, in fact, members of the southern population have created a growing whale-watching industry.

Though very large, right whales feed on tiny shrimplike creatures called copepods.

COURTESY OF MARINE MAMMAL COMMISSION / FISHERIES AND OCEANS CANADA

Trans# R5-4308(
RECORD
Card Type: VI
CHASE
233I
1325016
1413906
C2141390601
07-30-2019
15:35:41
$5.25
APPROVED - THANK YOU
this copy for your records
CUSTOMER COPY ***

); Endangered (IUCN)

sula and south along the North Ameri[...] to Baja California, some estimate the sea otter population in the early 1700s at between 150,000 and 300,000. Others believe there may have been as many as 600,000 sea otters throughout their range.

Almost all were doomed in the aftermath of Russian adventurer Vitus Bering's second Kamchatka expedition, between 1740 and 1742. Returning from Alaska, Bering's ship was wrecked on what became known as Bering Island. The captain and half the crew died, but the expedition's chief naturalist, German biologist Georg Wilhelm Steller, survived and used the terrible winter to study the region's fauna, among them the Steller's sea cow, Steller's (or northern) sea lion, the northern fur seal, and the sea otter.

Following his return home, Steller wrote about all these creatures in a book entitled *De Bestiis Marinis, (The Beasts of the Sea),* which was published in 1751 and soon translated into English. The Bering expedition sparked 150 years of what has been called The Great Hunt, the wholesale slaughter of all these animals, and led to the extinction of the small remnant of Steller's sea cow, the extirpation in many parts of its range of Steller's sea lion and the near demise of one of the most appealing creatures on Earth, the sea otter.

Sea otters were hunted for their fur – the densest on the planet, with 100,000 or more hairs per square centimeter or 650,000 hairs

LIBRARY AND ARCHIVES / ACC. NO. 1991-265-126

This drawing by John Woodhouse Audubon was created when he visited California at the beginning of the Gold Rush. The artist just once saw a live sea otter, in a river estuary. He therefore imagined the otter holding a large fish, which are rarely, if ever, eaten, and inaccurately posed it standing on a rocky shore.

per square inch. By 1911, the species was at the edge of extinction. Just 13 remnant populations, totalling between 1,000 and 2,000 animals, were believed to exist. Most were in the northernmost reaches of their range, from Russia's Kuril Islands and Kamchatka Peninsula through the Aleutian Islands and along the Alaska Peninsula. Tiny populations also clung to survival in the waters of Haida Gwaii, off the northern British Columbia coast; off the coast of central California and around the Islas San Benito off Mexico's Baja California. Though an international hunting ban was instituted, several of these remnant populations disappeared.

Sea otters were extirpated from Canada for 40 years, between 1929 and 1969, when they were reintroduced. Population estimates in 2007 indicate close to 3,500 sea otters living in Canadian waters along Vancouver Island and the mainland coast. Approximately 1,300 sea otters live off the Washington coast, along a 70-mile stretch from Destruction Island to Pillar Point, while Californian sea otter numbers have slowly grown; between 1999 and 2009, they hovered between 2,000 and 2,800 animals.

Currently, nearly 75 per cent of the world's 100,000 sea otters live off the coast of Alaska. In 1973, the population in Alaska was estimated at between 100,000 and 125,000. However, three decades later, the number of sea otters in Alaskan waters had fallen to an estimated 43,000, largely because of a huge decline in numbers in the Aleutian Islands. Orca predation is suspected as the main reason for the decline, perhaps the indirect result of a decline in harbor seal and sea lion populations. Sea otters may have been the most recent victims of the orcas need to replace whales in their diets. Other threats also exist. The *Exxon Valdez* oil spill, for example, killed between 2,650 and 4,000 sea otters in 1989, while nuclear testing in the Aleutians in 1965 and 1971 killed or injured many of these animals and other wildlife. In part to regain public goodwill, the American Atomic Energy Commission helped to transplant 89 Alaskan sea otters to the coast of Vancouver Island as well as Oregon, Washington and southeast Alaska.

The sea otter is both the world's smallest sea mammal and the heaviest member (at up to 100 pounds or 45 kilos) of the Mustelidae family, which includes 12 other otter species worldwide, as well as terrestrial animals such as weasels, badgers and mink. It is unique in other ways, for it is the only mustelid that does not make a den or burrow, has no anal scent gland and is able to live its entire life in the water.

As their Latin name suggests, sea otters likely descended from an otter-like creature known as *Enhydritherium* some five to seven million years ago. Believed to be the most recently evolved marine mammal, fossil evidence

RON WOLF

The world's smallest sea mammal, sea otters are delightful to watch while delicately eating, below opposite, or simply enjoying life, above.

RON WOLF

indicates that the Enhydra lineage became isolated in the North Pacific about two million years ago, which coincides with the beginning of the Pleistocene, Earth's current ice age.

Fossil evidence indicates that sea otters (and there were once at least two species) first evolved in the eastern North Pacific and then spread west. As research biologist Jim Estes writes, "Enhydra probably was confined subsequently to the North Pacific by the barriers of sea ice to the north and warm water to the south." A fossil of a second sea otter species – *Enhydra macrodonta* – found in northwestern California at the end of the last glaciation, has led to speculation that ancient human hunters may have been responsible for its extinction.

Their recent evolution is evident from their physical characteristics, for though their back feet are modified flippers, their front paws are still those of a terrestrial carnivore. They have large lungs that allow them to dive for minutes at a time while hunting for food; they can see well and forage in dim underwater conditions, and do not depend on land to mate, give birth, or raise their pups. By contrast, pinnipeds such as seals and sea lions must haul themselves out on land to give birth.

Unlike whales, dolphins, seals, walruses and other mammals that spend their lives in the oceans, sea otters do not have a layer of blubber to keep them warm. Instead, they have the thickest coat of any animal, with up to 800 million tightly packed hairs, beautiful dark brown hair that not only prevents frigid water from touching the otter's skin, but also absorbs sunlight to aid in warming.

Sea otters have few natural predators; they are hunted by killer whales, which may be responsible for a decline of more than 90 per cent in the Aleutian Islands since the 1980s, as well as sharks and bald eagles. Humans remain a significant threat to the otters' well-being, for the fur harvest shows a clear capacity for overharvesting. In addition to the devastation

Sea otters have strong back legs for paddling, and use their broad, flipper-like feet and flat tails as rudders as they swim on their backs. They spend most of their lives at sea, moving slowly and awkwardly when they do go ashore.

DISTRIBUTION

Today sea otters occupy most of their historic range from the Kuril Islands northeast of Japan to Prince William Sound in Alaska. An estimated 70 to 80 per cent of the world's sea otter population lives in Alaskan waters. At the southern end of its range, the California sea otter is found along the coast from just south of San Francisco Bay to Point Conception. There are also transplanted populations of sea otters in southeastern Alaska, British Columbia, Washington State, and San Nicolas Island, off the coast of Santa Barbara, California.

HABITAT

Sea otters live in shallow coastal waters. Their northern permanent range is limited by sea ice, while in the south their distribution coincides with the southern limit of cool water upwelling and distribution of giant kelp. Sea otters often wrap themselves in kelp when at rest to keep from drifting and use kelp canopies to hide in. The reintroduction of sea otters has helped to improve the marine environment for other species; otters keep the sea urchin population in check, allowing the kelp forests on which other species feed to flourish, which provides habitat for fish and reduces coastal erosion.

a great deal of body heat, as a result, they need to consume about a quarter of their body weight in food every day. So many sea urchins are consumed that botanists have found sea otter skeletons that are often purple in color from the pigment in the urchins.

BEHAVIOUR

Meticulous in their grooming habits, sea otters spend several hours a day cleaning and fluffing their fur to improve its insulation. They also conserve heat by keeping their furless paws and hindflippers out of the water when resting.

Sea otters socialize in groups called "rafts." Females gather in groups along with pups and sometimes a territorial male, while males of all ages raft together in larger groups. It seems that males tend to travel farther and more often than females.

MATING & BREEDING

Sea otters become sexually mature between two and five years of age. A territorial male will mate with several females throughout the year, while females often form a pair bond with one male for the three or four days that they are fertile. During this time, the couple will feed, groom and mate together, but they separate soon after the mating period. Female sea otters nearly always bear a single pup in water close to shore and care for it with great devotion, carrying it around on her chest or back and providing for all its needs and training until it can fend for itself at about six months of age.

VIEWING SEA OTTERS

Sea otters can be seen in a number of places from Alaska to California, mainly where the coastal waters are protected from strong ocean winds, such as barrier reefs, rocky coastlines or kelp forests. sA company on BC's Salt Spring Island offers sea kayaking to view otters, and captive populations are found at the Vancouver and Monterey Bay Aquariums. Both take orphaned or injured otters.

A "raft" of nearly a dozen sea otters – paws in the air to keep them warm – watches a photographer with great interest. They often link paws to keep from drifting apart.

DENNIS FAST www.dennisfast.smugmug.com

caused by oil spills, which mats their coat, causing it to lose its insulating properties, industrial runoff can pollute the near-shore environments preferred by sea otters, contaminate the food they eat and be ingested as they groom.

As efforts to determine the cause of the declining numbers grow, disease is increasingly recognized as taking a toll on sea otter populations. Scientists studying the California sea otter have found that since 1998 disease has been a factor in an astounding 40 per cent of sea otter deaths annually. The reasons the otters are developing intestinal worms, bacterial and fungal infections, and protozoal encephalitis are not clear. It has been speculated that organochlorine compounds, polychlorinated biphenyl (PCBs), heavy metals, and other types of pollution from agricultural run-off and industrial discharge are suppressing the animals' immune systems, causing organ failure, and possibly inhibiting normal reproduction.

Another threat to sea otters and other marine creatures such as seabirds, porpoises, sea lions and seals, are gill and trammel nets used by the fishing industry. Hanging like curtains in the water, they are designed to catch large fish, while letting smaller ones pass through, but other species can become entangled and drown.

Today, the worldwide sea otter population is estimated at about 95,000 animals. Sea otters are protected throughout their North American range; in Canada they are considered threatened by COSEWIC, while the California sea otter is listed as a threatened species under the federal Endangered Species Act. Since 2005, the southwestern population of sea otters in Alaska has been listed as threatened as well. Since 2006, the Sea Otter Bill has provided protection for sea otters and coastal ecosystems in California and direct funding for research, conservation and recovery programs. Nevertheless, several sea otter populations continue to decline and all remain vulnerable to pollution, environmental disasters, and other human-caused threats.

In 2008, the IUCN listed sea otters as endangered.

Vancouver Island Marmot

Marmota vancouverensis • Endangered

ONCE, in days of relative abundance, they were called "whistle pigs" and compared to overfed domestic house cats. In the late 19th century, a settler's diary referred to "swarms of ground hogs", and in 1922, a hunter boasted of killing a "brace" of marmots.

But since the Vancouver Island marmot was recognized as Canada's most endangered mammal, these mountain-dwelling vegetarians have won considerably more respect. In the mid-20th century, marmots occupied sites on 15 mountains on central Vancouver Island, and even in the mid-1980s, the wild population was estimated at more than 350. Over the last 15 years of the 20th century, however, marmots disappeared from two-thirds of their historical range. By 2004, it was estimated that fewer than 35 individuals were living on the high mountain meadows of central Vancouver Island, with another 93 animals in captivity.

It was once thought that Vancouver Island marmots, one of four marmot species in Canada and five in North America, were isolated from mainland populations relatively recently, perhaps 10,000 years ago, at the end of the last glaciation. But discoveries in the past few years have altered that idea. David Nagorsen, mammal curator at the Royal British Columbia Museum for two decades and now a member of the Vancouver Island Recovery Team, says that "remarkable new cave bone discoveries" include 16,000-year-old marmot bones from a sea cave on Port Eliza on the island's rugged west coast. Clearly, Vancouver Island marmots have been around longer than was previously believed. In fact, some biologists now believe that their isolation and subsequent evolution may have occurred prior to the Early Wisconsin or Penultimate glaciation, about 65,000 BP.

Above: A youngster gets to know its alpine home, while its elders, above left, keep watch.

JACK MOST: www.themostinphotography.com

JACK MOST: www.themostinphotography.com

Vancouver Island marmots require very specific conditions to thrive, including mountain meadows above 800 metres or 2,650 feet in elevation, where deep winter snow serves as an insulating blanket and winter avalanches help to keep trees from rooting. They also require deep soil to allow the creation of large burrows and support a lush growth of flowers and grasses, and large rocks for sun bathing and for use as lookout posts. Though an adult marmot ranges in weight from 3.5 to 6.5 kilos or seven to 14 pounds, these gregarious creatures are strictly vegan, dining on more than 50 species of flowers and grasses. Their weight is gained during the relatively brief alpine summers, for Vancouver Island marmots spend most of their lives in their burrows. From mid-September to late April each family or colony of families lives in large hibernacula a metre or more below the ground and extending up to four metres or about 13 feet in length. Families include an adult male, one or more adult females, along with subadults, yearlings and young.

Though studies have shown Vancouver Island marmots to be remarkably vigilant, quick to whistle alarms and reliably digging numerous escape burrows, the dramatic decline in numbers seems to have been largely due to increased predation by cougars, wolves and golden eagles, complicated by an increased use of unsuitable habitats. Unlike many endangered species, the habitat of Vancouver Island marmots has remained relatively intact, for their preferred sites are generally remote and treeless. In recent decades, however, they have taken to colonizing clearcut mountainsides, which initially resemble their natural habitat. This not only diverts the population from natural higher elevation meadows, but results in locations that are suitable for only about 15 years, for as the trees regrow, they provide ideal cover for predators. Since the marmots' average life span is less than 10 years in the wild, the colony's collective memory may not include colonies established on natural meadows. These changes have likely had an impact on survival rates in these new colonies.

To determine the cause of a steep decline of marmot populations, beginning in 1992, a total of 78 marmots were surgically implanted with tiny radio transmitters. These were tracked for a period of months or years by a team intent on determining the causes of death. Of the 37 that either died or disappeared (perhaps after the transmitter ceased to function), 38 per cent were preyed upon by wolves, 21 per cent by cougars and 10 per cent by eagles. Thirteen per cent died of unknown causes, 10 per cent of winter mortality and one apparently fell from a cliff. But since wolves, cougars and eagles had always preyed on marmots, why had the population crashed the way it did in the late 1980s and early '90s?

After looking at a variety of possibilities, it seemed the most likely answer lay not with the marmots themselves, but in numbers of black tailed deer, the prey most favored by wolves and cougars on Vancouver Island. Over the decades prior to the 1990s, deer populations had fluctuated greatly, increasing in the 1970s, possibly due to new food sources in clearcut areas, and then dramatically declining. Shadowing the increased number of deer, cougar populations also grew significantly in the late 1970s. And wolves, which had been so diminished in numbers through the early decades

To publicize its endangered status, the Vancouver Island marmot, "Mukmuk", was named a "sidekick" to the three official marmots of the Vancouver 2010 Winter Olympics: Miga, the sea bear, Sumi, the thunderbird, and Quatchi, the sasquatch.

IMAGES BOTH PAGES: JACK MOST: www.themostinphotography.com

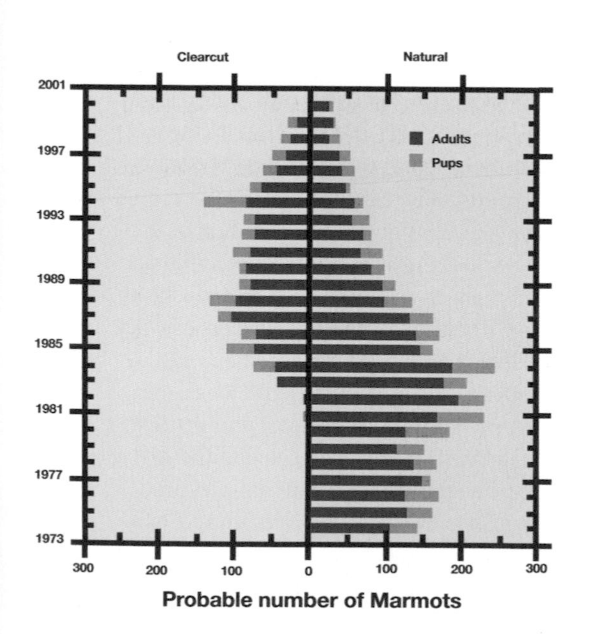

of the 1900s that in 1970 serious consideration was given to listing them as endangered, became so common by 1977 that legal wolf hunting was allowed. Then, in the early 1980s, just as deer populations were declining, and cougar and wolf populations were larger than they had been for decades, Vancouver Island marmots began to colonize clearcuts in significant numbers. Though the real problem was predation, and perhaps as marmot numbers declined, inbreeding and limited dispersal in isolated colonies, the result, though not evident at the time, may have been inevitable.

Natural meadows (and therefore marmots that inhabit them) can also be threatened by natural succession, which occurs when trees invade sub-alpine meadows. Colonies can be destroyed by fire, by ski hill development or by forestry, though there is no conclusive evidence that logging, per se, has affected marmot populations. However, both marmots and their

Above: A mountain biker sails past a marmot burrow, while below, a sign marks a favorite road crossing.

predators have regularly been seen using logging roads.

The researchers also found that the majority of deaths from predators occurred largely between early August and mid-September, with the greatest number of deaths concentrated in the last two weeks of August. The reason for this spike in mortality is not yet apparent.

Like many other species, marmots go to great lengths to avoid close interbreeding. To safeguard against it, approximately one in three male or female two-year-olds leaves the colony of its birth and travels up to 20 kilometres or 12 miles to join an established colony or to form a new colony with a mate. Clearly then, a growing population depends on the ability of these marmots to move safely from one alpine meadow to another. In other words, like elk, caribou and many other mammals, Vancouver Island marmots require safe migration corridors.

By 2009, the world total had risen to more 450 animals, 174 of which are in breeding facilities in Toronto, Calgary, Langley, BC, and on Vancouver Island's Mount Washington. In an attempt to restore and retain the population at between 400 and 600 animals, a multi-agency, multi-pronged program has been developed, focused on maintaining existing habitats and safeguarding wild populations, as well as on captive breeding and reintroduction programs. To enhance the odds of survival, a variety of experimental safeguards were tried, including fencing, netting and even human shepherds, to prevent predation in reintroduced colonies.

By late 2008, survival rates appeared to be increasing; with 74 per cent of marmots surviving through the year. By comparison, survival rates are close to 90 per cent for Colorado's yellow-bellied marmots and Europe's alpine marmots, according to Andrew Bryant, a private consultant who has worked closely with the recovery program.

Though government officials were determined to take a wait-and-see attitude before calling the recovery program a success, by 2009, the numbers were beginning to speak for themselves, with a wild population of between 230 and 280, a total of 233 captive-bred marmots released into the wild between 2003 and 2009 (just four were released the first year), and solid evidence that at least some of those were adapting extremely well to their age-old haunts. Among those is Haida, a captive-bred female released near Haley Lake in 2004. Over the next four years, she had two litters, in 2006 and 2008, and two of her pups from 2006 also had pups in 2008, making her an excellent candidate for mother (and grandmother) of the year. The year also marked a another milestone, for by summer's end, marmots were living on virtually every mountain where they had been historically documented.

Like the other 13 marmot species worldwide, Vancouver

Vancouver Island Marmot Fast Facts

Marmota vancouverensis

WEIGHT & LENGTH
The Vancouver Island marmot is the largest member of the squirrel family, with males averaging about five kilos or 12 pounds. Males can grow to be 70 centimetres or 28 inches long, while females are smaller.

LIFESPAN
Males generally live about six years, while females of eight or nine are not uncommon. Breeding every other year, a female might have between 12 and 15 pups during her lifetime.

CALL
Vancouver Island marmots are very vocal, with three distinct alarm calls and a unique "kee-ow" trilling sound.

ACTIVITY
When above ground, usually in the morning and evening hours, marmots spend their time eating, lounging, nose-touching and boxing.

RELATIVES
The Vancouver Island marmot's closest relative is the hoary marmot found in the northwestern US, Alberta and British Columbia; it is also related to Washington State's Olympic marmot and the yellow-bellied marmot of the mountainous regions of the northwestern US, southern Alberta and BC. North America's other marmot species include the Alaska marmot, which is found in mountains of Alaska and Yukon, and the woodchuck, which is found from the eastern US across Canada's aspen parkland to Alaska.

NAMES
Species of marmots in the eastern United States are often called ground hogs or woodchucks. Vancouver Island marmots were once known as whistle pigs, for their wide range of alarm calls and whistles used to communicate.

VIEWING VANCOUVER ISLAND MARMOTS
In the wild, marmots can be seen on the slopes of Mount Washington on Vancouver Island, as well as nearby mountains. Calgary and Toronto Zoos both breed Vancouver Island marmots for release, as does Mountain View Conservation and Breeding Centre just outside Langley, BC.. Guided tours at Mountain View can be arranged by email at visit@mtnviewconservation.org

A venerable marmot looks out on a ski hill, shown opposite left, perhaps satisfied that it has done its best to perpetuate its species.

IMAGES BOTH PAGES: JACK MOST: www.themostinphotography.com

Island marmots have beaver-like teeth, sharp claws and strong shoulder and leg muscles for burrowing. Closely related to hoary and Olympic marmots, Vancouver Island marmots differ in color and behavior. Their chocolate-brown coats have white patches on the nose, chin, forehead and chest, and this coloring can readily be seen as the marmots sun themselves on boulders or logs.

This propensity for sunbathing creates the need for relatively treeless conditions, for though blessed with acute hearing, even while basking on a log, a marmot must be continually on the lookout for predators. For most marmot colonies, these conditions are found in sub-alpine meadows between 1000 and 1400 metres in elevation. As discussed earlier, in recent decades, Vancouver Island marmots have taken to colonizing clearcut sites, which are usually located below 1000 metres.

Before hibernation, adult males can weigh up to seven kilograms or nearly 15 pounds, while females can be up to 5.5 kilograms or 11 pounds. Though their heart rate declines from between 110 and 200 beats per minute during the summer season to just three or four beats per minute in hibernation, marmots will lose about a third of their weight over the winter months. Despite this weight loss, marmots mate in early May, soon after emerging from their hibernacula. Both males and females begin breeding at three or four years of age and, following a gestation period of about a month, females usually give birth to three or four pups every other year, though litters of six have been recorded. Marmots are attentive parents and the well rounded pups can usually be seen above ground in July.

During their first winter, the pups hibernate in beds well padded with dried grass and moss, and stay with the colony until after the second winter.

Short-tailed Albatross

also called Stellers Albatross

Phoebastria albatrus • Endangered (US); Threatened (Canada)

FOR millennia, majestic birds with snowy white plumage and wingspans of seven feet, or more than two metres, owned the skies off North America's West Coast. Soaring on the wind, demonstrating a mastery of the air that was nothing short of awe-inspiring, short-tailed albatrosses were the largest and most numerous of the three albatross species that inhabited the North Pacific. As many as five million birds spent their lives gliding over the northern Pacific Ocean and the Bering Sea, hunting for food. But when the urge to mate overtook them, they headed west into the sunset, where they bred on a handful of islands off the south coast of Japan.

Then, beginning in the early 1880s, the Edwardian fashion for huge, broad-brimmed hats adorned with long feathers swept Europe and North America. Rsponding to the demand, in 1887, Japanese plume hunters headed for their nation's remote islands, where birds with long white feathers – birds that were remarkably unafraid of humans – were known to nest each year by the hundreds of thousands. Moving from one island to another, they slaughtered virtually every adult bird they could capture.

The demand for feathered hats and boas brought the short-tailed albatross, which *National Geographic* writer Carl Safina has called "the grandest living flying machine on Earth" to the brink of extinction.

To be fair, the Japanese plume hunters were likely unaware that the birds had come halfway across the world to nest on the islands, or that these few islands, just a dozen or so, were the breeding grounds for the Earth's entire population of short-tailed albatrosses. Nevertheless, in less than two decades, an estimated five million birds were slaughtered.

Fortunately, a number of juveniles were far across the Pacific and escaped the killing; after the killing ended, these young birds returned to nest on Torishima Island, the main breeding colony. However, in 1932, in anticipation of Japan's intention to establish the island as an albatross refuge, island settlers slaughtered almost all of the remaining 3,000 birds. The following year, fewer than 50 birds were seen, and, as if that was not bad enough, the tiny surviving population was further devastated in 1939 when a volcano on the island buried their breeding grounds under nearly 100 feet or 30 metres of lava. A decade later, no birds were seen on any of the known breeding colonies, and many believed the short-tailed albatross to be extinct.

Albatrosses have a very long history. A fossil seabird, known as *Tytthostonyx*, was found late Cretaceous rocks in New Jersey's Hornerstown Formation, formed about 65 million years ago. Though it doesn't resemble any living albatross forms, it may have been a distant ancestor. The earliest recognizable albatross fossil found thus far dates to between 30 and 35 mya. The North Pacific albatross appears to have diverged from the great albatross over the past 15 million years. And fossil evidence also shows that albatrosses once inhabited the North Atlantic. Though the region is not home to any varieties of the bird today, fossils show that the island of Bermuda once supported a colony of short-tailed albatrosses.

C. FRED ZEILLEMAKER / US FISH AND WILDLIFE SERVICE

Its dark plumage betraying its youth, a short-tailed albatross works at its flying lessons.

Despite all efforts to eliminate it, the species survived and in 1950, a few nesting birds were reported; by 1954, there were at least six breeding pairs on Torishima Island. And Japanese ornithologist Hiroshi Hasegawa began a campaign to bring the birds back from the brink.

To the south, reports in the early 1960s indicated that an albatross pair may have successfully nested on Midway Atoll, at the northeastern end of the Hawaiian archipelago, which had been set aside by US President Theodore Roosevelt in 1909 as a bird sanctuary in the wake of the Japanese slaughter. Since 1941, at least one and as many as several adults have been sighted annually on Midway. And following the American listing of the species as endangered in US waters, a recovery program was begun, with the goal of establishing a viable colony on the Pacific archipelago.

In Japan, meanwhile, mostly due to Hasegawa's efforts, numerous habitat improvements were undertaken on Torishima Island. To stabilize the volcanic ash and prevent it from being washed away (carrying the birds' nests, eggs and young with it), conservationists planted grasses on the breeding grounds. And strict protection for the island was initiated.

As the nesting population on Torishima increased, the colony there continued to be

The Ancient Mariner

As Samuel Taylor Coleridge's famous poem indicates, the albatross was often seen as a good omen by sailors, though this did not stop them from killing the birds for food or sport.

At length did cross an Albatross,
Thorough the fog it came;
As if it had been a Christian soul,
We hailed it in God's name.

It ate the food it ne'er had eat,
And round and round it flew.
The ice did split with a thunder fit;
The helmsman steered us through!

And a good south wind sprung up behind;
The Albatross did follow,
And every day, for food or play,
Came to the mariner's hollo!

– from *The Rime of the Ancient Mariner*
by Samuel Taylor Coleridge

threatened by volcanic eruptions. As a result, considerable effort has been put into trying to induce the birds to nest elsewhere. In 1971, 12 short-tailed albatrosses were discovered on Minami-Kojima in the Senkaku Islands, and in 1988, breeding was confirmed there.

Other land-based threats to the species include introduced predators such as rats and cats, against which albatrosses have not evolved any defences, and exotic vegetation that could reduce the suitable nesting habitat.

Out on the ocean, where the birds spend the vast part of their long lives, short-tailed albatrosses face other threats. They are vulnerable to what is known as bycatch, being accidentally killed by longline fishing gear, by becoming entangled in fishing nets, being befouled by oil or ingesting plastic debris.

In the late 1980s, Australian biologist Nigel Brothers raised the alarm that albatrosses and many other seabird species were declining primarily because they were being caught on longlines. Longliners deploy fishing lines up to 50 miles or 80 kilometres in length, with thousands of baited hooks. Drawn by the possibility of a free lunch, seabirds trail the boats, trying to steal the bait before the lines sink. However, many birds are hooked in the process and drown. Of the 22 recognized albatross species, 18 have declined in recent decades, with bycatch the main reason for the decline.

Working with fishermen in various parts of the world, Brothers and others are attempting to find solutions: setting the lines off the side of the boat, and weighting them so that they quickly sink beneath the hull, out of the birds' reach; dying the bait blue, or setting the lines at night, making them all but invisible.

As a result of all these efforts, the world population of short-tailed albatrosses is slowly growing. In 2008, it was estimated at 2,364 birds. And once again, these magnificent flying machines are being spotted on the coastal waters of Alaska, British Columbia and Oregon.

The largest of the North Pacific albatrosses, and the only one with a white body, the short-tailed albatross has narrow white wings with black tips. Almost three feet or 90

DARNA MICHIE, EAST ANGEL HARBOR HATS, www.eastangelharbor.com

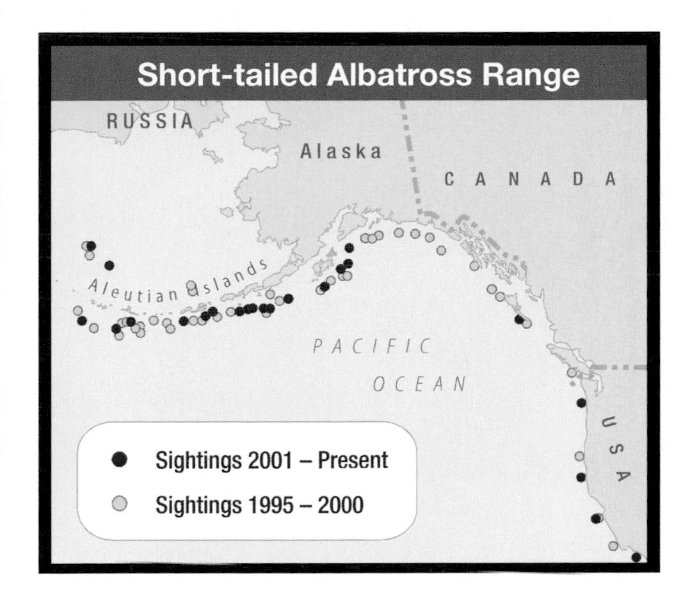

centimetres long, adults have a large, pink bill with a blue tip, a golden yellow cast to the head and neck, and black tail fringes. Other than their pale bills and legs, juveniles are almost entirely brown for several years.

Sexual maturity occurs at an average of six years of age, and short-tailed albatrosses mate for life, returning to approximately the same nest site each year. Building a nest with surrounding sand, volcanic debris and shrubbery, the female lays a single egg in October or November, which the pair takes turns

"ALBATROZ" WOODCUT FROM THE JOURNAL OF "O PANORAMA" 183, FROM THE DR. NUNO CARVALHO DE SOUSA PRIVATE COLLECTIONS / LISBON WIKIPEDIA COMMONS

The tail of the short-tailed albatross is actually no shorter than that of other species of North Pacific albatrosses.

Its unusual coloring, and its lovely rose and blue beak, set the short-tailed albatross apart, but along with its many relatives, it has intrigued humans for centuries.

STEVE MOORE / US FISH AND WILDLIFE SERVICE

incubating for 65 days. The egg hatches in late December or early January and for the first three or four weeks, one of the chick's attentive parents stays with it, while the other parent forages. After it's large enough to protect itself, they depart for days or even weeks to find food. And they go to enormous lengths to feed their young, travelling more than 1,850 miles or 3000 kilometres, to forage for food. Initially, the young chick is fed stomach oil, rich in calories and Vitamin A, which is regurgitated into its gaping maw in a meal that can equal a third of its weight. This injection of high protein food spurs enormous growth; at times the returning parents can only recognize their youngster by its sound and scent.

Soon, though, the parents bring squid and flying fish eggs. The chicks mature slowly and fledge between late May and early June, when the nests are at last abandoned.

Once they are airborne, life is precarious for all young albatrosses, as they must find food on their own, evade predators such as sharks and avoid being caught in longlines or drift nets. According to studies of other albatross species in the Indian Ocean, nearly 40 per cent of fledglings die in the first two months.

Short-tailed albatrosses spend the non-breeding season at sea feeding, and mate every second year. They are protected under the domestic laws of many countries, including the United States, Canada, Mexico, China, Japan, and Russia, as well as under international legislation.

IMAGES BOTH PAGES: JACK MOST: www.themostinphotography.com

Short-tailed Albatross Fast Facts

Phoebastria albatrus

WEIGHT & LENGTH

The largest of the North Pacific albatrosses, it is between 33 and 37 inches (84 and 94 centimetres) in length, with a wingspan of 85 to 91 inches (2.15 to 2.3 metres and weighs between 8 and 14.6 pounds or 3.7 and 6.6 kilos.

LIFESPAN

Short-tailed albatrosses mature slowly and can live for 60 years.

DISTRIBUTION

Though they travel vast distances across the North Pacific, populations today are concentrated in the Bering Sea, the Gulf of Alaska and the Aleutian Islands. However, in the last two decades, they have been increasingly observed off the shores of Alaska and British Columbia, and occasionally as far south as California, where they were once common during the winter months.

HABITAT

Short-tailed albatrosses spend most of their life at sea; beginning about age 10 they return in the fall to isolated islands off the coast of Japan, where there are no predatory mammals, to lay their eggs and raise their young.

DIET

Short-tailed albatrosses forage at sea, ususally during the day and possibly at night, surface feeding on squid, fish, flying fish eggs, shrimp and other crustaceans. They often follow fishing boats to feast on their waste.

BEHAVIOUR

This graceful seabird's long, narrow wings allows it to soar low over the ocean for long periods of time. Highly efficient in the air, it expends very little energy as it soars and glides on the wind. The birds have a sheet of tendon called a shoulder-lock, which keeps their wings fully extended without any muscular exertion. In fact, it has been found in another albatross species that its heart rate in flight is slower than when is at rest on land. It was once commonly thought that albatrosses actually slept while in flight, but biologists believe the birds likely sleep while resting on the ocean surface.

NAMES

It is also known as Stellers albatross, after George Wilhelm Steller, a German naturalist who collected skins during various 18th-century expeditions. North Pacific albatrosses have long been known as "gooney birds", by sailors.

VIEWING SHORT-TAILED ALBATROSSES

Because short-tailed albatrosses spend their lives at sea, they are generally seen by fishermen and other sailors, who have reported increasing sightings over the past 20 years.

The California Coast

Look closely at this image of a tidal pool at Salt Point State Park, and you will see harbor seals basking on the rocks. Aaaahhh, California.

COURTESY OF K GLAVIN / CREATIVE COMMONS SHAREALIKE 3.0 / HTTP://CREATIVE COMMONS.ORG/LICENSES

Sea lions, unbothered by the worries of the world, spend much of their time relaxing by the water.

DENNIS FAST www.dennisfast.smugmug.com

THE CALIFORNIA COAST

In 1510, Spanish writer Garci Ordóñez de Montalvo published *Las sergas de Esplandián* ("Exploits of Espladán"), a romantic adventure that described a fantastic realm brimming with gold and precious stones, peopled by Amazons, and ruled by Queen Califia.

When Spanish explorers first arrived on the southern tip of Baja California in the 1530s, they named their discovery after this fictional island paradise they must have thought it so resembled: California.

California features incredibly dramatic and varied geography, including both the highest and lowest points in the contiguous United States (Mount Whitney and Death Valley, respectively). Its coast is part of an ecological region known as Mediterranean California, which stretches for 800 miles or 1300 kilometres from Oregon in the north to northern Baja California, Mexico, in the south, and is bordered by the Pacific Ocean to the west and the Sierra Nevada and deserts to the east. It includes the mountains, collectively known as the Coast Ranges, that run northwest along the edge of the continent as well as the relatively flat Central Valley that lies east of them.

The Sacramento River, California's largest, runs through the northern half of the Central Valley, carrying water from the north-central part of the state to the Sacramento-San Joaquin River Delta and on to the Pacific through San Francisco Bay. The San Joaquin River begins high in the Sierra Nevada and drains the southern part of the valley, flowing north to the Delta. Together, the Sacramento and San Joaquin Valleys include more than seven million acres of irrigated farmland, and are among the most agriculturally productive lands in the United States.

The southern edge of this region includes the rugged Transverse Ranges, which run east-west forming the northern border of the Los Angeles Basin. A series of gentler hills, valleys and terraces run along the southern California coast and there are several islands close to the mainland, including five (San Miguel, Santa Rosa, Santa Cruz, Anacapa and Santa Barbara) that make up the remarkable Channel Islands National Park. This region is the only one in North America with a Mediterranean climate, with its hot, dry summers and mild winters. Average summer temperatures are above 64° F or 18° C , while average winter temperatures are above freezing; however, the coast has significantly warmer winters and cooler summers than the valley east of the mountains.

Precipitation occurs mainly in the winter when storms blow in off the Pacific, and the rainfall ranges between eight and 40 inches (200 and 1000 millimetres) annually, with more rainfall in the north than in the south. The region's west-facing slopes are particularly wet, because as the ocean winds meet the mountains and rise, they cool. Their excess moisture falls as precipitation. This creates particularly dry conditions in the southern Central Valley, but farther north the valley receives winter rains from storms travelling south from the Pacific Northwest.

Echoing the variety of its landscape, California is home to an amazing biological diversity. Its many different climates and habitats, separated by rivers, mountains or deserts, have given rise over thousands of years to a vast number of plant and animal species, many of which are not found anywhere else in the world. For example, of California's 3,488 native plant species, 60 per cent do not grow anywhere else on the planet.

The coastal plains and terraces and the offshore islands are dominated by sage shrub to the north and an increasing number of cacti and succulents farther south. Typical plants include coast sagebrush, California encelia, white, purple and black sages, California buckwheat, coyote bush, sawtoothed goldenbush, coast goldenbush and golden yarrow.

Colored much like the swift fox, the island fox has evolved several subspecies, including this one, on San Nicolas Island.

COURTESY OF THE NATIONAL PARK SERVICE

Chaparral, a distinctive vegetation primarily of evergreen shrubs with thick, waxy leaves or hard, thin needles to reduce water loss, grows at higher elevations. Chaparral plants are also well adapted to survive wildfires; some sprout from their bases after they are burned, while the seeds of others germinate only after fire has torn through the area. This dry, often harsh environment provides habitat for mammals large and small, including cactus, pocket and deer mice, kangaroo rats, coyotes and cougars. Coastal areas also include patches of cypress and oak woodlands, and some conifers; these areas are home to many species of birds and small mammals that make their homes in tree cavities.

Life on the offshore islands has evolved in isolation for thousands of years. Channel Islands National Park is home to more than 2,000 terrestrial plant and animal species, 145 of which are not found anywhere else on Earth. Island species arrived either by air or by drifting ashore on floating debris, and then adapted to the unique environment, often becoming distinct subspecies to their mainland cousins. Some, like the island fox, island deer mouse, Santa Cruz Island scrub jay, and the San Miguel Island song sparrow have evolved unique subspecies endemic to different islands. The islands are also an important breeding and resting area for a number of seabirds, and are home to colonies of ashy storm-petrels, western gulls, Xantus' murrelets and California brown pelicans.

The mountains and valleys of the Coast Ranges feature different vegetation and wildlife

Main image: The rugged south coast of Santa Cruz Island, one of only three islands where Pacific tree frogs are found (inset), and below, a visitor poses between two ancient sequoias.

MAIN IMAGE: COURTESY OF ROBERT SCHWEMMER, CIMNS, NOS, NOAA / NOAA PHOTO LIBRARY / DEPARTMENT OF COMMERCE. INSET: COURTESY OF THE NATIONAL PARK SERVICE

than coastal California. These low mountains, with summits between 3,280 and 5,250 feet or 1000 and 1600 metres, are heavily folded and faulted, the result of the North American and Pacific Plates slipping against one another along the San Andreas Fault. This active fault-line runs along the California coast from Cape Mendocino to the Mexican border. The Coast Mountains and the Transverse Ranges to the south were formed about 30 million years ago when the head-on collision between the two tectonic plates that had begun 220 million years earlier changed to a lateral slipping movement, crushing and crumpling the sea floor, and thrusting it skyward.

The ecology of the Coast Ranges changes dramatically from north to south. At the northernmost edge of the Mediterranean California region sit the state's famous redwood forests. Heavy winter rains, cool ocean air, and plenty of fog allow the giant conifers to grow on the mountain slopes and in the valleys from just north of the California-Oregon border to the Big Sur area of Central California. These are among the world's tallest and oldest trees, often growing more than 200 feet or 60 metres high, 15 feet or nearly five metres in diameter, and 2,000 or even 3,000 years of age. These old-growth forests support an abundance of plant and animal life. In addition to redwoods, the trees in this area include Douglas-fir, grand fir, western hemlock, Sitka spruce, western red cedar, tanoak, bigleaf maple, California bay, and Port Orford cedar.

Beneath, and even on their enormous branches, which soar high above the ground, a vast array of herbs, ferns, shrubs and fungi grow. Black bears, fishers, pine martens, Pacific giant salamanders, red-bellied newts, tailed frogs, and numerous bird species, including the endangered marbled murrelet and northern spotted owl (see page 217), all make their homes in these forests. Sadly, less than four per cent of virgin redwood forest remains, and just

PUBLIC DOMAIN / HTTP://CREATIVECOMMONS.ORG/LICENSES

COURTESY OF THE NATIONAL PARK SERVICE

A crash course in survival: Tiny Xantus murrelets like this one from Santa Barbara Island, home to the largest breeding population of murrelets, leave their steep hillside nests when only 48 hours old and drop into the surf below.

2.5 per cent of this is protected, despite the belief of the World Wildlife Fund that "Redwood National Park is the only hope for survival of functioning redwood ecosystems."

In the central Coast Ranges, grassy open woodlands dominate, including rare Garry oak woodlands (see page 112), but here are also patches of denser oak woodlands and coniferous forest; blue oak, digger pine, California buckeye and redbud are common tree species. Farther south, blue oak is replaced by coastal live oak and chaparral shrublands cover many of the slopes. This is home to mule deer, coyotes, cougars, California bobcats and gray foxes, as well as including three subspecies of kangaroo rat, the Sonoma chipmunk, the Suisun shrew, and the salt marsh harvest mouse.

Southern California's Transverse Ranges share many species with the southern Coast Ranges, but their higher elevations and hotter, drier climate allow a mixture of sage scrub, chaparral, pinon-juniper and oak woodlands, and pine forests to thrive at different altitudes. These habitats are home to reintroduced California condors (see page 171).

California faces critical threats. According to Peter Steinhart, author of *California's Wild Heritage: Threatened and Endangered Animals in the Golden State,* on average, more than 20 per cent of the state's naturally occurring species of amphibians, reptiles, birds and mammals are classified as endangered, threatened, or "of special concern" by agencies of the state and federal governments. There are many environmental challenges facing this region, including heavily polluted air and water, but the overarching reason for this sad situation is the huge, and ever increasing, human population. The landscape and climate that support such a wide variety of plant and animal life are also incredibly appealing to many people, drawing about a million more every year to a region that is already overcrowded.

Estimates vary widely, but prior to European settlement the Aboriginal population of California may have been between 310,000 and 705,000 people, but this number fell sharply through the 19th century to just over 15,000 by 1900. The non-Aboriginal population, in contrast, skyrocketed after the gold rush that began in 1848 and California's admission to the United States in 1850, jumping from no more than 15,000 at that time to an estimated 1.49 million by 1900. This number has continued to rise; California's population was 10.6 million by 1950, 20 million by 1970, nearly 30 million by 1990, and more than 38 million in 2008. This number is projected to grow to more than 44 million by 2020, and an astounding 59.5 million people by 2050.

As California's human population grows, the amount of wild space left will steadily decrease and unique and irreplaceable ecosystems will give way to human settlements, agriculture and industry. Urbanization not only reduces the amount of land available for wildlife, but also leads to habitat fragmentation, an increased invasion of alien species, and the alteration of natural fire regimes.

California's agricultural revenue in 2007 totaled $36.6 billion dollars, and accounted for about half of all American-grown fruits, nuts and vegetables. In addition to the amount of land that has been converted to crop production, the irrigation required to support these farms has created a tremendous demand for water – besides that exerted by the state's cities and industries. Agricultural runoff, along with storm runoff from city streets, effluent from industrial processes and discharge from sewage treatment plants is polluting California's water supply, despite the state's serious efforts to deal with the problem.

California is even more notorious for its air pollution. A 2008 study by researchers at California State University-Fullerton found that in 2006 more Californians were killed by air pollution than car crashes; 3,812 deaths were attributed to respiratory illness caused by particulate pollution in the San Joaquin Valley and South Coast Air Basin in 2006, as compared to 2,521 vehicular deaths.

Some would hold cars largely responsible for both types of fatality – there were almost 31 million licensed vehicles in California at the end of 2002 and motor vehicle emissions are responsible for the photochemical smog that many people instantly associate with the state, as well as about one-third of its total greenhouse gas emissions. California is the 12th largest carbon emitter in the world, despite the California Environmental Protection Agency's many programs aimed at improving air quality.

Ensuring the survival of California's distinctive terrestrial ecosystems will require not only cleaning up the water and air, but setting aside land for native plants and animals. As of 2006, California had protected 20 per cent of its land – a percentage second only to Alaska. And in March 2009, President Barack Obama signed The Omnibus Public Land Management Act, protecting another 700,000 acres of California's federal lands in the eastern Sierra Nevada, Sequoia and Kings Canyon National Park and Riverside County. As welcome as these measures are, unfortunately, most reserves are set aside for their scenic value, remote locations, and low economic impact and do not necessarily reflect the areas with the most biodiversity in need of protection.

California's coastal waters are some of the most productive and biologically diverse in the world. Often grouped into four zones – the inland watershed, the enclosed waters, the nearshore ocean, and the offshore ocean – these waters include many habitats and support a huge array of species. The inland watershed zone consists of approximately 7,800 miles of rivers, creeks and drainages that carry fresh water and sediment to the ocean. They provide critical migratory routes and spawning grounds for many species of anadromous fish, including the coho and chinook salmon (see page 229), steelhead trout, American shad, striped bass and white sturgeon (see page 244).

The enclosed waters zone is made up of California's bays and estuaries where fresh

The view from Serrano Ridge in Laguna Coast Wilderness Park; an orange cloud of pollution is clearly visible in the distance.

JENNIFER WOOD www.natureinorangecounty.com

Two male northern elephant seals fight for territory and mates in an explosive display of animal instinct.

COURTESY OF MIKE BAIRD / CREATIVE COMMONS ATTRIBUTION 2.0 / http://creativecommons.org/licenses/by/2.0

water mixes with ocean water, creating a wide variety of coastal wetlands, mudflats and seagrass meadows. These habitats provide spawning, nursery and feeding grounds for many species of fish and cover and crucial nesting sites for birds, including the endangered light-footed clapper rail and Belding's savannah sparrow. Even the superficially barren mudflats are home to many invertebrate species, such as clams, oysters and other mollusks, crustaceans, and worms, and provide important feeding grounds for many coastal birds.

The nearshore ocean zone extends from California's sandy beaches and rocky shores and outcroppings to an ocean depth of about 300 feet or 100 metres. The waters in this zone are rich in nutrients from the inflow of fresh water from the shore and upwelling currents from the deep offshore ocean. Shoreline, tidepool, sandy and muddy bottomed ocean habitats are all found in this zone, along with forests of giant kelp, the oceans's largest and fastest growing algae. This is an incredibly rich region of biodiversity, supporting a huge number of fish and invertebrate species (many of which are commercially important), providing feeding grounds for shorebirds, as well as forage and breeding grounds for marine mammals such as harbor seals, sea lions, elephant seals and California sea otters (see page 135).

California's offshore ocean zone, beginning at a depth of 320 feet or 100 metres, features two major ocean currents that add variety to its ocean life. The California Current, a cold water current, begins north of California and flows south along the coast, while the Davidson Current periodically brings warm waters from the tropics north into the near-shore waters of southern California. As a result, the southern offshore ocean supports temperate and warm water fish and invertebrate species, while the central and northern California coastal ocean regions are home to those species requiring colder water. Herring, mackerel and sardines live in the plankton-rich water of the shallower offshore ocean zone, and give way to tuna, swordfish, rockfish and other deepwater fish species farther from shore.

Large seabirds, such as albatrosses (see page 146), frigatebirds and a variety of gulls soar and dive above the open ocean, feeding on small fish and crustaceans and looking down on several species of dolphins and porpoises, gray and humpback whales and, more rarely, fin and blue whales (see page 164).

California's marine ecosystems may not be faring any better than those on land. According to a report published by The Resources Agency of California in 1997, agriculture and urbanization had resulted in the loss of more than 90 per cent of the state's historic freshwater wetlands – crucial ecosystems that help to filter and clean the water that passes through them. This report also stated that 75 per cent of the pollutants entering marine waters originated from land-based activities. California's water zones are intimately connected and activities in any one zone can have a negative impact on the others.

Realizing the importance of its coastal wetlands, in 1993, California adopted a policy of no net loss of wetland habitat, and preservation and restoration projects have met with some success.

In California's nearshore waters, the region's enormous coastal population, along with agricultural runoff has produced pollution deemed serious enough to harm marine life and endanger human health. Though pollution point sources, such as factories and sewage treatment plants, are easier to target, in large part this is a difficult problem to monitor.

Offshore events, such as oil spills, can be devastating; in November 2007, the Cosco Busan spill in the San Francisco Bay resulted in the death of more than 2,500 birds and the closure of several nearby beaches. It is estimated, however, that oil spills are responsible for only 12 per cent of the oil entering the ocean each year; three times this amount comes from polluted runoff.

Invasive non-native species in California waters have also had a negative impact on plant and animal life. Whether brought in unintentionally on ships arriving from foreign waters and discharging bilge wastes or on smaller commercial or recreational watercraft, or even intentionally introduced, many invasive species reproduce quickly in the absence of natural enemies, and compete with native species for food and habitat. California's coastal waters now have more than 300 invasive species; quagga mussels, New Zealand mudsnails, and dwarf eelgrass are three that cause particular concern.

The waters of California's offshore ocean are subject to many of the same environmental pressures as other parts of the world's oceans, including dredge disposal, oil and gas and shipping operations and military exercises. It's believed that these activities can disturb or harm marine life; however, our knowledge to date is limited, hampering the effectiveness of calls to action.

Finally, climate change is posing an unprecidented threat to California's land- and water-based ecosystems. The global sea level has risen by eight inches over the past century, and the 2009 California Climate Action Team Biennial Report projects that it may rise by another 55 inches or nearly 1.4 metres by the end of the century. This would result in significant flooding in coastal California, the destruction of bluffs, beaches and other coastal habitats, and possibly compromise some of California's water sources, as saltwater pushes farther into coastal rivers and aquifers.

Climate change is also likely to bring with it more frequent and severe storms and floods and a rise in ocean temperature and acidity. These changes could be devastating to California's plant and animal life, and are also predicted to also have a disproportionately negative impact on some of California's most vulnerable human populations.

On a brighter note, a number of organizations and networks have rescued and rehabilitated literally hundreds of injured, starving or entangled marine creatures (see Plastic Soup on page 162), as well as those that were simply lost. Over the past decade these have included seals, sea lions, whales, sea otters and sea turtles. Almost all are supported by donations as well as the time and efforts or many volunteers. And many are open to visitors.

OURTESY OF MIKE BAIRD / CREATIVE COMMONS SHARE-
LIKE 3.0 / HTTP://CREATIVECOMMONS.ORG/LICENSES

A California brown pelican in flight.

Plastic Soup

AN island in the middle of the Atlantic Ocean – fabled Atlantis – has long captured the human imagination, though no reliable source credits its existence. An island in the middle of Pacific Ocean, however, is a growing reality – literally. But this island, located between Hawaii and the continental US, is not composed of earth and sand and stone. It's composed of garbage, most of it plastic. Variously called the Great Pacific Garbage Patch, the Pacific Trash Vortex, or the Great Plastic Vortex, and estimated to be the size of Texas (or larger), this island threatens the wellbeing of marine life and birds and, in the long run, the health of other creatures higher up the food chain, including human beings.

You can't see this island as a consolidated mass. If you were passing through it by ship, you might notice a rogue soda bottle and piece of bubble wrap on the surface of the water, but you wouldn't be aware that you were passing through an enormous soup of very, very small pieces of plastic, much of which is invisible to the naked eye, suspended at or just below the water's surface. Once the pieces were shopping bags, milk jugs, fishing nets, children's toys or any of the thousands of other plastic products of consumer society, but under the desiccating effect of the sun's ultraviolet rays and the grinding effect of the waves, they break down into increasingly tiny fragments. While paper, for instance, biodegrades and eventually turns into plant nutrients, plastic only photodegrades, fragmenting into smaller and smaller editions of itself, never changing its essential nature. Seabirds, mistaking the plastic for food, dive down and swallow indigestible shards that eventually fill their stomachs, fail to pass, and cause them to starve to death. But the smaller the particles become, the more ingestible they are to sea life, until they are small enough to be eaten by aquatic organisms that live near the ocean's surface and which are themselves an abundant source of food for other forms of sea life.

While the Earth's seas brim with floating objects, the concentration in the northwest Pacific is due to the rotational pattern of the ocean currents, called the North Pacific Gyre, which draws debris from coastal waters off North America and east Asia into a large convergence zone. Waste dumped by ocean-going vessels forms part of it, but between 60 and 80 per cent is believed to be from land-based sources, whether from beaches or storm sewers or from rivers and streams. Some have compared the effect of the gyre to that of a giant toilet bowl that never completes its flush; the outer rim is constantly being swept by a clockwise current, dragging more and more floating material to the centre where it grows more and more concentrated. Oceanographer Charles Moore, who has studied the

J. THOMAS AND PROYECTO ANIDACION TORTUGAS MARINAS / REPUBLICA DOMINICA

Out of sight, out of mind: a stretch of neglected beach in the Dominican Republic, one of the most important leatherback turtle nesting sites, makes the scope of the problem clear.

The Great Pacific Garbage Patch

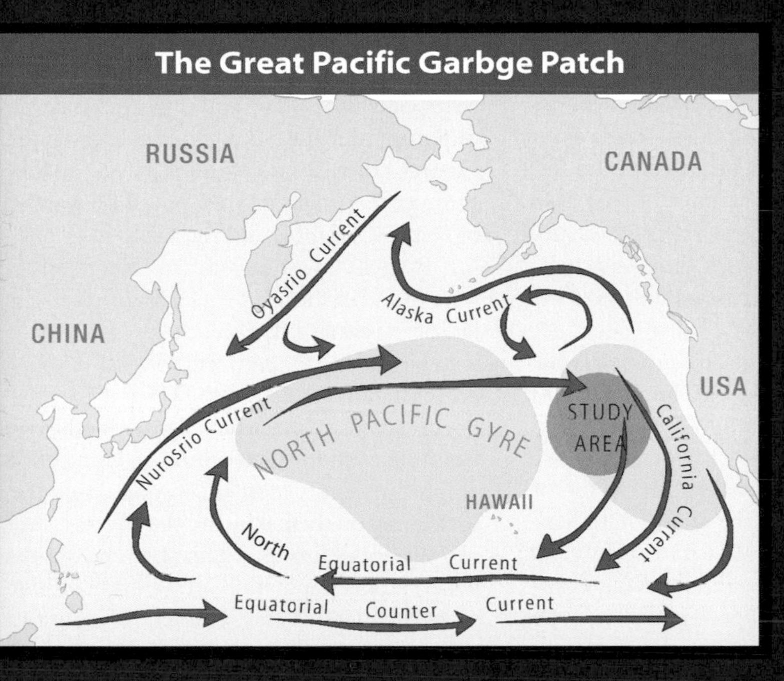

n early August 2009, two ships set sail from California with a team of scientists and environmentalists to tudy the marine debris collected in the North Pacific Gyre, as part of Project Kaisei (Japanese for "ocean lanet"), a multi-year campaign to address concerns related to ocean health. Among the expedition's goals re to capture ocean debris and experiment with ways to process and detoxify it without harming marine fe. For further information, including photos, videos and project blogs, visit <www.projectkaisei.org>.

reat Pacific Garbage Patch for more than a decade, uggests that at current rates of growth the island will ncrease in size by a factor of 10 every two to three years ntil it begins to resemble a genuine solid island.

Removing the floating garbage, estimated by some ources as weighing 3.5 million tons, doesn't appear to e a workable solution in the short term, though at least ne organization, Project Kaisei, is studying possibilities ee above). Because the patch lies in international aters, no single nation is likely to make cleaning it up its own responsibility. A coordinated multinational cleanup effort, meanwhile, would cost billions. Instead, environmentalists say, the long-term solution is on land: manufacture fewer plastic products, consume fewer plastic products, substitute biodegradable materials, increase recycling programs, and enforce litter laws. By some estimates, more than 60 billion tons of plastic are produced each year and less than five per cent is ever recycled. A good portion of it is cycling, though – round and round and round in the Great Pacific Garbage Patch.

Blue Whale

Balaenoptera musculus • Endangered (IUCN)

SPEEDING, submarine-like, through deep ocean waters, calling with low-frequency sounds that travel for miles, blue whales are the largest and likely the loudest creatures on Earth. In fact, scientists believe they are the largest creatures known to man, heavier than a *Tyrannosaurus rex* or even an *Argentinosaurus* and longer than a giant mosasaur. And their structured, low frequency sounds are louder than a jet engine.

Yet these gentle, solitary mammals, which dine on tiny crustaceans, were hunted almost to extinction. Always sought for their vast amounts of blubber, blue whales were too swift and powerful for most whalers prior to the 19th century. However, with the invention of harpoon cannons – which allowed whalers to kill the enormous animals without having to approach them in small boats – in the late 1800s, the great blue whale became a much sought-after species.

For a time in the early 20th century, this magnificent creature was the most hunted animal on Earth. The slaughter reached a peak in 1931, when 29,649 blue whales were taken and by 1966, an estimated 350,000 animals had been killed worldwide. Finally, with fewer than 2,000 blue whales remaining, the International Whaling Commission declared them protected and off limits to whalers everywhere. To deter renewed hunting, blue whales were listed as endangered worldwide in 1970 by the IUCN.

In the decade to follow, blue whales continued to be taken illegally by the USSR, and blue whale meat was available under other labels. The total number in the Antarctic, where an estimated 250,000 blue whales were once found, dropped to about 4,500 in 1980, and about 2,200 animals in 1998.

The situation in the North Pacific is more hopeful. Here, an estimated 10,000 blue whales once wintered, mating and calving in the warm waters off southern California, Mexico and Central America, and migrating north in the summer to feed in the cold, bountiful waters of British Columbia and Alaska.

By 1980, an estimated 1,700 blue whales were found in the temperate and tropical waters of the eastern Pacific, but very few were seen farther north. Between 1997 and 2009, however, biologists from the US and Canada documented 15 blue whale sightings off British Columbia and the Gulf of Alaska. Using photographs to identify individual whales from the patterning on their skin and the size and shape of their dorsal fins, they linked several of those whales to the population off California. "We speculate that this may represent a return to a migration pattern that has existed for earlier periods for [the] eastern North Pacific blue whale population," they wrote in the scientific journal *Marine Mammal Science.*

In 2009, the North Pacific population, estimated at nearly 2,000 whales, was considered to be the the largest population of a worldwide total of somewhere between 5,000 and 12,000. And there are hopeful signs elsewhere. Though the North Atlantic population is deemed by many to be less than 1,000, in 2009, the voice of a singing blue whale was positively identified in the waters off New York for the first time in living memory. And numbers are apparently growing in the waters off Iceland.

Blue whales are enormous even at birth, weighing between 5,000 and 6,000 pounds or 2.5 metric tons, and about 23 feet or seven metres long. Calves drink more than 100 US gallons, or 400 litres of milk a day and, not

This dead blue whale, a 70-foot-long male weighing more than 50 tons (see inset), was found floating in the Santa Barbara Channel. Riddled with broken bones, it was likely struck by a ship.

COURTESY OF TODD JACOBS / NOAA FISHERIES

surprisingly, grow very quickly. By the time they're weaned at six months of age, they have doubled in length.

Despite dining on krill, shimplike creatures the length of a human finger, they grow up to 200 tons or about 181,000 kilos over an estimated life span of between 35 and 80 years, with females growing larger than males. An adult blue whale heart has been compared in size to a small car.

Blue whales are found in all the world's oceans, and there are at least three subspecies: *B. m. musculus,* which lives in the North Pacific and North Atlantic; *B. m. intermedia,* found in the Southern or Antarctic Ocean, and *B. m. brevicauda,* also known as the pygmy blue whale, which lives in the South Pacific and Indian Ocean.

Blues are rorqual whales, a family of baleen whales with pleated throat grooves (see photo above); the word rorqual comes from the Norwegian for "furrow". In a blue whale, the ventral pouch runs from its chin to its navel and expands enormously as it takes in water when feeding. During the summer feeding season, blue whales eat an estimated 40 million krill, squid and small fish a day. To make that possible, they always feed in areas with high concentrations of krill, typically at depths of more than 100 metres during the day, sometimes feeding on the surface at night.

When feeding, a blue whale "headstands", lunging head first and open-mouthed into masses of krill, often lifting its tail flukes above the surface. With its ventral pouch greatly expanded to accommodate the tiny crustaceans and water, it then contracts the ventral pleats and, aided by its enormous tongue, squeezes the water through the 300 baleen plates that hang from the front part of its mouth, retaining

Tail flukes in the air, a young blue whale dives for food beside its mother in the Gulf of the Farallones.

DON SHAPIRO / NOAA PHOTO LIBRARY / DEPARTMENT OF COMMERCE

The rorquals (family Balaenopteridae) include the humpback, blue, fin, sei and minke whales. Recent genetic studies have shown that humpback and fin whales are closely related, which is surprising because the humpback whale possesses enlarged flippers and other morphological characteristics so distinctive that it is classified as belonging to a different genus than the other species of rorquals. The different species belonging to this family are in fact related closely enough that at least 11 cases of adult blue/fin whale hybrids have been documented in the wild and blue and humpback whales are believed to have mated to produce hybrid offspring on rare occasions; biologists compare their relationship to that between humans and gorillas.

and swallowing the krill. Blue whales submerge for an average of 10 minutes when feeding, but can dive for two or three times that long. When breathing on the surface, they produce a spectacular vertical spout of water up to 40 feet or 12 metres in height.

Long and slender, blue whales reach an average length of 80 feet or 25 metres, though females almost half again as long were reportedly taken by whalers in the early 20th century.

Whales are almost alone among mammals in that they spend their entire lives in the ocean and are completely at home there (sea otters, mammals of a very different nature, are also completely at home in the sea: see page 135) and their relationship to terrestrial mammals has long puzzled scientists. Did whales evolve before, or after mammals on land, they wondered? Even Charles Darwin weighed in on the question, in his 1859 edition of *On the Origin of Species.* Whales, he believed, descended from bears. But embarrassed by his critics following publication, Darwin removed the idea from subsequent editions.

Nor were the critics silent when English anatomist William H. Flower first proposed, in a lecture in 1883, that whales were evolutionary descendants of terrestrial ungulates and therefore related to even-toed mammals such as pigs and hippopotomases. However, the discovery of a host of ever more whale-like fossils over the past century, as well as new interpretations of fossils found before Flower's birth, have made it clear that he was well ahead of his time.

Today, thanks to the tireless work of

paleontologist Philip Gingerich and many others, it seems clear that the evolution of whales truly began about 53 million years ago, with Pakicetus, a wolf-sized carnivore first discovered in Pakistan in 1981. Initially, only a skull was uncovered on what was then the coast of the diminishing Tethys Sea. Though it was long muzzled, with sharp, serrated teeth for hunting fish, it was the inner ear that interested Gingerich, Neil Wells, Donald Russell and S.M. Ibrahim Shah, for its bones had features that are unique to whales. Though the complexities of the ear and nose that allow whales to hunt and communicate by echolocation were not present, the positioning of the ear bones and the folding of a bone in the middle ear, along with the long muzzle and the teeth definitely belonged to the whale or cetacean family. Complete skeletons of Pakicetus, discovered 20 years later, showed that it was a land mammal, but likely right at home in the water.

In 1992, another creature was found in the same part of the world by anatomy professor Johannes (or Hans) Thewissen, but in sediments more than 350 feet higher. Dating from about 50 million years ago, it was named *Ambulocetus natans,* "the walking whale that swims". Though it had functional legs capable of supporting its weight, it likely spent most of its time in the water, where it swam by paddling and dove with powerful kicks. When it came ashore, as it likely did to bear its young, it probably waddled more than walked. And its back legs were capped with small hooves, an indication of its ungulate ancestry; several scientists have speculated that it hunted rather like crocodiles do today, lying in wait at the water's edge and lunging out for its kill. Its skull, however, was significantly whale-like.

In 2009, Gingerich and his colleagues published a paper in the online journal *PloS One,* outlining the discovery of an eight-foot or 2.6-metre, pregnant protocetid female from the early part of the Middle Eocene, about 47.5 million years ago. *Maiacetus inuus,* or "mother whale", as the discovery was named, had died just before giving birth and both mother and infant were so perfectly preserved that it was clear that this species was another step along the road toward the whales we know today. Yet, as the group wrote, "The fetal skeleton is positioned for head-first delivery, which

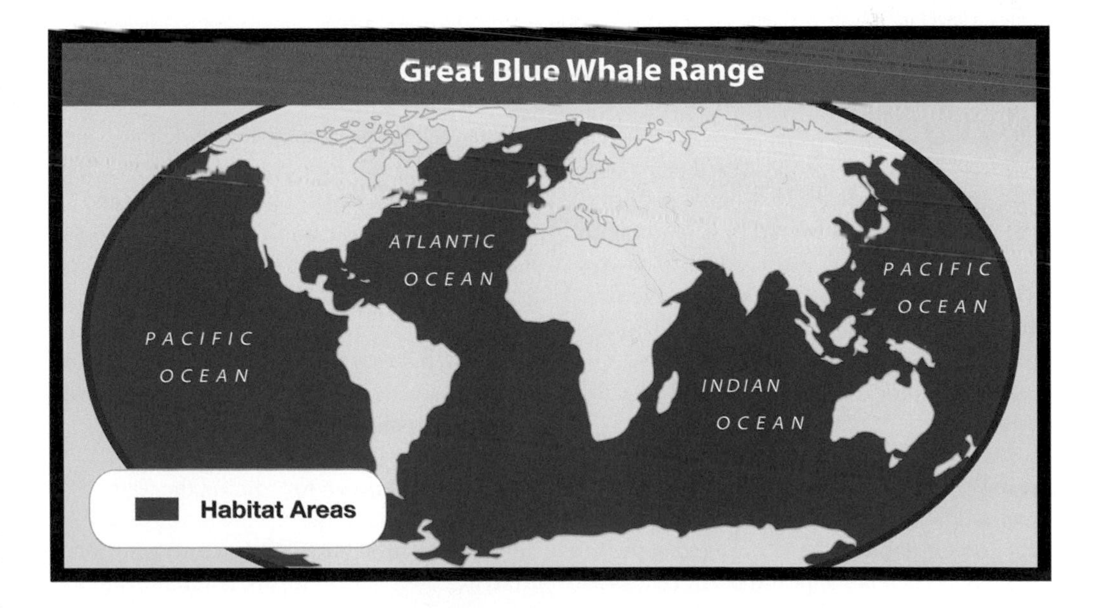

LIBRARY AND ARCHIVES CANADA / C-111498

Lashed to a sailing ship, a blue whale's blubber is cut from its carcass in a painting from 1850.

typifies land mammals but not extant whales, evidence that birth took place on land."

The fetus also had a well-developed set of teeth, suggesting it "would be able to get up and move shortly after birth, probably having to keep up with its mother, learning to feed and escape predators," Gingerich told Discovery News. The species was described as combining aspects of a cow, whale, shark, alligator and sea lion.

Moving along the evolutionary road are the huge, vicious looking *Basilosaurus,* and the much smaller *Dorudon,* both dating from the Late Eocene, about 41 to 33 mya and both with a markedly reptilian appearance. In fact, with its long, sepentine physique, *Basilosaurus* (the "king lizard") was long believed to be a reptile. Averaging 18 metres or 60 feet in length and found from Pakistan and Egypt's now desert-like "Valley of Whales" to the southeastern US, it was likely the largest animal of its period. So common were the fossils in Louisiana and Alabama that they were used as furniture, while Alabama and Mississippi chose *Basilosaurus* as their state fossil.

It was English anatomist Sir Richard Owen (better known for coining the term "dinosauria" and as the driving force behind London's Natural History Museum) who, after studying *Basilosaurus* fossils, realized it was a

Blue Whale Fast Facts

Balaenoptera musculus

DESCRIPTION

Blue whales look like no other animal on Earth, torpedo-shaped, with a flat head and a ridge that runs from the blowhole to the top of the upper lip. Small dorsal fins are located about three-quarters of the way back on the body and their flippers are between 10 and 13 feet or three to four metres long. Their skin is a mottled mixture of dark and light gray; each individual pattern is unique, allowing researchers to identify them and follow their activities over their 70- to 80-year lifespan. The blue whale's blow, a single vertical column that soars 40 feet or 12 metres into the air, can be seen from great distances.

HABITAT

Like humpback whales, blue whales ideally migrate annually from polar waters, where they feed during the summer months, to tropical-to-temperate waters where they mate and calve during the winter. Recent studies seem to show that while these migrations have been either missed or missing in recent decades, at least some blue whales in the Eastern Pacific are once again migrating.

DIET

The blue whale's diet consists mainly of krill, but also includes some small fish, crustaceans and squid taken in along with its main meal. This giant species can consume up to 8,000 pounds or 3,600 kilograms of krill in a single day.

BEHAVIOUR

Blue whales are solitary animals, usually living alone or in pairs, but like many other species of whale, small herds will occasionally migrate together to spend the winters in equatorial waters. Blue whales are the loudest animals on earth, communicating with sonic booms that can travel great distances underwater. About half these vocalizations, or songs, as they are sometimes called, are below the frequency of human hearing.

MATING & BREEDING

After reaching sexual maturity between the ages of six and 10, female blue whales begin mating every two or three years. The season begins in late autumn and lasts until the end of winter. Like other baleen whales, blue whales have a long gestation period of about 12 months. Females give birth in early winter to a 23-foot or seven-metre calf weighing about 2.5 tons. Young blue whales drink between 100 and 150 US gallons of milk a day until they are weaned at about six months.

On rare occasions, blue whales have been known to mate with fin whales.

Populations of whales and other large, long-lived mammals recover slowly due to their low rates of reproduction. The current status of the blue whale is uncertain; some populations seem to be increasing while others are declining or remaining stable.

THREATS

The blue whale's only natural predator is the orca (see page 124) and an estimated 25 per cent of adult blue whales have scars from orca attacks. Ocean traffic, fishing equipment, and noise pollution are also threatening blue whales in many parts of the world. In 2007, for example, when food was abundant in the Santa Barbara Channel off California, four adolescent blue whales were washed up to shore, apparently (given their injuries) killed by collisions with ships. In 2009, blue whales once again congregated in the channel, perhaps in response to an explosion of krill as a result of climate change.

NAMES

The species name *musculus* is Latin and could mean "muscular", but can also be interpreted as "little mouse". Linnaeus, who named the species in 1758, likely intended the irony. These majestic animals have also been called great blue whales and great northern rorquals.

VIEWING BLUE WHALES

Their vast bulk keeps them ocean bound and prevents the acrobatics for which orcas and humpbacks are famous. However, they can often be seen in Big Sur off the California coast from June through October, particularly in years when krill is abundant. The whales can be see from roadside turnouts overlooking the coast. Also, Monterey Bay Whalewatch sponsors trips out into the bay, which is also home to gray and humpback whales, seals and sea lions, as well as sea turtles (see page 176) and sea otters (see page 135), which often anchor themselves in the forest of giant kelp.

Surfacing with a spay of water, a blue whale mimics the ocean horizon.

DENNIS FAST www.dennisfast.smugmug.com

mammal. Moreover, its "whale-like yoke teeth" and the likely presence of tail flukes, convinced him it was a whale and one that swam.

With limbs smaller than a human's, no melon organ and a brain much smaller than today's toothed whales, it's quite clear that this ocean warrior lived in the upper reaches of the sea, but perhaps most importantly in terms of whale evolution, it had abandoned the land for good. There are those who believe that *Basilosaurus* still lives, perhaps explaining sea serpent sightings, but there is no evidence to support that.

Though much smaller, at about five metres or 16 feet in length, *Dorudon* was similar in shape and, like other basilosaurids, its nostrils were located halfway between the end of its snout and the top of its head. Gingerich has written that it represents the group most likely to be ancestral to modern whales.

Two things stand out about this chronology. First, with the exception of basilosaurids, which fanned out over the Earth's changing seascapes, whales evolved in a relatively small area around the Thethys Sea. Most of that ancient body of water is gone now, absorbed as the continents moved and India slammed onto the southern edge of Pakistan; its remnants include the Mediterranean, as well as the Black, Caspian and Aral Seas.

Second, Flower was right about his hunches. Genetic evidence has recently shown that whales' closest terrestrial relative is the omniverous hippopotamus.

Baleen whales, such as the mighty blue whales, evolved later. In fact, according to the Museum of New Zealand Te Papa Tongarewa, fossil evidence seems to suggest that for a brief time in whale history, mysticete or baleen whales had both teeth and baleen. The first fossil of this type was found in sandstone at Seal Rock State Park near Newport on the Oregon coast in 1964. Soon after, sometime between 28 and 23 mya, during the Late Oligocene, what are termed eomysticete whales (or primitive baleen) whales appeared.

California Condor

Gymnogyps californianus • Endangered

DESPITE its forbidding appearance and rather distasteful dietary habits, the California condor is the perfect poster child for endangered species. At the end of the last glaciation, California condors could be found almost everywhere along North America's coasts, from what is now British Columbia south to Baja California, as well as across the southern US to Florida and north along the Atlantic coast to New York State. Even 200 years ago, when Meriwether Lewis and William Clark arrived at the mouth of the Columbia River, they found condors feeding in large numbers on whale carcasses. More than a half-century later, in 1860, an estimated 300 of these huge birds were seen feeding along the California coast on the remains of sea lions that had been slaughtered for their oil.

Yet, despite being protected by the State of California since 1953, by the early 1980s the entire population of North America's largest land bird had been reduced to fewer than 25 individuals, all found in the mountains and foothills around the San Joaquin Valley. In a last-ditch attempt to avert extinction, all wild condors were captured and by 1987, all 23 birds then known to exist were in captivity.

What happened to California condors between 1860 and 1980? One must go much farther back to determine their fate. Fossil records of the genus *Gymnogyps* reveal that perhaps 100,000 years ago, condors ranged over much of the southern US, nesting in west Texas, Arizona and New Mexico. According to articles in a number of scientific journals, fossils, eggshells, unfossilized bones and feathers radiocarbon dated between 22,000 and 9,500 BP have all been found in caves in the Grand Canyon, indicating that is was an important nesting area for thousands of years during the height of the last glaciation.

The disappearance of condors (and other large scavengers) from the southwestern interior coincides with the Great Extinction of Pleistocene megafauna (see Introduction on page 10) – the rapid disappearance of mammoths, mastodons and *Bison antiquus*, the great ice age bison – soon after the glaciers retreated. Though there is some indication that condor numbers grew

COURTESY OF SCOTT NIKON FRIER / US FISH & WILDLIFE SERVICE

COURTESY OF RON GARRISON / SAN DIEGO ZOO / USFW SERVICE

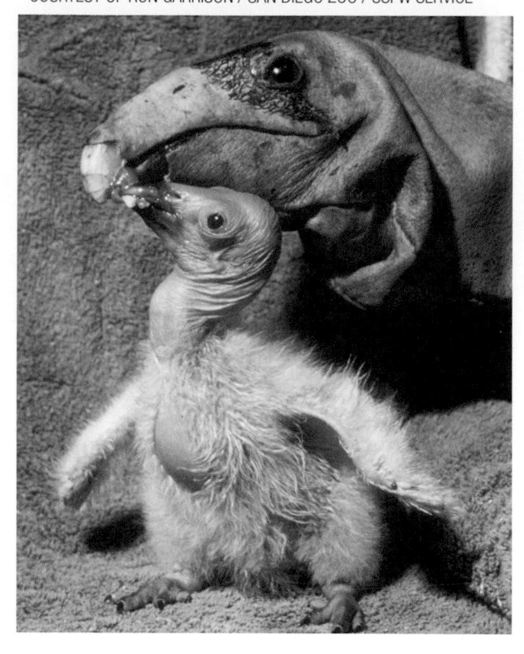

Above: Condors are bald for a reason; feathers would get in the way when they feed. Breeding facilities use hand puppets, above, to feed the chicks.

Aloft with its remarkable wings outstretched, below, or opposite, at rest as a trio contemplates dinner at Castle Crags, California, each bird is clearly identified.

COURTESY OF DAVID CLENDENEN / USFW SERVICE

again in the American southwest in the early 1700s when Spaniards and Mexicans introduced large herds of cattle, horses and sheep, by the 1930s condors were found exclusively along the California coast, where they dined on the carcasses of marine mammals, such as whales, sea lions and seals, as well as salmon that had died or been killed.

California condors, one of seven species of New World vultures and among the largest flying birds in the world, have always depended on a steady supply of large carrion. And following the extinction of the ice age megafauna, their range was substantially narrowed. Though there were scattered reports of condors in Arizona between the mid-1880s and the early 1920s, they thrived only on the Pacific Coast. Given the numbers witnessed by early Europeans, it seems they flourished on a diet that had changed from land mammals to marine mammals and bountiful runs of fish. Then the world's whalers arrived, slaughtering the great whales and elephant seals by the tens of thousands. As the oceans emptied, so did the skies above, for without marine mammal carcasses on which to feed, California condors began to disappear.

Their demise was hastened by predator poisoning campaigns, egg collecting and the capture and killing of specimens for museums, as well as by accidental lead poisoning from lead ammunition ingested as they dined on carrion, and by collisions with electrical wires. By 1982, with just 25 California condors left in the wild, it was decided that only captive breeding would save the species. In April 1987, the last wild condor was taken into captivity.

Over the next dozen years, more than 200 condors were raised in three breeding facilities in southern California and Idaho. Releases of condors back into the wild began in 1992 in southern California, in central California and northern Arizona in 1996 and 1997, and in northern Baja California in 2002. The first eggs laid in the wild failed to hatch, and the first three wild chicks, hatched in California nests in 2002, all died. However, since then, condor chicks have been raised and fledged in the wild and the annual mortality rate is dropping. Deaths today are usually from power line collisions, ingesting small pieces of trash or lead poisoning – all things Pleistocene condors didn't have to worry about.

COURTESY OF DAVID CLENDENEN / USFW SERVICE

Today, about 20 juveniles continue to be released into the wild each year and progress is slowly being made, though these are still deemed to be among the rarest birds in the world. All captive and wild condors were vaccinated with an avian vaccine for West Nile virus in 2003; hunters are being schooled on mechanisms to reduce lead poisoning and a poacher responsible for shooting a condor was convicted.

Condor deaths are investigated; when three of the magnificent birds died following collisions with a span of power lines on the Big Sur coast, the electric utility responded with marking devices on the lines and is considering other bird avoidance options.

Strategies are also being considered to provide enough food for reestablished condor populations. Though coastal regions lack carcasses of large land mammals, populations of northern elephant seals, which were hunted almost to extinction in the late 19th and early 20th centuries, have recovered substantially. Following a Mexican ban on hunting elephant seals in 1922 and an American ban shortly after, the worldwide population of northern elephant seals is estimated at more than 150,000, enough to repopulate many coastal areas and provide carrion for resident condors. Elsewhere, ranchers are assisting in providing food for reestablished condor populations in the western interior by donating stillborn calves and other stock that have died from natural causes.

Condors expend remarkably little energy as they soar high above the ground on wings that can span more than 9.5 feet or 2.9 metres. They prefer mountains, hillsides and gorges, for these land forms create the thermal updrafts necessary for soaring. One of the best places to view condors, particularly during the summer months, is the Grand Canyon, where members of Arizona's re-established condor population can be found during the summer months, soaring on the updrafts.

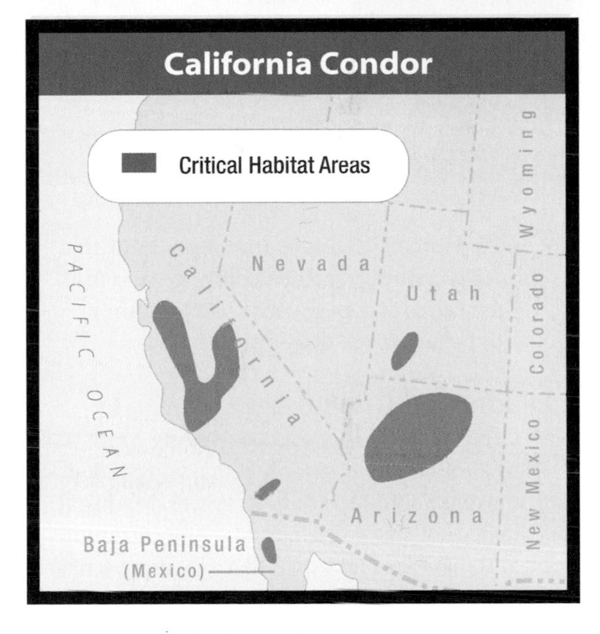

Condors have also been found in even closer proximity to human populations, according to field biologist Joanna Behrens. "As captive-bred birds return to the species ancestral roosting sites, they often find that people have taken over their habitat. With no wild parents to show them how to behave, these juveniles often settle in on roof tops and balconies, risking the dangers of close contact

COURTESY OF SCOTT NIKON FRIER / US FISH & WILDLIFE SERVICE

A condor feeds at a breeding site at Ventana Wilderness Centre prior to its release in 2000.

with humans." Behrens and others hope that as they mature, reach breeding age, and find themselves occupied with raising a family, this behavior will disappear.

The California condor is among the largest birds in the world, averaging 20 pounds or about nine kilograms, with a wingspan of more than 9.5 feet or three metres. Mating at the age of six or seven, they are monogamous and mate for life, a span that can approach 60 years. Normally, a single large, round egg is laid every second year between January and March. The eggs are incubated by both parents and hatch after two months. The downy gray chicks have bare patches on their heads, necks, bellies and underwings and grow quickly, reaching twenty pounds or nine kilos by six months, about the time they begin practice flights. Despite their size, youngsters stay with their parents for two years.

Adults have distinctively pinky-orange bald heads and blush when excited or upset, with the skin on their head and neck turning a darker red. During courtship, they puff out their throat sacks. With large white triangular patches under their wings, they are easy to spot when soaring aloft. While some confuse them with turkey vultures, identifying them is easy when the two are seen together; condors were described by author John Moir as "small airplanes". Condors also fly with their wings flat, rather than holding them in a "V" shape, as turkey vultures do, and their long flight feathers are splayed like fingers at their wingtips.

With claws that are blunt, rather than sharp like those of eagles, and lacking an opposing toe, condors are unable to carry their prey aloft. And though humans are virtually their only predators, they are bold and curious and often attracted to human activity.

VIEWING CALIFORNIA CONDORS

Since captive-bred condors began to be released in the 1990s, small populations have been established in central California, Arizona, Utah and Mexico's Baja California peninsula. Two of the best places to view condors flying free are in California, south of San Francisco at Big Sur, and at Vermilion Cliffs National Monument, on the Utah border in north-central Arizona just west of the Colorado River. Both regions offer remote cliffs with suitable caves and ledges to be used as nesting sites, as well as open savannahs and grasslands – important to condors while searching for food. In California, a number of sites near Big Sur offer good viewing, including Andrew Molera State Park; Bottcher's Gap; Julia Pfeiffer Burns State Park, Jack's Peak in Monterey, and pullouts through Bitter Creek National Wildlife Refuge in Maricopa, California. In Arizona, pullouts have been created along Highway 89A to allow motorists armed with binoculars to stop and scan the skies.

California Condor Fast Facts

Gymnogyps californianus

LENGTH & WEIGHT

Condors are between 3.5 and 4.6 feet (or 1.1 to 1.4 metres) long, with a wingspan of nine to 10 feet (or 2.7 to three metres). Their weight varies between 18 and 31 pounds (or eight and 14 kilograms).

LIFESPAN

They can live up to 60 years, and mature at between five and seven years.

HABITAT

Condors like to roost in noisy colonies on rocky cliffs, generally in areas where updrafts of warm air carry them aloft.

DIET

Condors feed on carrion, usually dead land or marine mammals and recently, stillborn calves donated by cattle and dairy producers. However, lead poisoning from lead fragments in unrecovered game or gutpiles that have been shot with lead ammunition has been deemed the most common cause of death in condors (lead shot game can be a risk to human and other wildlife populations, as well). Unfortunately, attempts to convince hunters to switch to copper and other non-lead ammunition have been met with considerable resistance from the members of the American gun lobby, including the National Rifle Association, which seems to view it as an attack on the right to hunt. Nevertheless, following decisive evidence that many wild condors had blood-lead levels far beyond the acceptable level, in July 2008, California Governor Arnold Schwarzenegger signed legislation banning lead bullets for big game hunting across much of his state.

Wild condors don't eat every day, so captive birds are also not fed every day.

BEHAVIOUR

Despite their dietary choices, condors are remarkably clean birds. After feeding, they clean their bare heads and necks by rubbing them on grass or branches, and also bathe frequently, spending hours preening and drying their feathers. And not surprisingly, they have hardy digestive systems and effective immune systems, which allow them to dine on animals in various states of decay.

MATING & BREEDING

Breeding begins between five and seven years of age, condors mate for life and their bald pink heads often turn dark red when mating. Pairs do not build nests, but lay a single large, round pale blue-green egg in a cave, crevice or cavity of a large tree, or on a ledge. Both parents share the care of the egg and chick and the large offspring usually stays with its parents for two years. As a result, pairs will generally breed once every two years.

RECOVERY PROGRAM

The California Recovery Program led by the US Fish and Wildlife Service includes partners that span the Pacific Coast from Oregon to Mexico's Baja California peninsula, and extend well inland through northern Arizona and southern Utah. One of the earliest and most active partner organizations is the Ventana Wildlife Society, which has been involved in California's only public releases of condors into the wild. Releases of condors in California take place in Pinnacles National Monument, Big Sur, Hopper Mountain National Wildlife Refuge and Bitter Creek National Wildlife Refuge. As of late 2009, the total population of California condors had risen to 350, with the majority living in the wild.

Birds raised in captivity (as many still are) go through a three- to-six-month pre-release program before being released into the wild. Lodged in a large flight pen, a group of chicks of similar age are placed with a "mentor bird", an experienced adult who serves as a role model for the inexperienced youngsters. Pre-release pens are even equipped with a mock utility pole capable of giving young condors a mild shock, to condition them against landing on electric poles and wires following their release.

Prior to release, each bird is fitted with a numbered radio transmitter and numbered wing tag and once free, is closely monitored. The radio signals will change to a distress signal if the bird fails to move for more than eight hours. Some birds are also fitted with GPS transmitters, which provide hourly locations and movement speeds during daylight hours via satellite.

NAMES

The scientific name comes from the Greek word *gymnast*, meaning "naked", referring to the head, *gyps* is Greek for "vulture" and the word condor comes from the Spanish *cuntur*, the Inca name for the Andean condor.

Leatherback Sea Turtle

Dermochelys coriacea • Endangered (US and Canada) Critically endangered (IUCN)

ALMOST everything about the leatherback sea turtle is unique: its enormous size; its speed; its appearance; its leathery shell; its vast migrations and ability to live in waters as far north as Alaska and Norway or as far south as the southernmost tip of New Zealand; its remarkable prowess as an extraordinarily deep diver, and its favorite food – jellyfish, not their floating parachutes, but rather their viciously stinging tentacles.

Unfortunately, there's one thing leatherbacks do share with the world's other six species of sea turtles – plummeting numbers worldwide. The devastation is particularly appalling among the three Pacific populations of leatherbacks.

When ocean currents and upwellings bring jellyfish by the millions to the Pacific coasts of the US and Canada, the sea turtles follow. Though they often feed along the coast of North America, these magnificent animals are classified as belonging to the western Pacific population, which nests 7,000 miles or 11,000 kilometres away on the beaches of New Guinea, Indonesia and the Solomon Islands.

The eastern Pacific population nests in Mexico and Costa Rica and forages south along the west coast of South America. And a third population, now all but gone, once nested in Malaysia and fed in the Indian Ocean. All three Pacific populations have been devastated over the past quarter-century (see below).

Leatherbacks are also found in the Atlantic Ocean, from the Gulf of St. Lawrence to the Cape of Good Hope, and recent population counts estimate that between 20,000 and 30,000 nesting females remain worldwide, a dramatic decline from the estimated 115,000 in 1980.

Though they live their lives almost totally out of sight – once they leave the nest and scoot into the sea, males never return to land, while females come ashore only to lay their eggs – we are slowly learning what remarkable animals leatherbacks are. They are not only the world's largest sea turtles, but also the last surviving members of their family.

The Pacific populations are somewhat smaller than their Atlantic counterparts – the world's largest recorded specimen, discovered on a beach on the west coast of Wales, meas-

Her eggs laid, disguised and hidden on a beach in French Guiana, a female leatherback heads for the sea.

COURTESY OF MATTHEW GODFREY

Marine photographer Herb Segars spotted this leatherback only because it was caught in a fishing line attached to a buoy on the surface. Puzzled as the buoy skittered across the water, he and a friend dove down to see what was going on. Apparently accustomed to the drag, the turtle swam forward, paused to let the buoy catch up, and only then moved on. After filming it, the divers cut the line, setting the animal free at last.

HERB SEGARS – gotosnapshot.com

ured 10 feet or just over three metres from head to tail and weighed 916 kilos or more than 2,000 pounds – nevertheless, Pacific leatherbacks often reach more than six feet or two metres in length and weigh about 1,500 pounds or 700 kilograms.

Not surprisingly, given their size, adult leatherbacks have few enemies. Yet they are in dire straits worldwide. And though large adults are sometimes injured by sharks, the plummeting numbers of leatherbacks everywhere are almost exclusively the result of human predation or irresponsible behavior.

The western Pacific population, which nests on the beaches of Irian Jaya, Indonesia, and feeds along coastal North America from California to Alaska, dropped from an estimated 13,000 nesting females in 1984 to 1,690 in 2000. Farther south, what was once perhaps the world's most important leatherback nesting region has been almost destroyed. As recently as 1980, an estimated 30,000 females laid their eggs on beaches in Mexico's coastal states of Michoacán, Guerrero and Oaxaca. In one of these rookeries, at Playon de Mexiquillo near the Rio de Manzanillo in Michoacán, the number of nests plunged from 4,796 in 1986–87 to just 70 in 1993. That same year, egg shells in the nests were tested and found to contain oil and grease, as well as copper, zinc, nickel and lead, contaminants that could have leached from the beach sand or been picked up by the female leatherbacks as they travelled.

And in Malaysia, where 10,155 nest clutches were found along one stretch of beach in 1956, there were fewer than 40 nests in 1997 and in 2008, just two nesting females returned; the eggs they laid were infertile.

Though a number of species consider turtle eggs a prized food, and hatchlings face a daunting natural world until they grow big enough to fend for themselves, in each of the places where Pacific leatherbacks nest, humans are very largely responsible for the steep decline in their numbers. The species is classified as critically endangered by the IUCN, yet harvesting eggs and killing nesting females (as well as hunting both males and females out in the open ocean) continues. Turtle eggs are used in traditional Asian medicines; Latin Americans consider them an aphrodisiac; Europeans have traditionally had a marked taste for turtle soup; leatherbacks continue to be killed in the Persian Gulf for calking boats and in Indonesia for use in oil lamps; Indonesian shops are still filled with turtleskin bags and other curios and in just one year, 1989, Japan imported 1.5 million pounds or 680 metric tones of shell, representing about 700,000 dead turtles. One hundred and sixteen countries have banned both the import and export of sea turtle products, yet the illegal trade continues.

Other threats impact the magnificent leatherbacks as well, most of them due to human carelessness or a complete lack of awareness. Until regulations were instituted to allow them to escape from the nets, an estimated 55,000 sea turtles were drowned annually by shrimp trawlers in US waters, and today thousands drown every year after being snagged on lines used by the long-line fishing industry.

Leatherbacks are also being killed by floating plastic garbage (see sidebar on page 162), particularly plastic bags, which they apparently mistake for their favorite food, jellyfish. Once ingested, the thousands of downward pointing spines that line a leatherback's throat make plastic bags virtually impossible to regurgitate and can partially or completely block the gastrointestinal tract, making digestion either difficult or impossible. Like albatrosses, the digestive systems of nearly half of leatherbacks recently studied have been reported containing everything from pieces of plastic milk jugs to bits of balloons, and some have died of starvation. And phthalates, a group of chemicals used as plasticisers, have been found in the yokes of leatherback eggs.

Nesting beaches are not only heavily exploited by local peoples around the world, but closer to home, waterfront beaches in Mexico are being transformed into oceanside condos and hotels, and industrial development erodes or contaminates beaches, resulting in habitat loss for all sea turtles as well as other species.

COURTESY OF NPS – CANAVERAL NATIONAL SEASHORE

A female leatherback digs a nest to lay her eggs at Canaveral National Seashore.

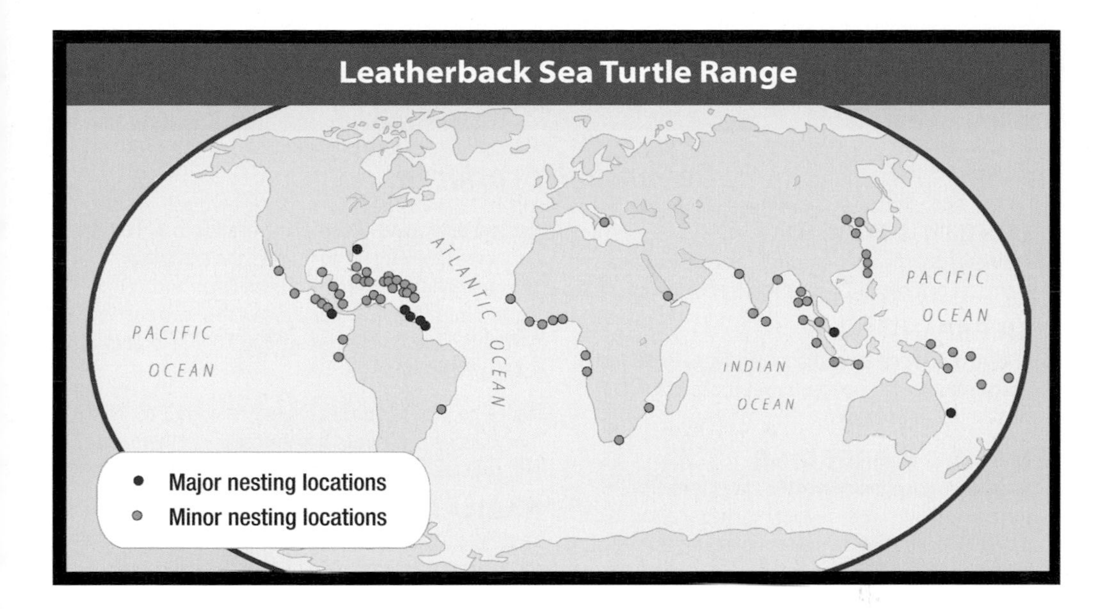

Oil spills around the North American coastlines have had negative results on several species of juvenile sea turtles, including loggerheads and green turtles, but perhaps because of their immense ranges, their nesting locations or a lack of information about oil spills, leatherbacks do not seem to be among them. However, the contamination of nesting beaches by accidental oil spills is an ongoing concern. In the past, when threatened by the blowout of an oil well off the coast of Mexico, US Fisheries and Wildlife officials assisted the Mexican Fisheries Department in moving 9,000 Kemps Ridley sea turtle hatchlings to a safe holding area, before releasing them into clean water offshore several days later.

Pathogens are also potentially deadly for sea turtle populations. Recent studies show that diseases emerging in wildlife populations are similar to those of humans. In sea turtles, Texas A & M biologist Adam Jones has found what he terms "an appalling disease", fibropapillomatosis. Once rare, this tumor-causing virus "likely caused by a herpes virus", according to Jones, causes tumors to grow throughout a turtle's body. In one recent sample from the Hawaiian Islands, he wrote in "Sea Turtles: Old viruses and new tricks", in *Current Biology*, "more than 90% of green turtles showed symptoms of the illness."

Though research continues, the early history of sea turtles is only beginning to be understood. Well-preserved fossils of a primitive ancestor from the Early Cretaceous found in Australia indicate that early sea turtles were both larger and more diverse than previously thought. At least five distinct sea turtle lineages existed about 105 million years ago. But their deep history goes back 200 million years farther, to the cotylosaurs, the earliest true reptiles or stem reptiles, for all vertebrate life leads back to them, just as a tree's branches converge in its trunk.

Cotylosaurs' hard scales and shelled eggs, marked an evolutionary turning point, for they "insured them against the age-old disaster of drying out both before birth and after," wrote the celebrated dean of sea turtles, Archie Carr, in his *Handbook of Turtles.* This was the beginning of a vast outpouring of reptilian forms, but the oldest true turtle fossil came later. Dating back some 220 million years, *Proganochelys*

Leatherback Sea Turtle Fast Facts

Dermochelys coriacea

LENGTH & WEIGHT
Leatherbacks have been known to reach a length of eight feet or more (or 2.5 metres) and weigh a ton.

LIFESPAN
In ideal conditions, leatherbacks can live to be 80 years of age. Most die far younger, however, of a long list of human-induced hazards.

MATING & BREEDING
Solitary creatures for most of their lives, leatherbacks mate at sea, either near nesting beaches or along migratory corridors. The female then returns to the tropical or subtropical beach of her birth to lay her eggs. Digging in the soft sand, she excavates a nest above the high-tide line with her flippers, lays her eggs and then carefully backfills the nest, scattering sand over it to try to disguise it from predators. The track she leaves as she returns to the sea is unmistakable, however.

REPRODUCTION
One female can lay as many as 10 clutches of eggs at eight- to 12-day intervals, with each clutch including about 70 or 80 eggs the size and shape of billiard balls, topped with 20 to 40 SAGs, smaller, yokeless (and therefore infertile) eggs that perhaps are meant to serve as decoys for the viable eggs below. The temperature in the nest determines whether the hatchlings will be male or female. After 60 to 65 days, the hatchlings emerge, each the length of a human finger and weighing between 40 and 50 grams or less than two ounces. Biologists estimate that perhaps 1 in 1,000 hatchlings survives to adulthood.

In ideal conditions, females might nest every three or four years, up to 30 times; however, many leatherbacks nest only once and are killed before they can return.

HABITAT
Leatherbacks are found in all the world's oceans, from the Arctic and Antarctic Oceans to tropical seas, though almost everywhere in diminishing numbers.

DIETS
Leatherbacks live largely on jellyfish, and have sharp-edge jaws and cusps (though no teath or chewing plates) perfectly designed to feed on man-of-war jellyfish, as well as on sea urchins, squid, small fish and floating seaweed.

NAMES
In Trinidad, leatherbacks are known as *caldon,* while in Latin America they are sometimes called *canal* or *las baulas.*

VIEWING LEATHERBACK SEA TURTLES

Given the vast distances they swim and the enormous area they cover, viewing leatherbacks seems to occur mainly by happenstance. In years when ocean currents and upwelling water brings krill and other nutrients to the surface, attracting jellyfish, the chance of seeing leatherbacks is greatly increased in the waters off the California and Oregon coasts. (In late September 2008, spotters in a NOAA twin otter plane reported seeing six leatherbacks, "surrounded by miles of jellyfish.")

Small nesting areas are located on the southeast Florida coasts, and the leatherbacks born here are slowly increasing in number. In the US Virgin Islands, Sandy Point Beach on the western end of St. Croix has been designated critical habitat.

Scientists travel to Costa Rica between October and February, for that is when leatherbacks nest in significant numbers in Las Baulas National Park on the northwest coast of Costa Rica. Earthwatch provides guides for tours of the park. For more information, contact www.leatherback.org and look under Las Baulas (Spanish for leatherback) National Park.

Leatherbacks also congregate north of the Hawaiian Islands.

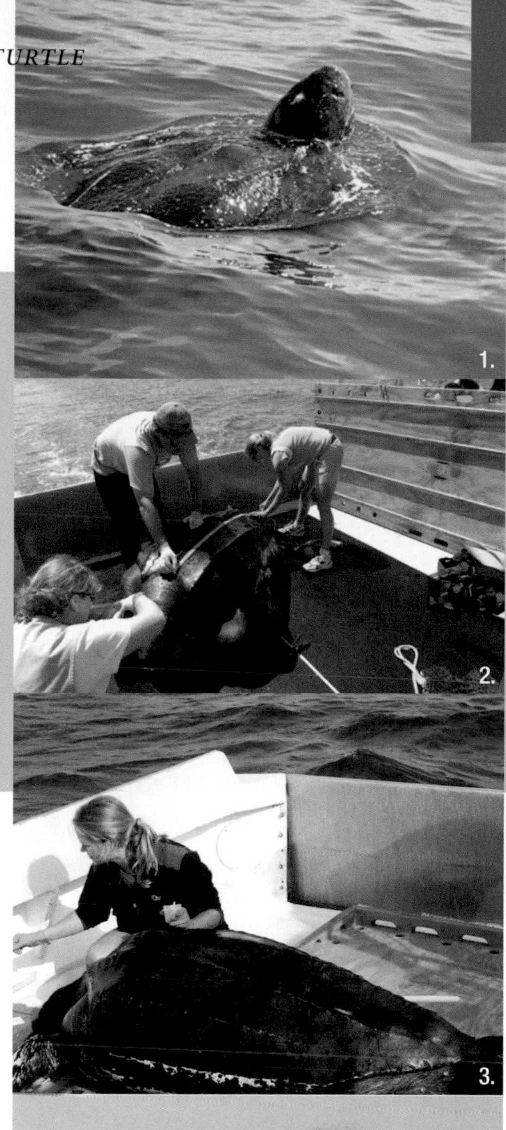

Scientists from the Large Pelagics Research Centre at the University of New Hampshire began satellite tagging leatherback turtles off the US East Coast in 2007 to learn about their movements, behavior and migratory routes. After tagging 22 leatherbacks (photos at right), scientist Kara Dodge writes that ". . . preliminary results show that the turtles spend up to three months in coastal and shelf habitats where they are vulnerable to interactions with various fisheries, vessel strikes, habitat degradation and pollution. Their migrations from northern to southern waters are impressive, showing clear orientation and navigation abilities over thousands of miles of open ocean. We hope our tagging data will help protect important leatherback habitats and reduce human impacts on this endangered, enigmatic species."

was a much larger version of modern snapping turtles. "By a cryptic series of changes, few of which are illustrated in the fossil record, there evolved a curious and improbable creature, which had a horny, toothless beak and a bent and twisted body encased in a bony box, the like of which had never been seen," wrote Carr.

The great mystery surrounding turtle evolution is its rapidity. In his beautiful, award-winning book, *Sea Turtles*, Drexel University's James Spotila theorizes that even a minute change in genetic composition could lead to the creation of a carapace and plastron, respectively the upper and lower parts of a turtle's shell, and that this small genetic change could have had huge evolutionary advantages.

Clearly, it was a model that worked. As the world evolved around them, as dinosaurs rose to dominate the land, plesiosaurs the seas and pterosaurs the skies, as comets rained down on the Earth, some turtles climbed through the swamps onto the land to become tortoises, while others stayed in the seas.

At the end of the Mesozoic, about 66 million years ago, there was a huge mass extinction that spelled the end of the dinosaurs and the Age of Reptiles gave way to the Age of Mammals. But in the oceans, four families of

PHOTOS COURTESY OF: 1. KARA DODGE Large Pelagics Research Center 2. MICHAEL DODGE 3. SCOTT LANDRY 4. NUNO FRAGOSO For images 2-4, Research conducted under NMFS Permit 1557-03

COURTESY OF SCOTT R. BENSON / NOAA FISHERIES

sea turtles survived, Toxochelyidae, Protostegidae, Cheloniidae and Dermochelyidae. As time passed, the first two disappeared, Toxochelyidae during the Eocene epoch between 55 and 34 million years ago (mya) and Protostegidae during the Oligocene between 34 and 24 mya.

The Earth continued its revolution around the sun, its elliptical path changing shape over time and the climate warmed and cooled. Mammals evolved wondrous shapes and, to quote Archie Carr again, "a mob of irresponsible and shifty-eyed little shrews swarmed down out of the trees to chip at stones, and fidget around fires, and build atom bombs." But turtles held fast to their remarkably successful structural plan.

Today, of the seven species of sea turtles that remain, six are in the Cheloniidae family, the other – the leatherback – is the last member of the Dermochelyidae family and the only one without a bony shell. Instead, its dark gray or black carapace is thick and leathery, with miniscule embedded bony plates and seven distinct ridges running from front to back. The underside is mottled pinkish-white and black.

In addition to being the largest, most distinctive and most often seen off North America's Pacific coast, leatherbacks are inimitable in many other ways. No other reptile swims as fast or dives as deeply; the *Guinness Book of World Records* reports that leatherbacks have achieved speeds of 21.9 miles or 35.3 kilometres per hour in the water (though they generally swim at a quarter that speed), and are known to dive to depths of almost 4,000 feet or 1200 metres.

No other sea turtles have the range that leatherbacks do. They have been found as far north as Alaska and Norway and as far south as New Zealand's south coast. They can do this because they have a metabolic rate at least four times higher than other large reptiles, an insulating layer of fat, a dark body color that absorbs heat and "countercurrent heat exchangers" in their huge flippers. These blood vessels, which are well insulated by their body size, are bundled in such a way that those carrying warmed blood from the heart are located immediately beside those carrying cooled blood back to the heart. The result is a core body temperature as much as 64° F or 18° C warmer than the surrounding water. Remarkably, only large adults are found in cold water environments; smaller, younger leatherbacks seem to know they do not have the necessary insulation and stay in waters with temperatures of 73° F or 26° C.

Leatherbacks travel vast distances each year. A transmitter attached to one female nesting in Indonesia has shown that in 2003, she travelled at least 12,774 miles to the coast of Oregon, out to the waters near Hawaii and halfway back before the transmitter batteries died.

As awareness of the dire future facing sea turtles grows, the US and Canada, as well as other nations around the world, are attempting a variety of methods to decrease bycatch, protect nesting beaches and assist sea turtle populations in other ways. Bycatch, the accidental take of non-targeted species, or species of an undesirable size or age have amounted to more than 90 per cent of trawl net catches in the past, devastating ocean life. Around the south and east coasts of the US, Turtle Excluder Devices or TEDs, with openings large enough to allow leatherbacks and loggerhead turtles to escape, have helped somewhat. However, as of 2009, only the US and Caribbean countries were routinely using TEDs, while some shrimp trawlers are reported to be wiring the nets closed. If used properly, the TEDs are reported to be 97 per cent effective for sea turtles.

The US is also working with other nations to develop teams of volunteers to patrol the beaches to protect nesting females, eggs and young when they appear. Though nest predation has declined in recent years, this not only continues to be important in major nesting areas where turtle eggs are still considered an important human food source, but to guard against animal predators such as raccoons and dogs.

Even when nesting beaches are protected, coastal developments have proven a threat to hatchlings for when they emerge from the nest, generally at night, they head toward the brightest area on the horizon. Historically, that was the ocean, for the moon and stars reflected off the water. Today, however, it is just as likely to be a coastal development. To combat the tiny turtles' confusion, many resorts and condo developments are dimming or turning off their exterior lights during the nesting season.

To counter the affects of all terrain vehicles, major nesting beaches are often restricted during the breeding season or, as indicated above, protected by an army of volunteers.

Deaths from longline and gillnet fisheries are a more serious matter, and affect not only sea turtles, but other marine creatures, including endangered great hammerhead sharks. The answer, the IUCN believes, might be conservation corridors or "swimways", free of longlines and nets, in regions known to harbor high concentrations of wildlife. The first such swimway, proposed at the World Conservation Congress in the fall of 2008 and supported by more than 8,000 scientists, government officials and environmental organizations, was the Cocos Ridge Marine Wildlife Corridor, a 360-mile or 600-kilometre geological hotspot that runs along the Pacific coast from Ecuador to Costa Rica. Scientists have discovered that western Pacific leatherbacks nesting in Costa Rica often head for the Galapagos Islands after nestings. Closing the area to fishing trawlers at appropriate times would protect them during the crucial mating and nesting season.

Headstarting, or assisting leatherback hatchlings (other than helping them find the ocean while emerging from their nests) has also been considered. While this has helped other marine turtle species by allowing them to be raised in a safe situation for six to 12 months until they are big enough to fend for themselves, leatherbacks have one key characteristic that makes this impossible. They have only one gear – forward. Unable to swim backwards, the tiny turtles have been found to swim relentlessly forward until they reach an impediment (such as a hatchery wall) and then repeatedly bump their tiny heads against it until they hurt themselves.

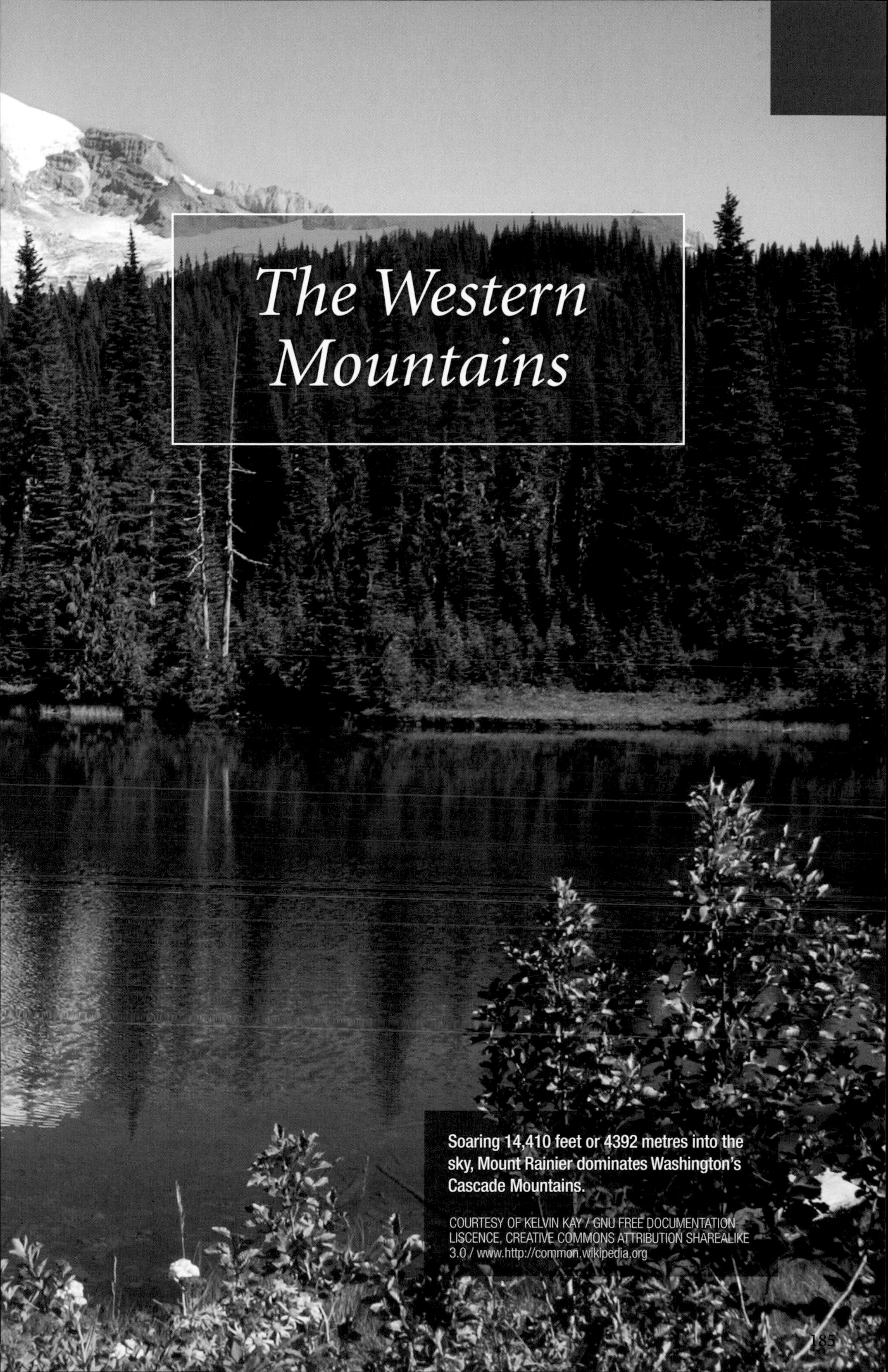

The Western Mountains

Soaring 14,410 feet or 4392 metres into the sky, Mount Rainier dominates Washington's Cascade Mountains.

COURTESY OF KELVIN KAY / GNU FREE DOCUMENTATION LISCENCE, CREATIVE COMMONS ATTRIBUTION SHAREALIKE 3.0 / www.http://common.wikipedia.org

Beautiful Upper Thornton Lake, at 5,040 feet or 1536 metres, and Middle Thornton Lake about 300 feet or 100 metres below it, are in Washington's North Cascades National Park.

COURTESY OF WALTER SIEMUND / GNU FREE DOCUMENTATION LICENSE, CREATIVE COMMONS ATTRIBUTION SHAREALIKE 3.0 / www.http://common.wikipedia.org

The Western Mountains

Almost no one who has ever visited North America's western mountains have come away unmoved. And many have described the experience as "otherworldly".

The **the magnificant** Rocky Mountains that link Canada and the United States, British Columbia's Columbia Mountains, the Cascade Range of the northwestern US and California's dramatic Sierra Nevadas provide a virtual textbook on many of the things that set North America's western mountains apart (see Building the Western Mountains, page 194). Take plate tectonics, for example. Both the Rockies and the Columbia Mountains, which with the Omineca and Cassier Mountains comprise the Omineca belt, were formed as a result of pressure from collisions with large island terranes farther west. Both are comprised of rock that has always been a part of the ancient North American continent. The Omineca belt mountains are older than the Rockies, however, and have been through at least two mountain building periods, while the Rockies were built in one major episode that began about 140 million years ago. Nevertheless, locals often refer to the Columbia Mountains as the "Kootenay Rockies".

While the Rockies and the Columbia Mountains include plenty of hot spots and hot springs, including Yellowstone, with its intense volcanism that continues today, they are not, on the whole, considered to be volcanic mountains like the Cascades and Sierra Nevadas to the west and south.

The Cascade Range, which extends from southern BC to northern California, is a mix of volcanic and non-volcanic mountains, as are the Sierra Nevadas farther south. Some of these are still active, including the volcanoes of the Cascade Volcanic Arc, which extends into the southern Coast Mountains.

Beautiful Mount Rainer, at 14,409 feet or 4392 metres the highest peak in the Cascade Range, is a volcano. And Canada's small portion of the Cascade Range includes Mount Garabaldi, which towers over the Vancouver skyline. Created in a long series of violent volcanic periods interspersed by stretches of dormancy, Garibaldi, which is quiet at present, is almost certain to erupt again someday. However, it and the other Cascade Mountains of southwestern British Columbia, as well as the North Cascades in Washington State are very different in character than the series of high volcanic cones from Mount Rainier southwards. Steeply rugged, the North Cascades show the dramatic effects of repeated glaciations.

All of this is part of the Pacific Ring of Fire, the volcanoes and mountains that circle the Pacific Ocean, testament to the power of the ever-expanding Pacific Plate. In the US, though the results of massive eruptions can be clearly seen in Yellowstone National Park, all the historically recorded volcanic eruptions have been from Cascade volcanoes. The most recent of these were the Lassen Peak volcano in north-central California, and Mount St. Helens in southwestern Washington. Lassen Peak – the highest of a series of 30 volcanic cones in the area – last blew its top with a vast explosion in May 1915, destroying Mount Tehama and raining volcanic ash as far as 200 miles or 320 kilometres to the east. The remaining cinder cone is 10,457 feet (3187 metres) high, about a third lower than the original mountain.

Mount St. Helens, which continues to build toward another eruption, is an active stratovolcano, or composite volcano. It has erupted many times during four long stages over its 37,000-year history. Its most recent, deadliest and most economically destructive eruption was on May 18, 1980, when it killed 57 people, destroyed 250 homes, and laid waste to surrounding bridges, railways and highways. The explosion reduced the mountain's height from 9,677 feet (or 2950 metres) to 8,365 feet (2550 metres) and created a crater almost a mile wide.

Though ash was found as far away as Edmonton, Alberta, and it created the largest debris avalanche in US recorded history, the Mount St. Helens eruption pales beside the massive eruption of Mount Mazama about 7,700 years ago, when more than half the 11,000-foot (or nearly 3500-metre) mountain blew away, spewing ash as far north as Alberta and Saskatchewan and as far east as western Montana, and creating a caldera that today is half-filled by Crater Lake, the deepest lake in the United States. And the enormous eruption that created the Huckleberry Ridge Caldera

Mount St Helens, above, on the day before the 1980 eruption and right, four months after the event. The photos were both taken from Johnston Ridge, six miles to the northwest.

BOTH IMAGES: HARRY GLICKEN / US GEOLOGIC SURVEY

Reflecting both the sky and the power of the Earth, Crater Lake is the deepest lake in the United States and seventh-deepest in the world.

(http://www.flikr.com/photos/75736915@N00900901338) / CREATIVE COMMONS ATTRIBUTION SHAREALIKE 3.0 / en.wikipedia.org creativecommons.org/licenses/by/3.0

(also known as the Island Park Caldera) at Yellowstone about 1.2 million years ago was estimated to have been 6,000 times the size of Mount St. Helens in 1980.

A great deal can be learned about Western North America's fascinating, though admittedly violent tectonic history by visiting Mount Ranier National Park, Lassen Volcanic National Park, Mount St. Helens National Volcanic Monument, and Crater Lake National Park. Each allows a view of a different stage of volcanic orogeny or mountain building. And Yellowstone has volumes of material on its longlived volcanoes, which, some believe, are fed by subterranean molten rock that rises to the surface where it finds areas of crustal weakness created by stretching and thinning that continues in the US Western Interior.

To the south of the Cascades is the Sierra Nevada, California's inland mountain spine, which runs from just south of Mount Lassen to the Mojave Desert, forming what was long an almost impenetrable barrier between the Central Valley and the Great Basin. Rising in height from north to south, the range includes Lake Tahoe, site of the 1968 Winter Olympics, and Yosemite National Park, with its magnificent waterfalls. It culminates in the highest point in the continental US, 14,505-foot (or 4421-metre) Mount Whitney, one of a jumble of soaring peaks. In July 1864, the region was described this way by Yale graduate William Brewer, a member of California's first geological survey team:

Such a landscape! A hundred peaks in sight over thirteen thousand feet – many very sharp – deep canyons, cliffs in every direction almost rival Yosemite, sharp ridges inaccessible to man, on which human foot has never trod – all combined to produce a view of sublimity of which is rarely equaled, one which few are privileged to behold.

This was the lofty stronghold of Sierra Nevada bighorn sheep, today listed as endangered. As indicated on page 331, their disappearance was due to a number of factors, but first among them was their unpreparedness for man's long-distance predation. As Robert McCracken Peck writes in *Land of the Eagle*, quoting English hunter George Ruxton following his introduction to Rocky Mountain bighorns in 1846:

As they love to resort to the highest and most unaccessible spots, whence a view can readily be had of approaching danger, and particularly as one of the band is always stationed on the most commanding pinnacle of rock as sentinel, whilst the others are feeding, it is no easy matter to get within rifleshot of the cautious animals. When alarmed, they ascend still higher up the mountain, halting now and then on some overhanging crag and looking down at the object which may have frightened them ... leaping from point to point and throwing down an avalanche of rocks and stones as they bound up the steep sides of the mountain.

COURTESY OFJEFFREY PANG / en.wikipedia.org / creativecommons.org/licenses/by/3.0

The southern Sierra Nevadas are also home to trees that rank among the largest on Earth, including the world's largest living tree. Dubbed General Sherman and estimated to be as tall as a 26-story building, this soaring beauty, along with four others determined to be among the largest 10 trees in the world, can be found at the north end of the aptly named Giant Forest in Sequoia National Park. All are between 1,500 and 2,700 years old.

This is a complex range of mountains, both geologically and physically, for while the northern Sierra is mainly volcanic, the southern reaches of the range are largely granitic, rock that was formed deep underground and forced upward millions of years ago. Beginning as an ocean plate, the rock was forced by tectonic forces deep beneath the North American plate, where it melted, forming molten rock or magma. Because molten rock is lighter than the cooler rock around it, the magma began to rise in plumes, forming plutons. As, one after another, island terranes slammed into what was then the west coast of North America between 140 and 90 mya, a second and a third wave of plutons formed. Buried deep underground, they comprised what is known as the Sierra Nevada batholith.

In those distant times, the Sierra Nevada was much closer to the Rocky Mountains than it is today. Then, between 30 and 17 million years ago, through a series of complex interactions between ocean plates and the continent's west coast, the tremendous pressure on North America eased, and an era of faulting began as the Earth's crust rebounded. Though this faulting created dramatic results, including the Rio Grande Rift, perhaps most remarkable is the Basin and Range Province, an area of the southwestern US and northwestern Mexico that spans almost 300,000 square miles or 800,000 square kilometres. As in other basin and range topography, the landscape was created by crustal extensions and rifts, in which the Earth's crust is pulled apart until it is thinned to the point of fracture. As each crustal block tilts into the rift, one side drops and the other is forced into the air. In the Great Basin, the blocks of rock tilting into the air resulted in a distinctively linear pattern of mountain ranges and valleys, each with a relatively gentle slope to the west and a steep escarpment on the east side. And as the long-buried rock containing the plutons rose in this dramatic fashion, it brought with it veins of copper, silver and gold, as well as other

COURTESY OF NASA / en.wikipedia.org

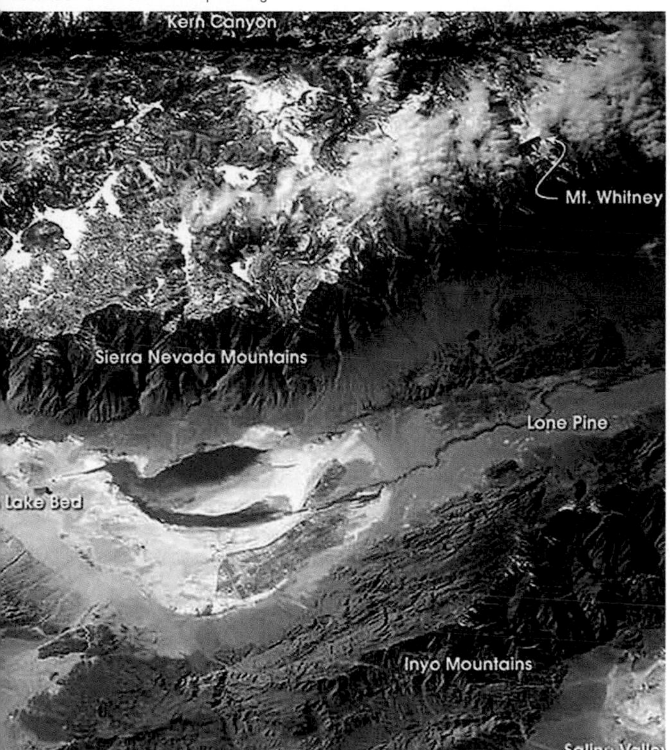

Opposite: The Sierra Nevada Range, photographed en route from Oakland to Las Vegas, makes clear the mountains' complex beginnings.

Left: The Owens Lake Bed, a deep rift in the Earth just east of Mount Whitney, seen in the upper right, was photographed from space on February 6, 2003.

minerals, which had risen as liquids and gases from deep within the Earth so long before.

These rare minerals might not have been discovered, however, had not repeated glacations and the rivers and streams they created eroded the rock until, at altitudes of between 1,000 and 2,700 feet (about 300 and 850 metres), there appeared a string of pockets of gold – rather like beads in a necklace – that stretched for 120 miles (nearly 200 kilometres). It was tiny bits of this gold – placer gold worn from the heights – that drew more than 100,000 people to California in the decade that followed their discovery in 1848. And it was the people who, in their rush for fortune, slashed the magnificent trees from the hillsides, fouled the streams, killed the mammals and birds and turned what had been a self-sustaining paradise into a state that has become too populous for its own good.

Crustal stretching can also be seen far to the north, in eastern British Columbia and northwestern Montana, where between 55 and 35 million years ago, the "Kootenay Rockies" were separated from the main Rocky Mountains by crustal stretching. This rebound effect, which occurred as pressure from the west diminished, created the Rocky Mountain Trench, a deep fissure that is one of the few geological features on Earth that can clearly be seen from the moon.

Along the trench south of Prince George, BC, the block faulting created "half-grabens", much like the faulted blocks in the Great Basin. These can be seen in several places along the trench, but nowhere quite as clearly as in the southeastern Kootenays, where the Rocky Mountain Trench separates into several valleys that run south of the American border. The Flathead River Valley, which is actually part of the Rocky Mountains and borders Montana's Glacier National Park, is an almost perfect half-graben valley, with the rock on the west side of the valley precisely matching that on the east. Geologists have determined that the Flathead Valley half-graben is almost 10 miles or six kilometres deep – about two-thirds as deep as Mount Everest is tall.

This faulting didn't happen all at once, of course; instead, as the Earth's crust stretched or rebounded as the western pressure relaxed, the rock slipped downward, causing regular earthquakes for millions of years.

Despite the beauty and volatility of many of the ranges farther west, when most visitors to North America think of its mountains, they imagine the Rockies. Stretching almost 5000 kilometres or 3,000 miles from northern British Columbia to New Mexico, the Rocky Mountains delineate the Continental Divide, the line at which water flows either to the Atlantic or the Pacific Oceans. And in the case of two mountain glaciers – Alberta's Snow Dome, the highest point of the Athabasca Glacier in the Columbia Icefield, and Triple Divide Peak in Montana's Glacier National Park – the melted ice crystals end up in three different oceans.

The loftiest peak in the Canadian Rockies is BC's Mount Robson, at 3954 metres or 12,972 feet, while in the US, Colorado's Mount Elbert is 14,440 feet or 4401 metres. (Mount McKinley, also known as Denali, in Alaska's Denali National Park, is hands-down the highest point in North America at 20,138 feet or 6138 metres.)

On both sides of the border, the Rockies contain significant wilderness areas, but with nine national parks, as well as much visited provincial and state parks, they are also magnets for tourists. The trick is going to be finding a balance for the animals and plants millions come to see each year.

As indicated in the section on grizzlies that begins on page 205, though Yellowstone National Park boasts that it has the highest concentration of mammals in the Lower 48 – 67 different species live in the park, including grizzly and black bears, gray wolves, wolverines, lynx, elk, bison and moose – there is consider-

US NATIONAL ARCHIVE 077-KS4-112

Members of Clarence King's Geological Exploration of the 40th Parallel (1867-1873) survey the land around Shoshone Canyon and Falls, which was then in Idaho Territory.

able concern in Canada's Rocky Mountain parks about their largest predator. Painstaking counts of grizzlies between 2003 and 2009, using lures and DNA testing to prevent counting individuals more than once, seemed to reveal that Alberta, long known for its watchable bears, has perhaps 700 grizzlies . . . or even fewer. Gord Stenhouse, chairman of the province's grizzly bear recovery team and head of the census, speculated in 2009 that the province has just 500 of the big bears "and maybe less". As a result – at least until the count was finalized – Alberta suspended its controversial spring bear hunt for a fourth year.

The Rockies, as well as other mountain ranges to the west and south, significantly impact both weather and climate. Caught in the rain shadow of the Rockies, the foothills and high plains immediately to the east are almost invariably dry grasslands, while to the south, the high barrier of California's Sierra Nevada helps to create the deserts to the east.

West of the Canadian Rockies, BC's Columbia Mountains divide the East Kootenays, centred around the broad Rocky Mountain Trench, and the West Kootenay region. The former has something akin to a continental climate, with hot summers and relatively dry, cold winters. The West Kootenays, by contrast, are a "coastal refugium", ecologically speaking, with mild wet winters, heavy snowfalls that allow typically coastal plants, including western redcedar, to thrive, and one of the world's best examples of inland temperate rainforest (see page 197). Yet among the coastal refugees, often on the same mountainside, one can also find outriders of the dry Interior – including the much beleaguered ponderosa pine. (Since 2000, a vast area of pine in the BC Interior has been devastated by the mountain pine beetle.)

Through these varied ecosystems, the Kootenay River is a waterway with a multitude of personalities. Beginning as a classic mountain stream, it tumbles southwest through BC's Kootenay National Park on what appears to be a collision course with the Columbia River system. But rather than spilling into Columbia Lake just north of Canal Flats, it curves south, missing the lake's southern shoreline by about two kilometres, and flows down the Rocky Mountain Trench into what is now northwestern Montana.

About 80 kilometres south of the 49th parallel, the Kootenai River (as it's known in the US) turns sharply north once more. Reentering Canada just south of Creston, BC, it meanders over the wide, fertile valley, renewing bountiful marshes that have long sustained both human and animal life, before spilling into a 100-kilometre-long fissure in the Earth known as Kootenay Lake.

Fed by rivers from south and north, the lake finds its outlet on the west. Flowing down the broad West Arm, the river at last joins the Columbia River at Castlegar. For millennia, the water thundered through this last stretch, roaring over Bonnington Falls just seven kilometres upstream from the confluence. These falls, which in historic times were 10 metres or 32 feet high, were created by isostatic rebound, an uplifting of the Earth's depressed crust after the great weight of the glaciers was removed. Before that, as soon as the ice was gone, high levels of glacial meltwater had allowed fish from the Columbia River to swim freely up the Kootenay. But eventually, as the water level fell and the land rose, the falls became too high for spawning fish to ascend, and it was this natural barrier that led to the evolution of kokanee salmon, the freshwater form of anadromous sockeye salmon, as well as the only naturally landlocked population of white sturgeon in North America (see page 244), an endangered population that now faces extinction.

Building the Western Mountains

ALMOST no one who has ever visited North America's western mountains have come away unmoved. Many have described the experience as "otherworldly". Citing the towering mountains, magnificent old-growth forests, great rivers and arid interior, they call it a place apart. And, geologically speaking, they're right. North America's mountainous West is a place apart. In fact, it's many places apart.

It should come as no surprise, then, that building this magnificent part of the continent was lengthy and complex. Even the geological descriptions of it are lengthy and complex, having to do with plate tectonics, accretionary complexes, magmatic arcs and crustal stretching. Geologists should not be blamed for their complicated jargon, however. Piecing together the formation of the West has been a convoluted process that has taken the past century and more, and involved the work of many individuals. Even today, there are still areas of the geological map that have question marks on them. Fortunately, the mountains, valleys, deserts and seashores have provided portals that, as our tools increase in sophistication, have increasingly assisted in piecing together much of the ancient past, not only of Western North America, but of our planet.

One of the best known of these windows on time is the Burgess Shale, an ancient undersea archive that today is far from any ocean and 2300 metres or almost 7,400 feet above sea level. Reached via a demanding alpine trail in Yoho National Park in the Canadian Rockies, this 513-million-year-old site offers an almost perfectly preserved picture of what may have been the Earth's most important eruption of life. The Cambrian Explosion, which according to zircon dating began 542 million years ago, was an orgy of evolution. Yet, in geological terms, it was the blink of an eye. In a period of about five million years, all the major body plans of animals evolved, including the ancestors of vertebrates or animals with backbones. Everything since, the late biologist Stephen Jay Gould once wrote, has been nothing but variations on a theme.

Like the Mount Stephen Trilobite Beds, located across BC's Kicking Horse Valley, the creatures of the Burgess Shale lay along the shifting shorelines of embryonic North America. A half-billion years ago, these shallows brimmed with life, in sharp contrast to the barren landscape of the continent. Without trees or grass or plants of any kind to anchor the soil, the wind and rain wore at the bones of the earth. Gushing rivers carried sand and silt from ancient mountains in the continental core to the sea. Layer upon layer, the coastal sediments grew, littered with the distinctive shells of dozens of trilobite species, as well as the many other wonderful

creatures that evolved over time.

By about 280 million years ago, sliding on the Earth's tectonic plates, the continents were again coming together to form a huge supercontinent that we know as Pangaea, "all lands". This was not the first congregation of continents in the Earth's long history, far from it. Geologists now believe that at least four other supercontinents had previously formed, but we know far more about Pangaea than we do about any of the others.

Ancient North America – Laurentia – occupied the northwest quadrant of Pangaea, which reclined on its side (compared to its modern orientation) touching the equator. The land that would one day be the mountainous West lay offshore, some of it not far away beneath a shallow continental sea, but most of it somewhere out in the Pacific, as embryonic island arcs or terranes far to the southwest, perhaps where Japan is today – as in the case of what would one day be parts of the Sea to Sky Region (see page 101). Other embryonic land, including what are now the stately volcanoes of the Cascades, lay deep within the Earth's molten core, waiting to be created. The Pacific Ocean covered most of the Earth's surface and the Atlantic was yet to be born.

Beginning about 270 million years ago, the islands far out in the Pacific began to coalesce into terranes – island arcs rather like the Philippines today. Gliding slowly northeast, some of them merged into a small continent, which by 200 million years ago lay just west of the giant supercontinent Pangaea.

This was the Jurassic, the beginning of the long ascendency of the dinosaurs and marine plesiosaurs. As the giant animals flowered worldwide, Pangaea was coming apart at the seams. Laurentia, ancient North America, slid northwest, creating the Atlantic Divide and colliding, between 185 and 170 million years ago, with the smaller continent that was riding northeast on the Pacific Plate. The smaller terrane slammed obliquely onto the edge of ancient North America, initiating the continent's second era of mountain building. (The first, which had taken place more than 1.7 billion years ago, created an arcing chain of lofty mountains in the heart of the North American continent. Over deep time, these wore down to become the Precambrian or Canadian Shield.)

Like a wedge, the edge of Laurentia peeled the lighter rock of the island terrane off the underlying oceanic plate, cementing it onto the continental margin. Today, a broad piece of territory, known as the Interior Plateau, covers a long swath of territory in central British Columbia. And far to the south, in California, beginning about 250 million years ago and continuing for about 100 million years into the earliest part of the Cretaceous, accreted island terranes were forming what would someday be the Sierra Nevadas, as well as most of the land to the west of them.

Coastal collisions like this continued over millions of years and inland, the new territory served as an enormous bulldozer. Rumpling the sedimentary layers like tinfoil, British Columbia's Columbia Mountains were created, while farther east, the pressure caused the land to buckle and jack-knife skyward as the main ranges of the Rockies were created.

To the north, the Pacific Plate, now once again on a

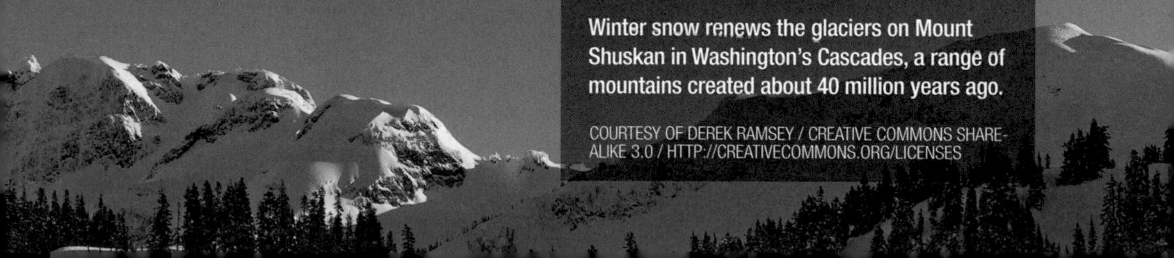

Winter snow renews the glaciers on Mount Shuskan in Washington's Cascades, a range of mountains created about 40 million years ago.

COURTESY OF DEREK RAMSEY / CREATIVE COMMONS SHARE-ALIKE 3.0 / HTTP://CREATIVECOMMONS.ORG/LICENSES

southwestern track, collided directly with Wrangellia, a huge terrane that today includes BC's Vancouver Island and Haida Gwaii. Inland, the melted crust of the collision zone caused the upheaval of the largest mass of granite in the world along the Coast Belt, and much farther east, in what is now Alberta, the front ranges and foothills of the Canadian Rockies.

Then, toward the end of the Cretaceous about 70 mya, the Laramide Orogeny (as geologists know it) produced a wide volcanic arc that created volcanoes in Colorado, Wyoming, northeastern Nevada and the interior Northwest. According to the University of California Museum of Paleontology, this " 'flat slab' episode profoundly shaped western North America," producing mountain ranges extending from the coast to the Cordillera.

In California, the "flat slab" lifted the northern Sierras well above their present elevation, but since the Basin and Range did not yet exist, they lay several hundred kilometres closer to the Rockies than they do today.

Then temperatures rose. Fossils in the northern Sierras, as well as in southern British Columbia far to the north, show that Western North America was tropical during the early Eocene epoch. About 55 million years ago, over a period estimated at just 5,000 years, the global temperature rose between four and seven degrees Celsius. By comparison, modern global temperatures have risen just over half a degree since the beginning of the Industrial Revolution. In the northern Sierras, in this tropical landscape, deep channels – known as the "ancestral Yuba River system" – were incised into the land, extending well into northwestern Nevada. Millions of years later, in gravel within these channels, placer deposits of gold would be found, sparking the California Gold Rush and profoundly impacting the forests and the wildlife of California (see page 155).

The glaciers of Mount Shuskan are reflected in the aptly-named Picture Lake in this photo taken from Mount Baker.

COURTESY OF SIRADIA / en.wikipedia.org / creativecommons.org/licenses/by/3.0

In Canada, most of the major mountain-building elements were in place by the Eocene, but in Oregon and Washington, the Cascades were still being constructed. Geologists believe they followed the breakoff of the "flat slab" discussed earlier; volcanic activity began about 40 mya in Oregon and Washington and built southward as the slab tore away. Then, perhaps 25 mya, complicated movements of the subducting Farallon plate, which collided with the coast of Southern California, created two "triple junctions", marked today by the San Andreas Fault. Rotating and dragging the solid Sierran crust northward, the Northern Sierras and Klamath Ranges were pulled north, slowly opening the Great Basin.

All these changes were accompanied by dramatic climate changes as the tropical and subtropical heat of the Eocene was replaced by drier warmth in the Southwest and winter cold that built from the north. Deciduous broadleaved and conifer vegetation was replaced by evergreen hardwood forests. By two million years ago, the western mountains had their great forests, interior grasslands and deserts, as well as a remarkable diversity of land and sea life.

Now, only the finishing touches remained. Most were courtesy of the Pleistocene, our (ongoing) ice age, which, with a series of lengthy glaciations, sculpted the mountain heights and the valley floors, and provided the causeway over which many new species of animals and, eventually, humans would populate a new world.

The Inland Temperate Rainforest

TRAVELLERS passing through Canada's Glacier National Park and along the southern edge of Mount Revelstoke National Park just to the west, or those driving through Glacier National Park's American counterpart in Montana and across the Idaho Panhandle to northwestern Washington may not realize that this is an environment found nowhere else on Earth. Here, in a region that includes the Columbia Mountains, as well as the Cariboo Mountains to the north and the western Rockies to the south, are the last old-growth remnants of the world's only inland temperate rainforest. Biologists calls these "antique forests", for they have been regenerating themselves since the last glaciation, and they exist thanks to a unique combination of geology and climate.

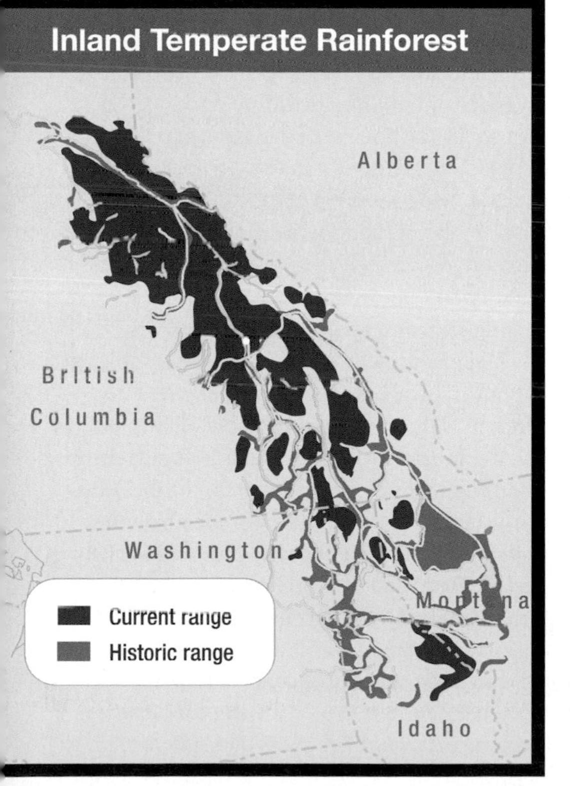

The Columbia Mountain ranges, which include the Purcells, Selkirks, Cariboos and Monashees, are geologically distinct from the Rockies, but closely related. Locals often refer to them as the "Kootenay Rockies". Underlain by beds of limestone more than eight kilometres thick, rock in this region, which geologists call the "Purcell Supergroup", had its beginnings as thick green mats of cyanobacteria - also widely, but erroneously, known as "blue-green algae". These primitive organisms grew along the shallow coastal seas at the edge of Rodinia, an ancient supercontinent that existed some 1.5 billion years ago. Living and dying in unimaginable numbers over hundreds of millions of years, washed by sand and silt from barren mountains inland – for there was no life on land – they laid the foundations for the lush mountains of the distant future.

But if the land was barren, in other ways the Earth was very much alive. About 750 billion years ago, Rodinia began to rift apart. A huge chunk of the continental shelf, from what are now the southern Purcells and Waterton/ Glacier National Park on the Alberta-Montana border almost 1,000 kilometres or 600 miles north to the Muskwa/Tuchodi region near Liard River in north-central BC, seemed to have disappeared.

For decades, geologists wondered what happened to that enormous chunk of the Purcell Supergroup. At last, with international co-

COURTESY OF FINPHISH / CREATIVE COMMONS SHAREALIKE 3.0 / HTTP://CREATIVECOMMONS.ORG/LICENSES

operation, the problem was solved. As Ben Gadd explains in *Handbook of the Canadian Rockies*, "Brace yourself: the rifted-off landmass is now most of Australia!"

What was left in nascent North America was both very thick – almost nine kilometres thick in some places – and very hard, providing a firm bed for the quartzite, limestone and marble that would be laid down later, as well as the tough plugs of granite, called plutons, that resulted from volcanic action as the mountains were formed between 175 and 150 million years ago. The outcome, particularly in the southern Selkirks and Purcells, was mountains that are rugged, with sharp, angular peaks and steep-sided canyons and higher than anything to the west. Since North America's weather almost invariably originates over the Pacific Ocean, year-round these mild wet, air masses flow east, unimpeded, until they hit the wall of the Columbias. The result is heavy rain in the summer and deep snow in the winter, particularly at higher elevations, where up to 23 metres or more than 70 feet may fall in a winter.

The natural effect of this copious year-round precipitation is a primeval forest crowned with huge western hemlocks and giant western red cedars, including some that are almost 2,000 years old. Here, too, are more tree species than are found together anywhere else on the West Coast, for this remarkable ecosystem brings together plants from the coastal, southern interior and boreal forests.

Not surprisingly, perhaps, these old-growth forests have long been home to a huge diversity of wildlife, including endangered mountain caribou (see page 200); grizzly bears, a species under review in the US and at risk in Canada (see page 205), gray wolves, mountain goats, bighorn sheep, wolverines, black bears and cougars, as well as many species of birds that depend on cavities in mature trees for nesting sites and raising their young.

An old-growth forest is also better able to survive natural disasters, including fires triggered by lightening, and invasions of insects, like the mountain pine beetle that is currently sweeping through the pine forests of the Western Interior. In fact, studies of old-growth forests demonstrate that invasions of bark beetles are a regular occurrence and one that a forest of different species and a variety of ages can often withstand. In Montana's Glacier NP, however, some of the oldest specimens of whitebark pine, a slow-growing timberline tree, were dated at 700 years of age before they recently succumbed to blister rust. And a whitebark pine in Idaho's Sawtooth range has been dated at more than 1,200 years old.

Though North America's old-growth inland temperate rainforest is unmatched anywhere on Earth, most of it has disappeared in the past half-century. Since World War II, enormous tracts have been logged and at the lower elevations, only three per cent of old-growth forest remains in Canada and less in the US.

Within the boundaries of Mount Revelstoke National Parks, the forests can no longer be logged, but even there, large swathes are maturing second-growth forest, mainly the result of fires over the past 150 years. Prime examples of old-growth forest do exist, however, and accessible trails have been created (see Viewing the inland temperate rainforest). But none of these areas is large enough to provide any real protection for the forest ecosystem as a whole, or for many of the animals at risk; particularly those, like mountain caribou and grizzlies, that must range widely for food.

Even at midday, the meandering Giant Cedars Boardwalk Trail in Mount Revelstoke National Park, British Columbia, is dark, for towering trees obscure the light.

Over the past thirty years, the plight of disappearing old-growth rainforests on BC's Pacific coast has been spotlighted with positive results on the part of the provincial government. But the province's even rarer inland rainforest has been virtually ignored.

The national parks mentioned opposite encompass examples of the best of the Columbia Mountains. Mount Revelstoke NP, which is entirely within the Selkirk Range, stretches north and east from the outskirts of Revelstoke, BC, and provides a snapshot of the "Rainforest, Snowforest, No Forest" ecosystem of the region. Driving up into the park on the Meadows-in-the-Sky Parkway, the rainforest, described above, gradually gives way to the "snowforest", dominated by Englemann spruce and fir. This forest thins near the treeline, creating the open meadows that explode into a brief blaze of colour in August, with paintbrush, lupines and monkeyflowers. Fully half the park is above the treeline and much of this is rock, permanent ice and snow or tundra. Grizzlies, which also are found at the lower elevations during berry season, mountain goats, white-tailed ptarmigan and the mountain caribou ecotype are among the many animals that summer above the tree line.

Montana's Glacier National Park encompasses more than a million acres of forest, including ancient Pacific cedar-hemlock forest in the southwest region. Here, western hemlock and western red cedar reach their extreme eastern limits in the US and some trees in the lower Avalanche Creek area are almost a half-century old.

PETER ST. JOHN

Mountain Caribou

Rangifer tarandus caribou • (Endangered (US); Threatened (Canada)

DRIFTING like silent shadows through ancient forests, dancing on snow drifts as much as four metres deep, feeding on veils of tree lichen that hang from the branches far above the ground, mountain caribou were likely among the first animals to return to British Columbia following the last glaciation. Since then, they have evolved to live in circumstances most large mammals find impossible. Yet today, in most places, they are rapidly disappearing.

Larger than mule deer, smaller than elk, mountain caribou are uniquely adapted for one of North America's most difficult winter environments. Their hooves are their chief asset, for they spread like snowshoes, allowing the animals to walk on the surface of deep, hard-packed snow, rather than sinking into it as a deer or an elk might. According to a paper produced by the BC Ministry of Environment, Lands and Parks, the hoof imprint of a mature caribou is about the size of a moose, yet the caribou weighs only half as much. Caribou also have exceptional coats, with hollow kinked hairs that trap a layer of warm air against their bodies.

A specialized type of woodland caribou, mountain caribou are so uniquely adapted to life in the old-growth forests of the interior wet belt of south-central British Columbia and the adjoining US states of Washington and Idaho – the world's only temperate inland rainforest – that ecologists have termed them a "flagship species". In short, mountain caribou might be found almost anywhere in their chosen environment, but not beyond it, for they have evolved a lifestyle quite different than that of their relatives, the boreal caribou of Canada's northern forests, where the snow cover is much thinner.

The keys to their survival are the mountain caribou's twice-yearly vertical migrations. In early spring, the caribou head for the valleys, where they fatten up on spring's early greenery. When the snow melts on the alpine meadows, they return to the heights, with pregnant cows seeking lofty, secluded locations that allow them to avoid predators during calving.

In late fall, when the winter's heavy snows begin, the caribou again migrate to the lower elevations where the closed canopy of old-growth forests shields the ground, allowing the animals to find forage, particularly mountain boxwood, a much favored evergreen shrub.

Eventually the snowpack on the lower slopes deepens to the point that digging for food becomes impossible. About the same time, the deep snow on the upper slopes becomes firm enough to hold their weight and once again the caribou climb to the high mountain forests, where they spend the balance of the winter moving about on snow three or even four metres (or nine to 12 feet) deep, and feeding on lichen that hangs from the branches of the oldest trees or blows off the branches during winter storms. Lichens alone provide the food they eat for the balance of the winter.

Not only does this high mountain environment provide the food they need, according to provincial government biologists it also creates a safety zone, because "the deep snow drives elk, deer and moose down to valley bottoms where there is less snow, and most predators follow them, leaving the caribou safely isolated."

It follows, therefore, that harvesting old-growth forests in the Columbia and Selkirk Mountains has had a disastrous effect on the

His magnificent antlers cloaked in velvet, a bull caribou greets spring.

MARK BRADLEY

numbers and range of mountain caribou.

Almost 98 per cent of the world's estimated 1,900 mountain caribou – down from an estimated 2,500 just a decade ago – live in British Columbia. The other two per cent, about 46 animals according to an aerial census taken in 2008, are in northern Idaho and northeastern Washington, and their rapid decline has caused great concern. Finally, after intense pressure from environmental groups on both sides of the border, in 2007, judicial rulings in the US and government regulations in BC were put in place to preserve the heart of the mountain caribou's habitat.

In Idaho and Washington, about 10 per cent of the Idaho Panhandle National Forest (or approximately 322,473 acres) has been declared off-limits to snowmobilers, while in British Columbia, the provincial government has put 2.2 million hectares (or almost 8,500 square miles) of old-growth forest off limits to logging and road building, and has also limited recreational use.

Environmental groups in both regions were delighted. "This ruling demonstrates that Idaho is big enough for both snowmobiles and mountain caribou," said Mike Peterson of The Lands Council in Spokane. "Once a species goes extinct, there is no bringing it back."

In British Columbia, Bev Ramey, president

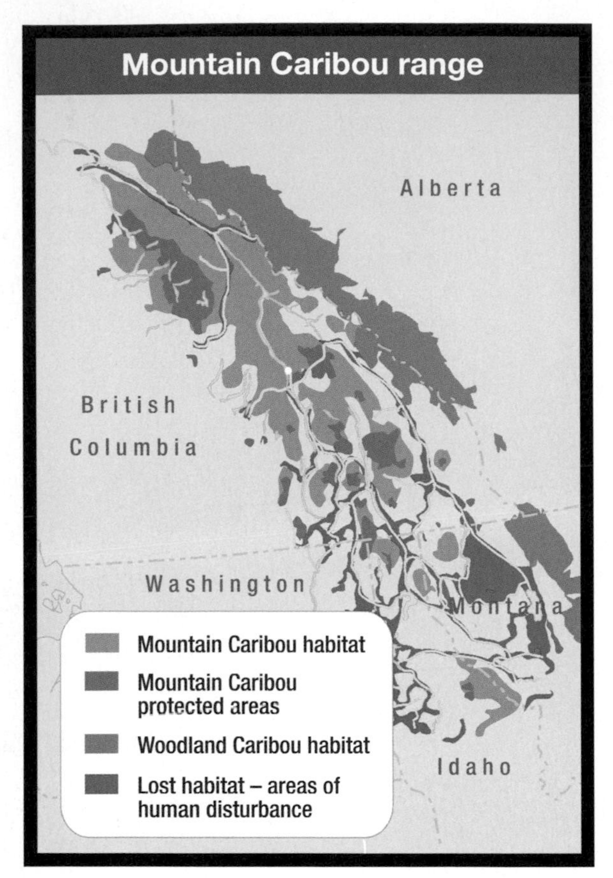

of BC Nature said, "This Mountain Caribou Recovery Implementation Plan is a really positive step towards habitat protection. We're also pleased to see the commitment to better manage motorized recreation . . ."

Just 150 years ago, mountain caribou inhabited the old-growth forests from northeast of Prince George, BC, south to Idaho's Salmon River and east into central Montana. As logging fragmented their crucial habitat, not only destroying the lichen-draped forests on which they depend, but opening corridors that allowed the penetration of elk and deer, along with their predators, wolves and cougars (or mountain lions, as they're also known), caribou numbers fell dramatically. Suddenly, researchers studying caribou mortality found that the highest predation rate was not in spring and early summer (when fawns and young calves are typically vulnerable to predators), or in late winter, when starving ungulates are easy prey, but in late summer and fall, when caribou should be in peak condition.

"[Cougars and wolves] were having a tremendous impact on bighorn sheep, mountain goats and caribou," Tim Layser, a wildlife biologist with the Idaho Panhandle National Forests was quoted as saying in the Winter 2009 issue of *Forest Magazine.* It didn't take long to find out why. "Changes in the habitat altered the historical predator-prey complex," wrote author Jim Yuskavitch. "Open areas created by the increase in logging attracted deer to the caribou's high elevation range, and mountain lions followed the expanded deer population."

Something similar, though with different species, was happening in British Columbia, where tracts of logged terrain attracted moose to the Revelstoke area, leading to a dramatic increase in the numbers of moose – from 400 to 1,600 in the decade after the mid-1990s, with a corresponding increase in wolves, which put new pressures on caribou.

Logging roads and areas of clearcut forest have also lured snowmobilers in ever increasing numbers over the past decade. In both countries, the open, mountainous terrain has become a major attraction for thrill-seeking snowmobilers; and in BC, heli-skiers and back-country skiers are often transported to their chosen slopes and back-country lodges by snowmobile. Even when their operators caught nary a glance of the elusive mountain caribou, snowmobiles have been proven to cause animals as far as a half-mile away to flee, inadvertently harassing them during a time when they are already stressed and need to conserve energy.

Mountain Caribou Fast Facts

Rangifer tarandus caribou

HEIGHT & WEIGHT
Mountain caribou are bigger than mule deer and smaller than elk; most bulls weigh about 200 kilograms or 420 pounds, while cows weigh about two-thirds as much.

DESCRIPTION
Their body hair changes color with the seasons, but with the exception of the nearly white Peary caribou of the far north, it is medium to dark brown in the summer and fall, with white around the rump and, in bulls, under the throat. The head and neck have a mix of gray, white and brown hair, but during the winter the darker guard hairs break off and the animals appear lighter in color. Unlike other species of the deer family, almost all caribou cows have antlers, which help in clearing snow to get at lichen. They are generally not shed until summer, allowing them to defend their young. The antlers of bulls, particularly those in their prime, can reach two metres or six feet in width and are shed in the late fall and winter, when the mating season is over. Caribou have a keen sense of smell and specialized nostrils that warm frigid incoming air during the winter and help to condense and capture the moisture from it before it's expelled. Their coats are made for winter, with a dense woolly undercoat and a longer-haired overcoat consisting of hollow, air-filled hairs.

LIFESPAN
Caribou can live up to 15 years, but their average lifespan is just five or six.

MATING & BREEDING
Mating takes place between late September and early November, when males battle for access to females. Dominant males collect as many females as possible; a barren-ground caribou bull's harem may number 25 or 30 females, but mountain caribou harems are much smaller. During the rut, males are so preoccupied they almost completely stop eating and lose much of their reserve weight. Calves are born the following May or June, and though they continue to nurse until fall, within a month, the calves are able to graze with the rest of the herd.

NAMES
The word caribou comes from the Mi'kmaq *qalipu,* meaning "one who paws", referring to its habit of pawing through the snow – known as cratering – to reach buried lichens, moss, sedges or grasses below. In Europe and Asia, caribou are known as reindeer and are domesticated in many places.

VIEWING MOUNTAIN CARIBOU
Mountain caribou are elusive and few in number and in February 2009, the British Columbia government announced new snowmobile closures covering nearly one million hectares in order to protect mountain caribou on public lands. These regions are marked by informational signs.

PETER ST. JOHN

Treading lightly on a deep snowpack, mountain caribou are elevated to a height where they can feed off the veils of lichen that hang from old-growth trees.

Mountain caribou are listed as threatened by the Committee on the Status of Endangered Wildlife in Canada (COSEWIC), as are boreal caribou. In February 2009, the BC government announced additions to snowmobile closures within large areas of the mountain caribou recovery region. Information about these closures can be found online on the government's website.

In the US, where mountain caribou are often cited as the most endangered mammal in the lower 48 states (recently surpassing the black-footed ferret), they are protected under the Endangered Species Act. Not only have they been extirpated from 43 per cent of their historic range, of the 14 local populations (13 in Canada and one in the US) identified in censuses gathered during aerial surveys in late winter, several are in steep decline, and five herds, including two transnational herds close to the international border, are at critically low levels with between five and thirty-five animals. A hard winter, avalanches, heavy predation, forest fires, disease, further harvesting of old-growth forest: any or all of these could spell the end of one or more of these herds and further endanger the mountain ecotype as a whole.

Particularly at risk are mature bulls. Not only are they key to the survival of each herd, but because of the autumn rut, when they expend most of their time and energy on competing for mates, they often go into winter undernourished and weakened.

Canadians carry the caribou in their pockets and wallets, for these beautiful animals grace their 25-cent coins. It might be argued that inflation has greatly devalued the Canadian quarter since artist Emmanuel Hahn created the design in 1937, but surely the real thing is even more worth protecting.

Grizzly Bear

Ursus arctos horribilis • Species of special concern (Canada)
Populations under review (Lower 48 States and Alberta)

IN September 2009, British Columbia's "stream walkers" who were working the province's Great Bear Rainforest (see map on page 120), began to warn of an impending – or perhaps on-going – ecological disaster. These volunteers, who annually walk the salmon streams searching for spawning salmon and the grizzly, black and Kermode bears that depend on them, found the streams almost empty and the surrounding forests and riverbanks ominously silent.

"I've never experienced anything like this," Doug Neasloss, a bear-viewing guide with the Kitasoo-Xaixais tribes told Mark Hume from *The Globe and Mail* newspaper. "I've been doing this for 11 years and this is the worst I've seen it. Last year on [one river] I saw 27 bears. This year it's six. That's an indication of what it's like everywhere."

Others agreed, saying that not only had the bears disappeared, but what were once dependable runs of large, fat chum salmon – the species that bears fill up on before heading into hibernation – had also vanished. Ian McAllister, conservation director of Pacific Wild, a BC non-profit group, told Hume, "River systems that in the past had 50,000 to 60,000 chum now have 10 fish."

In fact, chum salmon runs have been declining for a number of years. And, as Neasloss said of the bears, "It's scary. I think a lot are dead. I think they died in their dens [during the winter of 2008-09]."

It's certainly a possibility, for in 2008, just 3,008 chum salmon were caught by gill-netters in an area off the central coast that had a 10-year average of well over 100,000 fish.

Others have suggested that after assessing the dearth of salmon in the streams, the bears may have headed into the forests and mountains to try to fill up on berries and other foods. Almost nothing, however, has the fat and protein content that chum salmon do.

It's a scenario rather like the one that faced plains grizzlies as the plains bison were hunted to near-extinction in the 1860s and '70s. Almost solely dependant on the massive herds for their relatively effortless way of life, plains grizzlies disappeared in less than 50 years (aided, of course, by the blazing guns of European traders and settlers).

The collapse of the chum salmon fishery, following hard

Grizzly sows are particularly protective when nursing.

DENNIS FAST www.dennisfast.smugmug.com

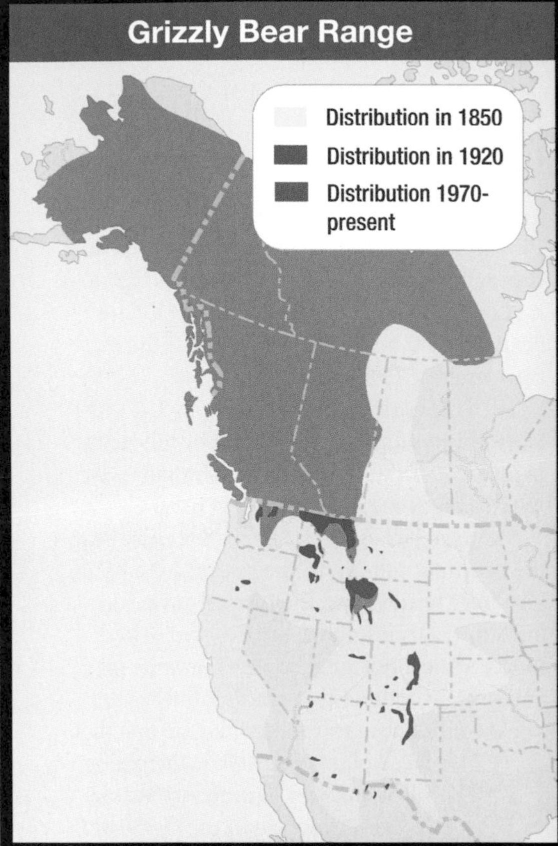

Above: The distinctive signs of a big male; lower left: remarkably human in appearance, twins wrestle in a river; below: an old grizzly drinks at waterside.

ABOVE AND BELOW: DENNIS FAST www.dennisfast.smugmug.com

STEVE HILLEBRAND / US FISH AND WILDLIFE SERVICE

on the heels of a similar decline of the Fraser River's world-famous sockeye salmon fishery, affects not only bears, of course, but also orcas, wolves, otters and eagles, among many other species.

All this came at the same time that Alberta – following a comprehensive six-year census of grizzlies in the province – was in the process of realizing that it has far fewer bears than previously believed. The study involved placing smelly lures of a mixture of cow blood and fish, surrounded by a strategically placed string of barbed wire, in areas known to be frequented by bears. When the grizzlies came to investigate, the wire scraped a few hairs off their backs. Studying the DNA in the hairs, researchers could determine how many individual bears had visited the site. As the numbers began to come in – the region west and south of Calgary, including the enormously popular Banff National Park, for example, an area of approximately 7600 square kilometres (or just under 3,000 square miles) was discovered to hold fewer than 100 bears – the province put a hold on its long-controversial annual bear hunt.

Though the numbers were still to be finalized, in 2009 Gord Stenhouse, who led the provincial census, said that if the total population in Alberta is less than 500, the big bears should be considered endangered in the province. In 2002, when it was believed that the province had about 1,000 bears, the Endangered Species Conservation Committee recommended that it be classified as threatened.

In British Columbia, meanwhile, though the government had maintained in 2008 that overall grizzly numbers were increasing, in the spring of 2009, it closed additional areas to hunting along the central and north coasts. The closures brought the total area closed to hunters to 1.9 million hectares (or 7,330 square miles), about one-third of the bear's known range in BC. However, in the rest of the province, fall trophy hunting began as usual.

In the United States, grizzlies were listed as threatened in 1975, when the population in Yellowstone National Park – the bears' last real stronghold in the lower 48 states – was believed to be 136. With care and conservation, that population increased over the following 30 years to more than 500 in 2006 and the following year, grizzlies in Yellowstone were delisted.

Elsewhere, small populations exist in Montana's Glacier National Park; in what is termed the Cabinet-Yaak recovery zone in northwestern Montana and northern Idaho; in the Bitterroot Mountains of Idaho and Montana, and in the Selkirk Mountains in northern Idaho. An area in Washington's North Cascades, directly across the border from south-central British Columbia, is also being studied as a recovery zone. Though the area has not been repopulated with grizzlies, bears that have crossed the border from Canada are sometimes seen in the mountain valleys. The total population in the lower 48 states is estimated to be about 1,100, less than two per cent of the pre-contact population.

Alaska, however, has a large grizzly or Alaskan brown bear population, estimated by some to be about 30,000 bears, about 98 per cent of the entire US population.

North America's second-largest land carnivore behind its cousin, the polar bear (see page 28), grizzlies are distinctive in appearance, with a large hump over their shoulders formed by their powerful foreleg muscles, and long claws on their front feet. Unlike black bears, which have a straight or "Roman" nose, grizzlies have concave faces with upturned snouts. The bears' coats can range in color from cream to black, and in some areas are tipped with silver giving the bears a "grizzled" appearance.

Except for the mating season and mother bears with cubs, grizzlies are solitary animals. They have no natural predators, and some argue that because of this they are not naturally aggressive, but when threatened, a grizzly's first reaction is to stand its ground and

JOHN JAMES AUDUBON / LIBRARY AND ARCHIVES CANADA / C-041933 / LAC

fight rather than to try to escape. Nevertheless, grizzlies generally avoid people; mother bears with cubs are most likely to attack humans.

These magnificent animals once roamed North America from the Pacific Coast to Labrador in Canada's eastern Arctic, and from the northern tundra through the prairies and open plains south into California and Mexico.

Biologists believe bears diverged from canids – the ancestors of dogs, as well as raccoons, weasels and cats, among others – perhaps 40 million years ago. According to San Francisco State University, the bear family has been traced back to a small, doglike animal that lived in Europe. Dispersing around the world, the Ursidae family eventually evolved into three genera: Tremarctos (South America's speckled bear and North America's extinct giant short-faced bear), Ailuropoda (Asia's giant and red pandas) and Ursus, which includes the Asiatic black bear, the Malaysian sun bear, the sloth bear of India and the bears we know in North America – the black, polar and brown or grizzly bear.

An African species of Ursus was driven to extinction in the 19th century.

For hundreds of thousands of years, most of North America was the preserve of the giant short-faced (or running) bear and the lesser short-faced bear of the eastern US and Mexico. The history of the giant short-faced bear on the continent goes back almost 800,000 years, while, according to Minnesota's North American Bear Centre, the lesser short-faced bear "may have died out due to competition with a large Pleistocene subspecies of black bear *(Ursus americanus amplidens)* and due to brown/grizzly bears *(Ursus arctos)* invading from the west near the end of the Ice Age."

Perhaps because of these two species, it was believed until recently that grizzlies did not migrate into the heart of North America until about 13,000 years ago (shortly before both short-faced bears disappeared).

Recently, however, scientists have found evidence that grizzlies had travelled south much earlier than once believed. A 26,000-year-old bear skull found near Edmonton, Alberta, suggests they had migrated south from Beringia before two massive glaciers met in the foothills about 24,000 years ago to block the route.

Grizzly Bear Fast Facts

Ursus arctos horribilis

HEIGHT & WEIGHT

Adult male grizzlies stand about a metre or over three feet at the shoulder and more than two metres (or between six and seven feet) when standing upright, as they sometimes do to investigate a situation. They usually weigh between 135 and 385 kilograms or 300 to 850 pounds, though much larger bears have been found, particularly on the Alaskan Peninsula. Females are between half and two-thirds of this size.

DESCRIPTION

Grizzlies can live in a variety of habitats, providing that they are large, unspoiled and isolated areas. Grizzlies cover large territories in search of food and shelter; an average male's home range can be anywhere from 27 to over nearly 1,000 square kilometres or 10.5 to more than 500 square miles, depending on the available resources available. This range will usually overlap that of two or three female bears.

DIET

Grizzly bears are considered carnivores, but are actually highly omnivorous. Their meals are determined by what's available at any given time, and plants such as fruit, berries, nuts and bulbs can make up 80 to 90 per cent of the bear's diet. Their massive shoulder muscles and long front claws allow them to dig for insects and burrowing mammals. They will also feed on carrion, small mammals and very young elk, moose, and caribou. Coastal populations will gorge themselves on salmon during spawning season.

MATING & BREEDING

Female grizzlies do not become sexually mature until they are five or six years of age. Cubs are born in the mother's winter den in January or February, usually two at a time. Tiny at birth, they weigh about 400 grams or just under a pound, but grow quickly. Young bears usually stay with their mother until their third summer, at which point they strike out on their own and she is free to have another litter. As a result, the reproductive rate of grizzlies is very low, for females may only have four or perhaps five litters of cubs in a lifetime; their litters are very small, and only about half of cubs survive their first year. As a result, even where both bears and habitat are protected, grizzly populations grow slowly and small populations take many years to recover.

NAMES

The word "grizzly" in its name refers to "grizzled" or gray guardhairs, but when naturalist George Ord named the bear in 1815, he misunderstood the word as "grisly", and therefore added the biological Latin specific name "horribilis". Grizzlies are also called silvertip bears.

VIEWING GRIZZLY BEARS

Though numbers are dwindling, it is still possible to see grizzlies in Banff and Jasper National Parks. Protected by high fences, provided with safe underground or overhead highway crossings, grizzlies are much less afraid of humans than they are in many other places and therefore often allow themselves (and sometimes their cubs) to be viewed.

Grizzlies also range throughout Yellowstone but are most often seen in and around the Dunraven Pass area and just past the turn-off to Mount Washburn. Other good areas are across the Yellowstone River in Hayden Valley and in the Fishing Bridge area.

Many lodges and touring companies on BC's west coast and in Alaska offer bear-watching outings; participants are often taken by boat to watch grizzly, black or Kermode bears along coastal areas.

But remember, even when protected, grizzlies are powerful wild animals, and never to be approached. A good pair of binoculars or a powerful lens on your camera is a must.

Trailed by her cub, a grizzly mother heads to the river for a fishing lesson. Opposite, she is always on alert for danger.

IMAGES BOTH PAGES: DENNIS FAST www.dennisfast.smugmug.com

To assign the Alberta fossil to a specific genetic group, Paul Matheus, a paleontologist with the Alaska Quarternary Centre, and Jim Burns, curator of the Provincial Museum of Alberta, which held the fossil, asked colleagues at the University of Oxford and Germany's Max Planck Institute to study the fossil's mitochondrial DNA. The result showed that it belonged to the same genetic group as do modern grizzlies in what is now the southernmost extent of their range – southern BC and Alberta, as well as Montana and Idaho – according to the November 2004 issue of the journal *Science.* And these bears, previous studies had shown, belong to a genetic population thought to be extinct in North America for as much as 35,000 years.

By contrast, fossils of grizzlies or brown bear fossils from Alaska, Yukon and other parts of the north belong to a different and more recent genetic population.

Coexisting for thousands of years with native North Americans, grizzlies did not begin to seriously decline in number until the arrival of European traders and settlers. The first written account of the grizzly comes from Hudson's Bay Company trader and traveller Henry Kelsey, who wrote in the rather awkward doggerel he employed for making notes:

His skin to get I have used all ye ways I can
He is mans food & he makes food of man
His hide they would not me it preserve
But said it was a god & they should Starve.

As Kelsey indicated, the indigenous peoples of North America held grizzlies in high esteem; if it was necessary to kill one, its teeth and long claws were highly prized.

Beginning in the mid-19th century, grizzlies were commercially trapped and hunted in large numbers for their meat and hides. The US government also encouraged the slaughter of the estimated 30 million bison as part of a deliberate campaign to force Plains peoples into submission and onto reserves. This slaughter removed the plains grizzlies' main prey, for they followed the great herds of bison and pronghorn, feeding on carcasses of animals that had died, or on the young.

As settlement rolled west, grizzlies were

also shot for sport and for the threat they posed to crops, domestic livestock and occasionally humans. They were most often seen along river valleys, in part because early Europeans used the rivers as travel corridors, but also because the valleys were particularly bountiful in terms of both the prey and vegetation favored by both grizzly and black bears.

By 1880, the great bison herds were all but gone and the prairie grizzly soon followed. The unregulated hunting of grizzlies continued, however, through the 1950s until the only populations left in the contiguous US were in remote and inaccessible mountain areas.

In the last 50 years, human development has often cut through remaining grizzly ranges, leaving them with less unspoiled wilderness than they need to survive. Unlike black bears, grizzlies have been unable to adapt to drastic changes to their habitat. Because of this, dwindling grizzly populations say a lot about the quantity and quality of natural space left in an area; when humans move in, the bear populations inevitably dwindle. And though they continue to be hunted, both legally and illegally, loss of habitat is the greatest threat to the grizzly bear's survival.

In Canada, grizzlies are designated a species of Special Concern. There have been ongoing arguments by conservation groups for habitat protection, as well as for the creation of natural corridors in the Northern Rockies to allow existing populations of bears to migrate from place to place. These have met with some success in the past few years. One prominent example is the creation of the Khutzeymateen Grizzly Bear Sanctuary, an isolated 44,300-hectare (or 110,000-acre) area of rainforest on the BC coast dedicated specifically to the preservation of the bears and their habitat. Preserving vast tracts of wilderness specifically for one large, highly visible animal such as the grizzly bear may seem like a waste of effort, or a way of cultivating political good-will, but it is actually a legitimate conservation strategy because of the "umbrella" of protection that the grizzly's habitat provides. Protecting an ecosystem to provide habitat for the grizzly will also result in the conservation of the bears' prey, along with smaller carnivores and many other native plants and animals.

American Badger

Taxidea Taxus jeffersonii • Endangered (British Columbia)

SOLITARY, sensitive to change and stalked for centuries, it's not surprising that badgers are today listed as endangered in British Columbia as well as Ontario. Yet it is only now, thanks to studies being done in both provinces on two subspecies of this ancient immigrant to North America, that the contributions this heavy-set, short-legged carnivore makes to Canada are at last being realized.

When European trappers and settlers arrived in North America, the continent's four subspecies of badgers – *Taxidea taxus* of the Great Plains; *T. t. jeffersonii* of British Columbia and the western United States; *T. t. jacksoni* of Ontario and the north-central US, and *T. t. berlandieri,* found from Mexico, California and Texas north to Colorado and Oklahoma – were widespread. In fact, as indicated below, during their very long tenure on the continent, ancestral badgers once lived much farther north than they did in historic times.

But fur traders, farmers and ranchers changed all that. The Hudson's Bay Company's records show that between 1843 and 1908, 96,000 badger pelts were trapped and traded and between 1919 and 1984, more than 122,000 pelts were traded in Alberta alone. That, according to a 2002 report by Dave Scobie for the Alberta Fish and Wildlife Division, was 39 per cent of the Canadian harvest. In other words, over the 65-year period, more than 300,000 badgers were killed in Canada, an average of almost 5,000 a year.

Most were killed for profit. In 1928, Scobie writes, a pelt averaged just under $50, a tidy sum of money at the time. But as settlement marched west, many were also poisoned or shot by farmers after the same fine-grained, well-drained soil that badgers find appealing, or by ranchers who feared that their stock would break a leg by falling into a badger hole. For decades, until badger numbers suddenly collapsed in the late 1940s, almost everyone overlooked the good badgers do, keeping rodent numbers down, providing burrows for other animals and birds, including endangered burrowing owls, and assisting in aerating native plant communities.

In British Columbia, where *T. t. jeffersonii* subspecies of badgers are separated (largely by geographical obstacles) into regional populations, badger trapping and hunting was discontinued in 1967, after fewer than 10 pelts were produced annually for nearly two decades. But the numbers continued to decline and badgers were extirpated from the Upper Columbia Valley in the mid-1990s. Young badgers migrating to new territories are often killed by highway traffic or farm implements and a certain percentage of young are lost each year to natural predators – coyotes, wolves, bobcats, cougars and even ravens. Very few adult badgers are preyed upon, for they are quick and have a fearsome reputation for self-defence.

Particularly over the past two decades, urban sprawl has also contributed to their decline. A British Columbia study released in 2008 estimated that between 25 and 30 animals remained in the ever more popular Okanagan Valley, and only eight or 10 in the region along the US border. The largest population, between 100 and 160 animals, seems to be in the East Kootenays, and even there, according to counts of animals captured (and released) by researchers, just 13 of 34 animals captured were juveniles. By contrast, in stable or increasing populations, more than 50 per cent of captured animals are juveniles.

Rarely seen above ground during the day, as juveniles, badgers travel long distances to establish themselves in new territories.

DENNIS FAST www.dennisfast.smugmug.com

The relatively small percentage of females in several populations is another concern. In the Thompson study region, only two of 13 study animals were females, while in the East Kootenays the numbers of males and females were about even, and in a study conducted in Idaho, there were 1.2 males per female.

On a more positive note, according to the BC study, "recent research in the [north-central] Cariboo region suggests the Badger population there may be increasing, as Badgers are now being observed in areas where they have not been seen in decades."

American badgers are mustelids, a family that also includes weasels, fishers, wolverines, polecats, martens and otters. Among the most diverse of carnivore families, mustelids evolved from ancestral canines about 25 million years ago as the global climate cooled. With a cooler climate, forests thinned and grasslands expanded, allowing burrowing herbivorous mammals to spread, and the mustelids adapted to exploit this new environment.

In Europe and Asia, they evolved into long, low-slung hunters that could follow prey into their burrows and large bear-sized diggers from which modern badgers and wolverines descended. Then, sometime between eight million and four million years ago, lower global sea levels created land bridges between Asia and North America, allowing the badger's Eurasian ancestors to cross to a new continent.

It seems that there were several waves of ancestral badgers, for a fossilized skull of a primitive American badger has been recovered from late Miocene (about 11.6 to 5.3 mya) sediments in northern New Mexico, while a species clearly related to today's Eurasian badgers, Arctomeles, was found in a peat deposit of what was, between four and five million years ago, a beaver pond on the southern tip of Canada's Ellesmere Island. Today, Ellesmere Island has an artic tundra climate, but then, as this and many other fossils – including ancient horses, musk deer, a primitive bear and a large wolverine – show, winter temperatures were nearly 15° C (or 59° F) higher than they are today, and summer temperatures about 10° C (or 50° F) higher.

In appearance, the American badger has yellowish-gray, grizzled hair on its body and a dark head with distinctive black facial "badges" or flashes, and a white stripe that runs from the nose to the shoulders. With long, strong front claws for digging, and short, flat back claws to scoop loosened soil away, badgers are built for burrowing. Much of their lives are spent underground in dens where they rest during the day, raise their young and store their food.

The needs of a badger are few. It needs earth for digging and small burrowing mammals, particularly ground squirrels, as prey. In fact, badgers are the only carnivores that burrow after and eat other tunnelling animals.

Though it will stand and fight when attacked, a badger's first response when threatened is to hide. It can dig itself into a hole remarkably quickly, all the while flinging dirt

JOHN JAMES AUDUBON / LIBRARY AND ARCHIVES CANADA / C-041774 / LAC

American Badger Fast Facts

Taxidea Taxus jeffersonii

HEIGHT & WEIGHT

From nose to tail, badgers range from 52 to 77 centimetres (or about 20 to 30 inches) and adults can weigh anything from four to 12 kilograms or between about nine and 26 pounds.

LIFESPAN

The oldest wild badger on record lived to be 14, while one badger in captivity reached 26 years of age. An average lifespan is about seven years.

DISTRIBUTION & POPULATION

Badgers have simple needs: friable soil that's easy to dig and abundant burrowing prey. As a result, they live in open areas across North America from the northern prairies to Baja California and the Mexican highlands. However, the *jeffersonii* subspecies is genetically isolated from the badgers on the prairies and is only found west of the Continental Divide, from California west to central Colorado and north to southern British Columbia, where it is only found in the Interior. Based on counts in 2004, the population in Canada has been estimated at between 240 and 360 animals.

DIET

Badgers are carnivorous, feeding mainly on small burrowing mammals such as ground squirrels, but supplementing their diet with birds, eggs, reptiles, amphibians, insects and sometimes plants.

BEHAVIOUR

Badgers are generally nocturnal, but can be active during the day, particularly in the mornings. Their home ranges, which largely depend on suitable habitat, as well as the availability of prey and mates, vary from two to 500 square kilometres (or between under a square mile to more than 190 square miles). Males have much larger home ranges than females do.

Burrows, where a badger sleeps, stores food and raises young, are at the centre of their daily lives, though unlike several species of European badgers that create one or two elaborate dens, American badgers often build new burrows on a regular basis, for brief stays as they hunt for food. The exception is dens used for raising young, which are usually more complex, with a larger mound of soil at the entrance. During the winter, badgers sleep most of the time, but do not enter a complete state of hibernation. Some individuals are more active than others.

MATING & BREEDING

Though badgers are loners, they are also polygamous and during the summer mating season both males and females may mate with several partners. However, implantation is delayed until February and one to five young are born in late March or early April. Looked after by their mother, the youngsters develop quickly and generally disperse from the den in June or July. Juveniles often travel long distances (up to 50 kilometres or 30 miles for females and twice that for males) to find a suitable territory. Males try to avoid territories that overlap with those of other males.

Females can breed at a year, but males often wait two or three years before successfully mating, particularly in areas where the female/male ratio is low.

THREATS

Suitable habitat in British Columbia's Interior is becoming scarce, as urbanization and intensive agriculture in the interior spread and open areas are invaded by trees due to fire suppression. As adults, badgers are well able to defend themselves and have few predators – other than humans – but are frequently killed by farming operations and cars.

RECOVERY EFFORTS

In an attempt to help BC's badger population recover, measures such as openings in road barriers and underpasses are being created to reduce the number of badgers that end up as road kill. Public outreach activities are also raising awareness of the need for badger habitat and landowners are being encouraged to maintain areas on their property where badgers and their prey can live. Efforts are also being made to relocate unwanted animals in suitable terrain.

NAMES

The American badger is also known as the New World or silver badger. In Mexico, it is called *tejón americano.*

VIEWING AMERICAN BADGERS

Given their nocturnal feeding habits and underground living arrangements, badgers are not easily viewed, except by the very determined. However, the British Columbia Wildlife Park, located on the TransCanada Highway just east of Kamloops in south-central BC (and not far from Badger Drive), offers regular badger viewing tours between May and September. The park operates a wildlife rehabilitation centre and has cared for many species from orphaned baby wolves and bobcats to injured deer and rattlesnakes. The park also operates a captive breeding program for burrowing owls (see page 282), which had been extirpated from British Columbia. The program raises about 100 owls each year for release in selected areas in the Thompson-Okanagan region. Its burrowing owl ambassador, Quetemie7 (pronounced KEH-tem-mee) travels the region promoting preservation of and information about BC's grasslands. For more information about these and many other programs, contact the park at info@bczoo.org

This beautifully marked youngster's facial "badges" are clear to see.

COURTESY OF US FISH AND WILDLIFE SERVICE

into the face of its attacker.

Though usually found in grassland regions and open valley bottoms, badgers have also been found in other environments, including open canopied forests and alpine slopes. However, its preference for soil that is also good for growing grapes or other crops, has put it at odds with farmers in several areas of southern BC, as well as with those seeking winter warmth.

Northern Spotted Owl

Strix occidentalis caurina • Endangered (Canada); Threatened (US)

IN the early 1990s, this appealing owl, the most endangered bird in Canada, sparked a prolonged battle with logging companies on both sides of the border. It was (and indeed still is) also losing territory to a more adaptable look-alike, the barred owl. And on all sides, it seemed, it was losing.

Fearing that designating old-growth forest as spotted owl habitat would lead to further declines for their industry, some loggers made their feelings known with protests, industry pressure groups and bumper stickers, including one that read "I Like Spotted Owls – Fried".

Meanwhile, larger barred owls (*Strix varia*), which, in part because they fare far better in young forest stands, have extended their range dramatically over the past 50 years, are not only competing for food and nesting locations in spotted owl territory, but may also be bringing with them a host of blood parasites. Once known only in Eastern North America, barred owls expanded across the continent and south down the Pacific Coast in the wake of logging and development. Now, as a 2008 study by San Francisco State biologists has discovered, nearly half of the spotted owls studied – both birds from the northern subspecies and their closely related California spotted owl cousins – were found to be carrying blood pathogens. Some spotted owls had as many as 17 different strains of parasites and for the first time, avian malaria was found in a wild spotted owl captured in an Oregon forest.

"While Plasmodium parasites have been found in thriving owl species, the detection in a spotted owl could further challenge the threatened species' survival," said Heather Ishak, an SF State graduate biology student who performed the research with biology professor Ravinder Sehgal and others.

Beset on all sides, spotted owls are declining in number. In 2006, the BC government estimated that there were 22 northern spotted owls, including six known breeding pairs, in the province, its only Canadian home, down from about 200 in 1991. Employees of the Mountain View Breeding Centre in Langley, BC – who use captured wild owls, as well as captive birds for their breeding stock – put the total even lower.

In neighboring Washington State, populations declined by 70 per cent between 1992 and 2003 in the central Cascade region, while across its entire US West Coast range the spotted owl population is declining at about four per cent annually.

The problem is that northern spotted owls are very selective about their environment, and will only nest in cavities in large, old trees. Old-growth forest also provides the needed

This juvenile spotted owl was photographed in Doe Creek, Oregon.

COURTESY OF THE BUREAU OF LAND MANAGEMENT / OREGON

These fluffy chicks were spotted in an old-growth forest near Eugene, Oregon.

COURTESY OF THE BUREAU OF LAND MANAGEMENT / OREGON

prey and protective cover. The owls are virtually non-existent in clearcut regions or forests of younger, similarly-aged trees.

In the US, the Fish and Wildlife Service called for proposals for developing a recovery plan in 2006, 16 years after adding the owl to its burgeoning list of threatened species. Despite that, in 2008, the Bush administration cut the bird's protected habitat by 23 per cent. However, in April 2009, the Obama administration indicated it would review those cuts and possibly rewrite the recovery plan.

As a 2004 paper on Managing Identified Wildlife and a 2006 editorial in the *Wall Street Journal* both indicate, the US plan depends on forest management (meaning limited logging in a given area), suggesting it can go hand-in-hand with a healthy population of spotted owls. Citing a privately-funded study, the article said, " ... we know little about the relationship between harvesting and owl populations. One [study] infers an inverse relationship between harvesting and owls. In other words, in areas where some harvesting has occurred, owl numbers are increasing a bit, or at least holding their own, while numbers are declining in areas where no harvesting has occurred.

"... Why is this?," the paper asks. "... Could it be that spotted owls are more resourceful than we think?"

This was the line of thinking that had been followed in British Columbia, where the government determined that the primary goal should be to set aside suitable habitat, while still allowing selected timber harvesting.

The Spotted Owl Recovery Action Plan, announced in 1997, mandated that 159,000 hectares (or more than 390,000 acres) in parks and protected areas would be fully protected, while another 204,000 hectares (more than 500,000 acres) of provincial forest will be managed as "spotted owl range". Timber harvesting was allowed, but only until a minimum of 67 per cent of suitable habitat remains and never within the 80-hectare core area.

However, unlike California spotted owls, which "prey on different species and survive in different habitats", according to BC Ministry of Environment provincial bird specialist Michael Chutter, northern spotted owls did not respond well to timber harvesting.

"There are some areas of California where the SPOWs [spotted owls] have responded positively to timber harvest; in these areas, the main prey is a species of wood rat that increases as the forests are opened," Chutter explained. "To my knowledge, this has not been the case for the northern spotted owl."

As a result, in 2006, the BC government changed its focus to fully protecting old-growth owl habitat, "by giving up [for timber harvest] a few areas that have not been occupied [by owls] for many years," Chutter wrote in an email.

In addition, spotted owls are being captive bred and released in British Columbia, and an effort will be made to manage competing barred owls.

At the time of publication, Oregon had a

plan that allows some logging, though it is aimed at private landholders. While timber harvesting generally takes place on 40- to 60-year rotations, those who sign a safe harbor agreement on behalf of spotted owls are asked to lengthen the period between harvests, in return for payment of 50 per cent of costs for restoring and enhancing a forest ecosystem.

The goals everywhere, in addition to protecting the nesting and roosting areas, are to minimize disturbance around those sites and create, enhance or maintain suitable multi-layered, multi-species forests with canopies dominated by trees at least 100 years old and with deadfall and standing dead trees that attracted prey species such as flying squirrels, tree voles, wood rats and deer mice.

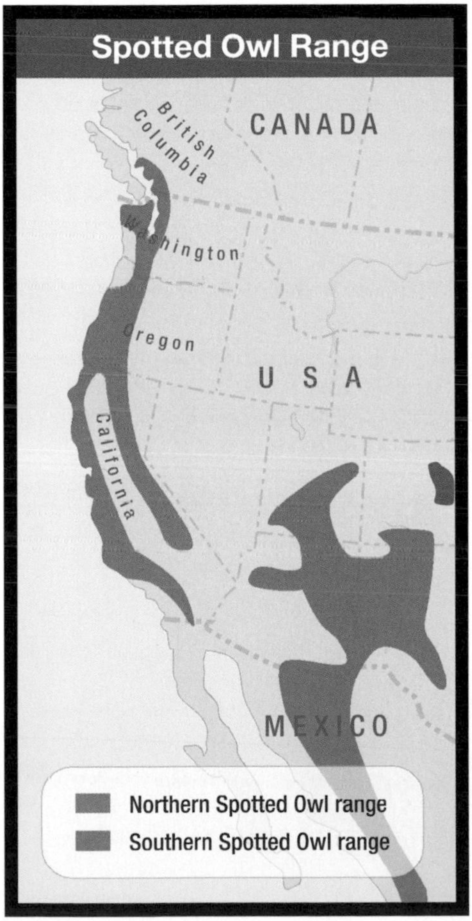

Nevertheless, while loss of habitat and increasing fragmentation of suitable forests due to logging, urbanization, agriculture or mining is a significant threat to spotted owls, the main problem is competition from barred owls. In addition, problems such as disease, decreasing genetic variability in very small populations or even the possibility that the number of spotted owls in a given region could decline to the point that individuals would be unable to find mates, are also threats to declining populations.

Spotted owls are medium-sized, averaging about 45 centimetres or about 16 inches inches in height, with an average wingspan of twice that. Their brown eyes are surrounded by tawny facial disks and they have no ear tufts. Their dark brown body feathers are patterned with round or elliptical spots, while their chest and tail have white horizontal bars. The barred owl, by contrast, has horizontal markings on its neck and breast that give way to vertical white and brown stripes on its chest and belly.

Nocturnal by nature, northern spotted owls hunt by selecting a perch and then waiting for their prey to appear. Since old-growth forest was widely available on the Pacific Coast and Pacific Interior until the past century, spotted owls not only became very selective about their hunting methods, but had large areas of climax forest in which to hunt.

This is particularly important since flying squirrels, for example, are typically more abundant in old-growth forests, yet are relatively small in number. To forage successfully for these, as well as tree voles and wood rats, spotted owls need large areas of old-growth forest.

Barred owls, by comparison, need less forest cover. Long at home in the marshlands and dense forests of the southern US, they moved north to the boreal forest and then west as old-growth forests everywhere gave way to clearings and younger stands.

Northern Spotted Owl Fast Facts

Strix occidentalis caurina

HEIGHT & WEIGHT

The owls are 40 to nearly 50 centimetres or 16 to 19 inches in height, with a wingspan of more than 100 centimetres or about 42 inches. Females weigh about 644 grams or 23 ounces, the slightly smaller males about 588 grams or 21 ounces.

LIFESPAN

Owls 22 years of age have been documented in the wild, but 10 years is an average lifespan. Spotted owls mate for life and remain in the same geographical areas year after year.

HABITAT

Spotted owls live in old-growth forests from southwestern British Columbia to northern California, where they nest in cavities in large trees or in abandoned nests of other species, mainly on Canadian provincial or US state lands, but also on private and tribal properties in the US.

DIET

Primarily nocturnal, spotted owls mainly hunt wood rats and flying squirrels, but will also eat other small mammals, birds, reptiles and insects. They often swallow their catch whole and then regurgitate indigestible hair, feathers or bones.

MATING & BREEDING

Though ready to reproduce at two years of age or even before, they typically breed at three, mating in February or March and laying one to three eggs in March or April. Incubated for 30 days, the young owls stay with the mother, while their father hunts and feeds them. Fledging in 34 to 36 days, they stay with their parents until early fall, then form their own winter feeding range, generally between three and 38 kilometres (or two and 24 miles) from their parents.

THREATS

The loss of old-growth forests has been the main reason for the plummeting numbers of spotted owls. More recently, barred owls are believed to be displacing spotted owls, and perhaps communicating mosquito-borne blood diseases, including malaria and West Nile virus.

NAMES

Also called the western barred owl and wood owl, the northern spotted owl is related to the California spotted owl (*Stix occidentalis occidentalis*) and the Mexican spotted owl (*Stix occidentalis lucida)*.

VIEWING NORTHERN SPOTTED OWLS

The best places to be certain of seeing Canada's most endangered bird is at the Donald M. Kerr Birds of Prey Center at the High Desert Museum south of Bend, Oregon. With facilities to breed spotted owls, three healthy chicks were born to parents Polka and Dot at the High Desert Museum over the past two years. When they are old enough to travel, the Oregon chicks are sent to Langley, which had a total of 10 spotted owls in 2009. Owls raised at the Mountain View Centre learn to hunt live mice and other rodents in large flyway pens, but are not open to public viewing. Ultimately they will be released into the wild.

Clear-cutting in northern spotted owl habitat has contributed to sharp decreases in their numbers in Oregon.

COURTESY OF STEVE HILLEBRAND / US FISH AND WILDLIFE SERVICE

This lovely photograph of an adult northern spotted owl was taken on the McKenzie River, a tributary of Oregon's Willamette River.

COURTESY OF JOHN AND KAREN HOLLINGSWORTH / US FISH AND WILDLIFE SERVICE / W05490

Pacific Salmon

WHEN Europeans first arrived on the Pacific Coast they were awed by the sight of rivers all but overflowing with fish, so thick in some places that it was possible to imagine walking across the water on their backs. Undoubtedly the best known of the many species of fish were salmon, which have played a major role in the lives of coastal peoples from central California to Alaska for more than 5,000 years. In fact, dozens of First Nations built entire cultures around the coming of the salmon.

The enormous and dependable runs allowed the evolution of sophisticated societies and the creation of internationally recognizable art. It is a sad irony, therefore, that both the cultures who produced these distinctive art forms and the species on which they depended are today greatly threatened.

Though paleontologists and archaeologists believe the enormous spawning runs – salmon returning in numbers large enough to reliably support societies for months at a time – go back five or even six millennia, salmon as members of a species are far older. The ancestors of today's salmon date back to the Early Cenozoic, between 50 and 60 million years ago. Fossils of a very early relative of the family *Salmonidae*, which includes salmon, trout, char, grayling and whitefish, have been found in fossil beds along Driftwood Creek near Smithers, in north central British Columbia, where until recently huge numbers of spawning salmon could still be found. *Eosalmo driftwoodensis*, the "dawn salmon", observes University of Alberta professor Mark Wilson, is an almost perfect morphological intermediate – a missing link, as it were, – between two salmon subfamilies – the *Salmonidae* (salmon, trout and char) and the *Thymallinae* (grayling). Though distinct from them today, Wilson believes that salmon evolved from a more grayling-like form, and also that they were originally freshwater fish.

Over the next 20 million years, descendants or relatives of the tiny dawn salmon, established themselves in North America, Greenland and Europe. Some of these species were enormous, including *Oncorhynchus rastrosus* (also known as *Smilodonichthys rastrosus*) one of two tusk-tooth or sabre-tooth

JOSEPH DRAYTON / OREGON HISTORICAL SOCIETY / OrHi 969

salmon species. Twelve-million-year-old fossils of this giant, which grew to be 10 feet or more than three metres long, weighed an estimated 500 pounds or 227 kilograms, and developed a pair of small spawning fangs from the tip of its snout, were identified by renowned paleontologist Robert Rush Miller in 1972. Fossils of *O. rastrosus* have been found in Washington, Oregon and California.

Salmon thrive in cold waters and recently, their fossilized remains have both provided answers and prompted questions about salmon during the Earth's Pleistocene era, or Ice Age. For example, in 2000, fossils almost indistinguishable from today's sockeye were found in slabs of sediment in Washington's southeastern Olympic Peninsula. Not only were the fossils judged to be at least 900,000 years old, they were found at what was then the toe of an ancient glacier. Today, the site is not far from Frigid Creek, which likely describes the temperature of the water in which the ancient fish were spawning during what was clearly an ancient glaciation. Much younger fossils, perhaps 15,000 to 18,000 years old have also been found on the south shore of Lake Kamloops in south-central British Columbia, raising interesting questions about the extent and timing of Earth's last glaciation.

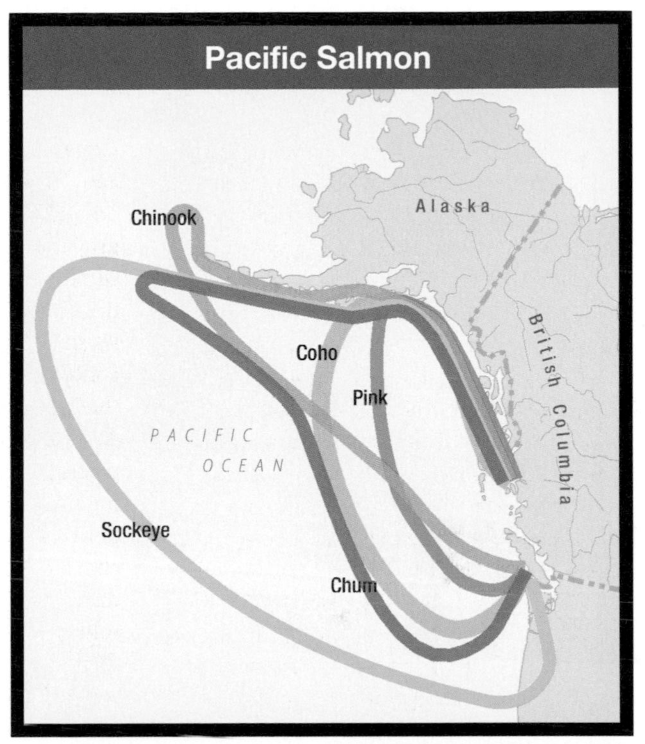

COLUMBIA GORGE DISCOVERY CENTRE / 1999.06.04

Opposite: Joseph Drayton's 1841 painting shows Native Americans salmon fishing at Willamette Falls; within 10 years, the falls had been harnessed for power for what became Oregon City. Above: Cleaning salmon beside the Columbia River.

Today, biologists recognize seven distinct types of anadromous Pacific salmon, ranging in size from the huge chinook or spring salmon, which can weigh in at 50 kilograms or more than 120 pounds and lives between five and seven years (see page 229), to the little pink, which weighs 2.3 kilos or less and lives only two years. Five of these – the sockeye (see page 237), chum (also known as keta and dog salmon), coho, pink and chinook – are found in North American waters. Two others, the masu and omago, are found only in Asian

Autumn colors in Alaska's Togiak National Wildlife Refuge, home to 48 species of mammals, several threatened bird species and salmon, which have long been so abundant that the Salmon River was named for them. Inset: A glacial valley in the refuge, and opposite, a big chinook salmon powers up a waterfall to spawn.

IMAGES COURTESY OF US FISH AND WILDLIFE SERVICE

waters. There is also a freshwater form of sockeye, known as kokanee salmon.

In addition, cutthroat and steelhead, once thought to be members of the trout family, have recently been reclassified as salmon. The freshwater version of the steelhead is the rainbow trout. While other Pacific salmon species invariably spawn and die, and rarely live longer than six or seven years (in the case of chinook salmon), steelhead may live nine years and return to the ocean following spawning, where about 20 per cent survive to reproduce again. On North America's eastern shores, more than half of Atlantic salmon also survive for at least a year after spawning.

Though all anadromous (the term comes from an ancient Greek word meaning "to run up") Pacific salmon begin and end their lives on the gravelly beds of cold mountain streams, they are also creatures of the ocean. And despite years of effort, scientists know relatively little about their lives at sea. Ocean tagging has produced some remarkable insights, however; a sockeye tagged at 177 degrees West, a longitudinal parallel that runs through eastern Siberia and south through the Pacific east of Kamchatka and Japan, was found later in the same year in the Nass River, a journey of at least 2000 kilometres or 1,250 miles. Known for its prodigious freshwater travel, it seems the sockeye also clocks the greatest distances in the three or four years it spends at sea.

Though the fossils found on the shore of Lake Kamloops may prove otherwise, it has long been believed that all salmon stocks disappeared from British Columbia during the last glaciation, when great sheets of ice covered virtually the entire province. But it is not really known where they went. Fossils of a large, extinct species of Pacific salmon, *Salmo australis*, found in 1972 along the northern shore of Mexico's Lake Chapala, seem to indicate that during global cycles of deep cold, salmon may have simply migrated south. Lake Chapala, Mexico's largest natural lake, is just northwest of Mexico City and more than 400 miles (or nearly 700 kilometres) south of the previously known most southerly salmon range.

Moreover, as scientists have recently discovered, it was at least 4,000 years after the glaciers melted before salmon repopulated the rivers of Canada and the northern US in appreciable numbers, and even longer before the stocks had recovered enough to constitute a significant part of the diet of early peoples. The ice began to disappear about 11,000 years ago but, according to painstaking years of work by archaeologists, even in more southerly regions such as Hells Canyon, on the Snake River along the Idaho-Oregon border, the earliest remains of chinook salmon yet found date to 7,290 BP. These were found at Bernard Creek Rockshelter, indicating not only were salmon beginning to supplement a diet that was largely big game and plants, but also that the huge fish were able to climb the river up to the natural barrier at Shoshone Falls, something that is no longer possible as a result of the plethora of dams on the Columbia and lower Snake Rivers.

It was even longer before salmon repopulated the rivers in numbers that could sustain large populations, something that archaeologists studying the Fraser River in British Columbia believe happened between 5,000 and 6,000 years ago.

The story of Pacific salmon, and particularly the tales of the endangered sockeye and chinook, are nothing short of remarkable. Hatched in the early spring from gravel beds in clean, cold, oxygen-rich tributary streams, small rivers or lakeside beaches, each tiny fry has a pendulous yolk sac that sustains it for its first two or three weeks of life. Then it's on its own. Depending on the species, salmon young, called smolts, spend between six months and three years in freshwater lakes and rivers before the travelling to the ocean. Once there, again depending on the species, they migrate hundreds or thousands of kilometres west into the Pacific and north to the Gulf of Alaska. Scientists are not yet sure where each population goes. Increasingly however, thanks to genetic testing, scale growth rings, parasites and other identifying markers, they do know where each

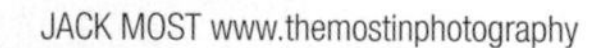
JACK MOST www.themostinphotography

salmon caught in the open ocean originates.

In Canada, scientific testing now allows Fisheries and Oceans patrol boats to not only apprehend ships fishing beyond their national limits, but to assess the origins (and therefore the "citizenship" of salmon found, generally in huge freezers, on board. This is done through genetic tests and assessment of the parasites each fish contains. In recent years, Chinese, Japanese and Russian ships have all been caught with between 50 and 80 per cent of their catch originating in the US or Canada.

Most North Americans are only now beginning to realize how fragile our salmon populations really are, despite the their great antiquity and historically enormous numbers. Scientists estimate that Western North America was once home to close to 1,400 Pacific salmon populations, groups of fish in a geographic area that do not spawn with any others, are genetically distinct, and have adapted to a particular environment. Today, more than one-quarter of these populations are extinct. Of the remaining salmon populations, more than a third are considered threatened or endangered.

Populations in interior watersheds have been particularly hard hit. Migrating chinook salmon have virtually disappeared from the Columbia River headwaters in BC and the sockeye that once gave Redfish Lake, at Stanley, Idaho, its name, have been reduced to numbers that can be counted on two hands. On the lower Columbia, the once enormous late spring and early summer run of huge chinook salmon, the "June hogs", as they were often called, has dwindled to three per cent or less. The fall chinook run on the Sacramento River has completely disappeared. And more than half the salmon populations once found in California's Central Valley have vanished. Coho that once crowded the Snake River are extirpated; the steelhead are all but gone and farther north, the enormous sockeye, coho salmon migrations that were a spectacular annual event in British Columbia's Fraser and Skeena Rivers are but a shadow of what they once were. And in 2008 and 2009, the chum salmon runs that have sustained the Pacific coast grizzlies for millennia have declined so dramatically that the grizzlies themselves are in danger of disappearing.

In short, almost everywhere, Pacific salmon are in a dire state of emergency. They face increasing threats on their long journeys to the sea and back upstream to their spawning grounds.

JERRY KAUTZ

Washington's Columbia River Gorge National Scenic Area was the first such designated region in the US. However, though still lovely, the Columbia is no longer the magnificent producer of salmon that it once was.

The main obstacles to the survival and rehabilitation are the "4 Hs": habitat destruction, hydroelectric dams on migratory rivers, over-harvesting of rare stocks and competition with hatchery fish. One could also add two other H's: human hubris, and our capacity for creating havoc, for we are responsible for every problem faced by these ancient and magnificent endangered and threatened species.

Salmon depend on clean, cold, well-oxygenated water to spawn and the resulting tiny fry depend on it for their very lives. They are remarkably vulnerable to changes in water quality. Yet for the past century, salmon spawning habitat has been degraded by human activities without any thought for the consequences. Development, pollution, agricultural and recreational use all have had an impact, but spawning streams are also vulnerable to invasive plant and predator species as well as to climate change. Climate change is something Earth has undergone since its very beginnings, and both the impact of human activities and the repercussions of our current cycle of change are not yet known. However, based on past experience and current observation, some of the effects global warming will have on salmon populations can be inferred. On the Atlantic coast, some salmon stocks are already migrating farther north, as far as Baffin Island, and on the Pacific coast, some runs in BC's Fraser River have taken to returning early. It's conceivable that the ranges of other species of fish will also shift north, bringing tuna, mackerel and Pacific sardines into increased competition with salmon, and increasing the predation of young fish. Changes in river levels and temperatures will also affect the timing and ability of salmon to return to their spawning grounds.

As will be shown on page 232, the creation of huge dams over the past century to meet the demand for hydroelectric power, along with irrigation schemes to make the desert bloom, have put additional pressures on salmon in the major spawning rivers. And despite ever-increasing expenditures on hatchery programs, as discussed above, the past half-century has seen declines in salmon numbers virtually everywhere in the Western US as well as the extirpation of salmon, particularly in upper reaches of many rivers.

However, the primary reason that Pacific salmon populations have declined so dramatically is plain and simple: overfishing. And while habitat degradation is an important concern, mixed-stock harvesting of wild and "enhanced" stocks is an immediate threat to salmon numbers. Enhanced salmon stocks, those supplemented by artificial propagation in hatcheries and spawning channels, cause a number of problems. Hatchery fish increase competition for resources with wild salmon stocks and artificially inflate numbers, creating an illusion of abundance. This supports increased commercial catches and allows what Arizona State historian Paul Hirt has called "a conspiracy of optimism", the idea that human ingenuity can overcome the need for natural processes.

Farming fish, which has been widely practiced in Canadian waters, not only demonstrably spreads diseases to natural stocks, but also causes a loss of genetic diversity. Finally, by increasing overall salmon numbers, these enhanced stocks support increased commercial catches. Most commercial salmon fishing takes place in coastal waters before individual stocks separate to return to their individual spawning streams; this not only ensures that the breeding salmon will never reach their spawning streams, but because harvest rates are based on maximum sustainable yields – at least in theory – the impact on less productive stocks migrating at the same time can be devastating. The less productive stocks are over-fished and salmon biodiversity is diminished.

Despite these negative ramifications, ocean fishing is preferred for two reasons; it is not only economically expedient, but also catches salmon when they are more commercially valuable,

> **Canadian farmed salmon spread diseases and sea lice that have been proven to be lethal for up to 95 per cent of tiny ocean-bound wild salmon smolts.**

when they are sleek and fat before they begin their long, exhausting upstream migrations and before physiological changes take place to ready them for spawning.

In 2008, the disastrous results made headlines across the continent. "Chinook salmon Vanish Without a Trace" reported *The New York Times* on March 17, 2008, over a story on the almost complete absence of fish in the late fall run on the Sacramento River, "the richest and most dependable source of Chinook salmon south of Alaska," according to the paper. The result was the unprecedented closure of the Pacific salmon fishery from northern Oregon to the Mexican border.

Less than a month later, on April 12th, Canadian fisheries officials announced the Fraser River sockeye fishery would be closed to both commercial and sport fishers for a second year in a row. And at the end of April, the Canadian government asked British Columbia's First Nations to ration their salmon catch. These, of course, are the same Aboriginal nations that had created entire societies based on huge, dependable salmon harvests and yet, for millennia, had managed to sustain enormous spawning runs – at least prior to the arrival of Europeans on the Pacific Coast. First Nations were among those who, in 2004 and 2005, predicted the collapse of the Fraser River sockeye run based on assessments of the number of young salmon descending the river to the sea, complicated by unusually warm ocean water temperatures, which can be lethal for salmon.

Though in the past, both Canadian and American governments had at times condemned First Nations for the decline of their fisheries, history and science have proven those accusations to be both irrational and fatuous.

Having climbed Deep Creek in Peachland, BC, and successfully spawned, a kokanee salmon will provide food for other species.

PETER ST JOHN

Chinook Salmon

Onchorhynchus tshawytsha • Endangered

THE Columbia River travels more than 1,200 miles or 1,900 kilometres to the sea, and aided by its major tributaries – among them the Snake, Okanogan/Okanagan, Spokane, Pend Oreille, Canoe, Blaeberry and Kicking Horse Rivers – drains an area the size of France. Where it greets the Pacific Ocean, its tidal estuary is so wide that the opposite shore is often lost in seaborne mists. Travellers crossing from Oregon to Washington State over the Astoria-Megler Bridge sometimes feel the span they're traversing is endless, as it disappears fore and aft into the swirling fog.

COURTESY OF GLENN K. YOUNG / US FISH AND WILDLIFE SERVICE

The Columbia is not the longest river in North America – the Missouri holds that claim – but for thousands of years, it produced more chinook and coho salmon, and more steelhead, than any other river in the world. All five North American species of Pacific salmon could be found in the Columbia, but the mighty chinook were most plentiful by far. *In Salmon Without Rivers*, authors Jim Lichatowich and James A. Lichatowich estimate that eight to 10 million chinook once journeyed up the river each year And many, especially the "'June hogs' of the summer run, weighed 50 to 60 pounds each."

For millennia, this magnificent bounty sustained the peoples of the river. Gathering at Cascade Rapids and Celilo Falls on the lower Columbia and at Kettle Falls and Priest Rapids farther up (and at dozens of cascades along the river's many tributaries), they used spears, dip nets and eventually sophisticated fishing platforms, basket traps and harvesting channels to reap huge quantities of fish in a short time during peak runs. With entire communities working to gut, clean and splint the fish, they smoked them over fires in the temperate rainforest of the lower reaches of the river and dried them in the hot sun and wind east of the Cascade Mountains. The bounty was enough to feed entire communities for most of the year.

Some archaeologists believe human habitation along the lower Columbia, which was south of the ice sheets during the last glaciation, goes back nearly 15,000 years. Certainly, a host of archaeological sites indicate great antiquity. These include Marmes Rockshelter, which was inhabited virtually continuously for 11,000 years, but is now drowned near the confluence of the Palouse and Snake Rivers in eastern Washington; Clovis points found along the Willamette Valley south of Portland, Oregon also indicate an 11,000-year history, while a record of human habitation going back at least 10,000 years has been found along the shoreline of The Dalles, and the almost complete skeleton of Kennewick Man, who had been carefully interred along the river bank near Kennewick,

Washington, has been radiocarbon dated at more than 9,300 BP.

How long salmon provided a significant portion of the diets of Columbia River people is as yet undetermined. The annual gatherings at the Five-Mile Rapids site near The Dalles, as well as salmon bones dating to 8,500 BP found there, indicate that the fish were being caught and consumed for many millennia. However, scientists with the National Research Council believe that salmon have made up the major part of the diet of Columbia River peoples for only the past 3,500 or 4,000 years.

Yet, despite those thousands of years of harvests, the bounty of the Columbia seemed undiminished when Europeans arrived on its shores. In the summer of 1847, Irish-born artist Paul Kane travelled through what was then called the "Oregon Country", intent on recording the lives and landscapes of native Americans. In early August, he arrived at Fort Colville, just above Kettle Falls, the highest falls on the Columbia. Though he described the cataract as "exceedingly picturesque and grand", it was the salmon that left Kane spellbound. As authors Diane Eaton and Sheila Urbanek write in Paul Kane's *Great Nor-West*, "The salmon were running, arriving in incredible numbers below Kettle Falls, where they gathered their strength to leap up the rushing waters to the spawning beds above."

Then, quoting Kane himself, " 'There is one continuous body of them, more resembling a flock of birds than anything else in their extraordinary leap of the falls, beginning at sunrise and ceasing at the approach of night.' "

Many of the huge fish were unequal to the challenge. Exhausted by a journey of more than 700 miles or 1,100 kilometres up the river, Kane watched as salmon by the thousand were battered on the rocks and ledges and fell back, casualties of the river that had given them life.

As they had for thousands of years, people had gathered at Kettle Falls to harvest the bounty; Kane believed there were 500 of what he termed Chualpays [today known as the

HENRY JAMES WARRE / ALBERTYPE COMPANY / LIBRARY AND ARCHIVES CANADA / 1845 C-047017

"Salmon packing above the Cascades" on the Columbia River was painted by Henry James Warre in 1845.

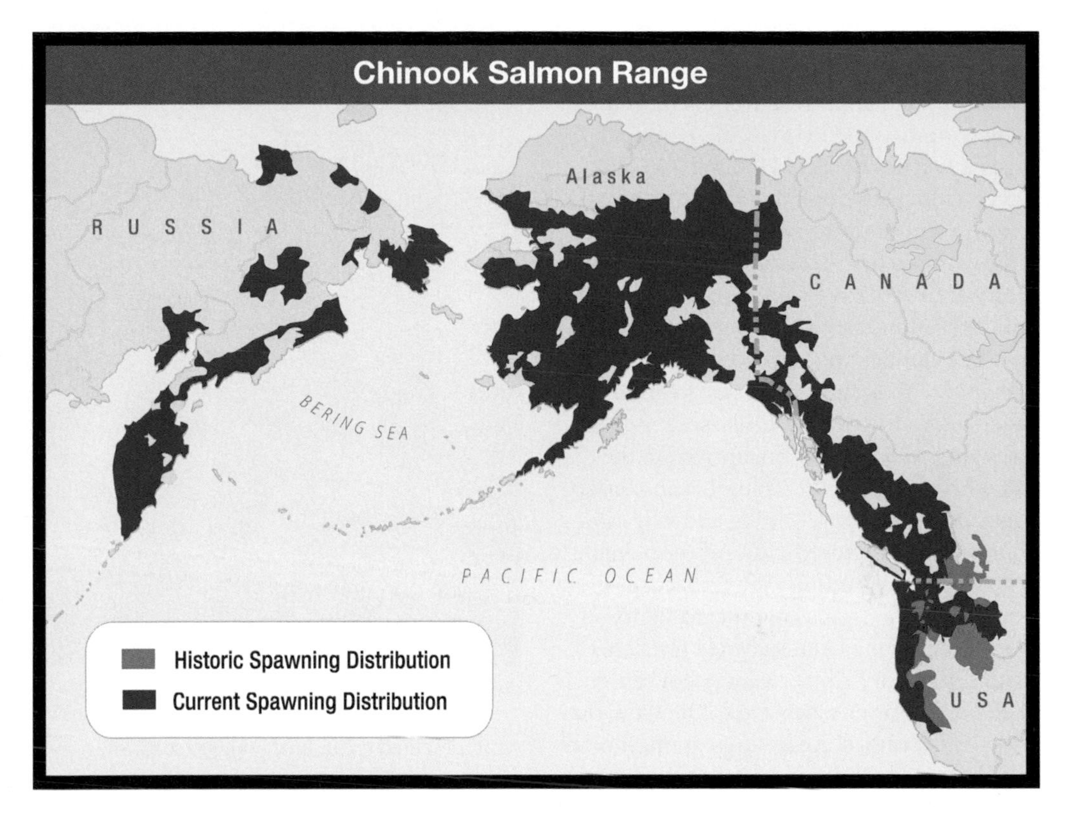

Colville] in 1847. Using a large wicker trap, the "Chief of the Waters", See-Pays, controlled the harvest, which Kane estimated at about 400 chinook salmon a day, and distributed an equal share to each of his people, "even to the smallest child".

" 'Infinitely greater numbers of salmon,' " write Eaton and Urbanek, quoting from Kane's diaries, " 'could be readily taken here, if it were desired; but as the chief considerately remarked to me, if he were to take all that came up, there would be none left for the [people] on the upper part of the river; so that they content themselves with supplying their own wants.' "

It was a lesson that Europeans arriving on the West Coast seemed not to understand. William and George Hume, who had been selling fresh and salted salmon on the Sacramento River since the 1850s, arrived in 1866 on the Columbia with Andrew Hapgood, a tinsmith from their home state of Maine. That year, according to the Washington State Department of Archaeology and Historic Preservation, the trio opened the first salmon cannery on the Columbia, at Eagle Cliff east of Cathlamet, upstream from the river's mouth.

Located on a scow, the cannery was initially a trial-and-error business. Nevertheless, at the end of the first year, it had managed to produce 4,000 cases of salmon, with each case holding 48 one-pound cans. Sold as far away as Australia, canned salmon created something akin to a gold rush, as canneries opened up and down the lower Columbia.

In 1883, they produced 43 million pounds of salmon, the largest harvest ever and the beginning of the end. For the next 35 years, harvests averaged about 25 million pounds, an average that was maintained by switching from spring and summer chinook to the fall run. By 1920, however, as the Lichatowiches write, the combination of overfishing and habitat destruction led even die-hard optimists to see reality. "Said Hollister McGuire, Oregon's first State Fish and Game Protector, 'Oregon has drawn her wealth from her streams, but now, by reason of her wastefulness . . . the source of that wealth is disappearing and is threatened with complete annihilation.'"

The chinook were particularly hard hit, not only because their size made them so much desired, but because, like sockeye, their spawning streams were far upriver, in the upper tributaries of the Columbia and Snake Rivers, where logging, mining and irrigation channels had destroyed much of the habitat.

McGuire's warning went unheeded, however, and a decade later the drought and Depression of the 1930s spawned not better management or habitat cleanup, but rather Franklin Roosevelt's New Deal. The West, the president reasoned, could support many more people, if only the dry interiors of Washington and Oregon could be irrigated. And so began the construction of the Bonneville and Grand Coulee Dams. From the beginning, the Lichatowiches write, "biologists could see that the dams, even when equipped with fish ladders, would destroy the remaining stocks, but felt helpless in the face of huge government programs." The answer, they hoped, might lay in artificial propagation and the transport of tiny smolts to the lower Columbia. In short, they believed that technology could solve the problems created by the destruction of natural environments. Or, as the Lichatowiches put it, they believed "the myth that humans could have salmon without healthy rivers."

The idea was first put into practice in 1946, when an investment of two million dollars led to the harvest of between 1.5 and two million salmon. Almost 50 years later, in 1994, the government invested $425 million in hatchery and transportation programs,

LIBRARY AND ARCHIVES CANADA / PA-031664 / ACC. NO. 1988-250-26

which yielded a catch of 500,000 salmon.

In the years since, the downward spiral, not just on the Columbia, but on the Sacramento and many other spawning rivers up and down the Pacific Coast, has become a tailspin. In 2001, for example, irrigation water from the Klamath River, which begins in southwestern Oregon and flows through northwestern California, was shut off in response to orders issued under the Endangered Species Act, in order to provide enough water in the tributaries for spawning salmon. The farmers affected by the ensuing water shortages went to President George W. Bush, who, despite spirited opposition from conservation groups and native Americans, restored the irrigation water in 2002. The result: 70,000 returning salmon died that year, before they could spawn, in the dried up creeks and streams.

With the images of tens of thousands of dead and dying salmon still fresh in mind, in November 2008, the Bush administration suggested that all parties involved should work cooperatively to remove four hydroelectric dams on the Klamath by 2020. If successful, this might be the beginning of a desperately needed collaboration. But will it be in time?

Dams are not the only problems both spawning salmon and young smolts face as they navigate the rivers of the West Coast. In fact, an international study, released in late 2008, seemed to indicate that ocean-bound chinook and steelhead smolts had higher survival rates in rivers with many dams than in those that were dam free. The rivers involved in the study were the Snake and Columbia Rivers in the US, and the Thompson and Fraser Rivers in Canada. Sound-emitting tags about the size of an almond were implanted in 1,000 chinook and steelhead smolts – each juvenile salmon was about the length and half the weight of a hot dog – prior to their journeys down to the Pacific and then north along the continental shelf to Alaska. The tiny salmon were followed using a system of underwater detectors.

The results were surprising (and encouraging for at least one of the study's funders – the Bonneville Power Administration), for the salmon in the Columbia system, which had to traverse eight dams and had farther to go, seemed to survive the trip as well or better than the salmon that traveled the undammed Thompson-Fraser system, leading researchers to wonder whether the complaints of environmentalists about the state of the water in the lower Fraser may well be justified (see Sockeye on page 237). On the other hand, because of the limited range of the tracking devices, counting tiny salmon in a free-flowing river, or in the ocean for that matter, is much more difficult than it is in rivers that are dammed, where the fish are funneled through structures that allow their tags to be read by the detectors. It may be, therefore, that the fish on the Fraser were simply missed, rather than non-existent. Electronic technology does hold considerable promise for tracking other species, however, including highly endangered white sturgeon (see page 244).

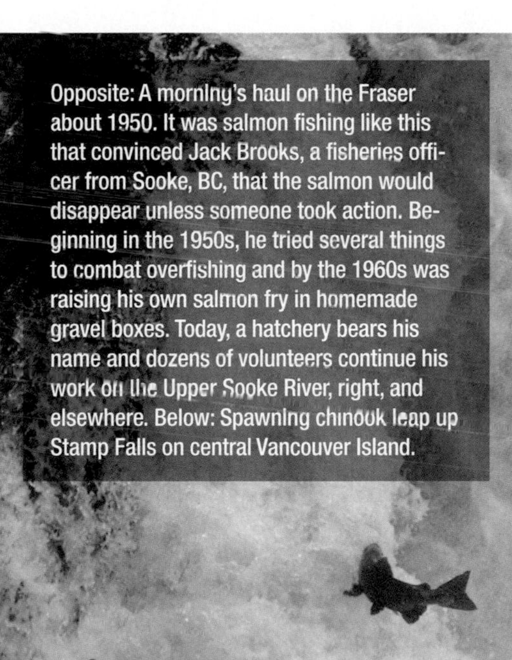

Opposite: A morning's haul on the Fraser about 1950. It was salmon fishing like this that convinced Jack Brooks, a fisheries officer from Sooke, BC, that the salmon would disappear unless someone took action. Beginning in the 1950s, he tried several things to combat overfishing and by the 1960s was raising his own salmon fry in homemade gravel boxes. Today, a hatchery bears his name and dozens of volunteers continue his work on the Upper Sooke River, right, and elsewhere. Below: Spawning chinook leap up Stamp Falls on central Vancouver Island.

BOTH IMAGES: JACK MOST www.themostinphotography

Above: Each autumn, dozens of dedicated volunteers take just a few large, healthy salmon from Vancouver Island's Sooke River run and, handling them with great care, "milk" them to capture the eggs and sperm that will produce a new generation. The eggs then go through a complicated process to establish the Accumulated Thermal Units or ATUs, which in turn determines whether the eggs will be viable or not. At exactly the right temperature, all infertile eggs are removed by hand; the rest are returned to the hatching towers to continue the process of producing tiny fry and, ultimately, smolts that are released into the river to begin the ancient cycle once again. The volunteers' motto: "We hatch them; you catch them."

But if salmon stocks are a priority, as discussed in the introduction on page 223, technology must be accompanied by a continued moratorium on both commercial and recreational salmon fishing in the Pacific, followed – should the stocks be revived – by far stricter regulations on the numbers of chinook and other threatened and endangered salmon species that can harvested. Moreover, stocks are unlikely to be revived without decisive and determined action to restore hundreds of degraded spawning streams from central California to British Columbia.

Though our own concerns are usually uppermost, humans are not the only species affected by dwindling and disappearing

VIEWING CHINOOK SALMON

A number of websites offer information on viewing spawning salmon. Among them are nearly a dozen spawning streams in King County, which includes Seattle. Best viewing times are mid- to late September through late October and early November. Chittenden Locks in Seattle offers salmon talks, sponsored by the US Army Corps of Engineers, twice a day through September. The adjacent recreational area is open year-round.

Spawning chinook salmon can also be seen at Stamp Falls on the Stamp River in central Vancouver Island, as well many in as other rivers on the island. Unfortunately, salmon spawning in the Comox River on the island's east shore, along with the smolts from its hatchery, must run a gauntlet of seals that take advantage of the deeply dredged river. Dredged to allow pleasure boats to go inland., the river is now deep enough to allow seals to swim far upstream, cutting significantly into the salmon numbers.

IMAGES BOTH PAGES: JACK MOST www.themostinphotography WITH ASSISTANCE FROM THE JACK BROOKS HATCHERY AND THE SOOKE SALMON ENHANCEMENT SOCIETY.

salmon stocks. In the fall of 2008, scientists at the Centre for Whale Research in Friday Harbor, Washington, linked the disappearance of chinook salmon to the rapid decline in the number of Puget Sound orcas (or killer whales, as they are sometimes called). The endangered Puget Sound population, a small resident group that numbered about 100 in the early 1990s, had slipped to fewer than 80 animals in 2005 and though the whales are inventoried annually, seven of those, including two breeding females, could not be found in 2008.

The reason for the decline, said scientist Ken Balcomb seems obvious. "People say they could eat hake, or they could eat cake. It's just arrogance, or actually ignorance. If chinook aren't doing well, the whales aren't either."

Added Dr. Rich Osborne, a research associate with The Whale Museum in Friday Harbor, "Restoring Columbia River chinook salmon is the single most important thing we can do to ensure the future survival of the southern resident community of killer whales. We cannot hope to restore the killer whale population without also restoring the salmon upon which these whales have depended for thousands of years. Their futures are [inextricably] linked."

RECOVERY PROGRAMS

A recovery program must be multi-pronged and involve a moratorium on commercial and recreational fishing, the cleanup and restoration of spawning streams, and in some cases the removal or alteration of dams to allow sufficient water to flow into upper tributaries.

In Canada, though the once-numerous Okanagan River population has dwindled to an estimated 50 spawning adults, the stock was downgraded from endangered to threatened in 2005. However, it is now listed under Canada's Species at Risk Act (or SARA) which includes prohibitions against "killing, harming, harassing, capturing or taking species at risk, and against destroying critical habitats." Much farther north, bilateral cooperation between the US and Canada involving quotas in the ocean fishery of the Alaska coast (in 2008, the Alaskan fishery quota was 170,000 fish, a decrease of 159,000 fish from the 2007 quota and the lowest harvest level since 2000) along with regulations based on the number of fish crossing into Canada have meant relatively steady numbers of chinook salmon reaching their spawning headwaters of the Yukon River in Yukon and northern British Columbia. On the Canadian border a count of 54,000 or more fish means a green light for fishing; between 19,000 and 54,000 means closure of the commercial and recreational fisheries, while, depending on the number, First Nations will be allowed to fish for personal consumption, and less than 19,000 fish crossing the border means a complete closure of the fishery. Chinook spawning in the upper tributaries of the Yukon River can travel more than 1,500 miles or 2400 kilometres.

Chinook Salmon Fast Facts

Onchorhynchus tshawytsha

DESCRIPTION

Chinooks stay in their home streams for more than a year, longer than many other species, before heading down river to the Pacific, where they live for between three and six years. During their years in salt water, the fish are greenish blue with black spots and a silver underside; their sides and back darken during the summer months. As the fish head upstream to spawn – something that happens as early as three years of age for males and as late as six or even eight years for females in northern waters, their body color changes to red or copper and can even appear black. Males are darker than females.

LENGTH & WEIGHT

The largest of the Pacific salmon, adult chinook salmon can grow to five feet or 1.5 metres long. and weigh upwards of 100 pounds or 45 kilograms. The largest chinook on record was 126 pounds or 57.27 kilos; the fish was caught in a fish trap near Petersburg, Alaska.

HABITAT

Chinook begin life in the clear, cold water of mountain streams;baby salmon, or alevins, and small juvenile salmon, or fry, live in fresh water for more than a year. Then, the juvenile salmon, now called smolts, travel downstream and spend from one to three months in tidal estuaries, before heading north and west into the Pacific. With a life span of between three and six years, chinook are vulnerable to environmental threats to their spawning streams, to river dams and diversions, to urban development and infilling, industrial, urban and agricultural pollution as well as climate change and El Niño upwelling in the ocean, which warms the water and contributes to high mortality numbers.

DIET

After hatching, alevins live in the stream-bed gravel for several weeks, slowly absorbing the nutrients in their attached yolk sak. Juvenile fry and smolts feed on plankton and insects, but once in the ocean chinook dine on other fish, squid and crustaceans, often doubling their weight each year.

BEHAVIOUR

Chinook travel huge distances. Fish from Washington, Oregon and California are believed to travel north to Alaska and then west to the mid-Pacific, where their migration route may overlap with chinook from Asia.

Though largely independent, chinook form loose schools, congregating in areas with underwater structures such as reefs or rocks while resting during their long journey home to spawn. They also cover remarkable distances; studies have shown that even juvenile fish can travel more than 1,500 miles or 2500 kilometres in just three months.

MATING & BREEDING

Spawning chinook do not feed during their upstream journey, which can take several weeks. Using stored body reserves, their condition deteriorates until at last they arrive at the stream of their birth. Once there, the large females build large nests or redds, moving rocks as large as a cantaloupe in relatively deep, fast-running water. The eggs are also larger than those of other species and a large female can lay up to 7,000 eggs, compared to an average of 2,500-5,000 for other types of salmon. Most breeding females are five or six years old; the males are typically a year or more younger. Spawning has been called the "final hurrah", for the fish die shortly after and float downstream again, providing food for many other species.

NAMES

Chinook are also called king or spring salmon, tyee, blackmouth and chubb salmon. The name "chinook" is named for the Chinook people, noted traders, salmon fishers and hunters, who lived along the Columbia River when Europeans first arrived in the region.

JACK MOST www.themostinphotography

A narrow canyon near Stamp Falls on central Vancouver Island draws spawning chinook.

Sockeye Salmon

Onchorhynchus nerka • Endangered (Snake River population, US; Cultus Lake population, COSEWIC); threatened (Lake Ozette population, US)

THINK of British Columbia's Fraser River, and magnificent runs of sockeye salmon almost invariably come to mind. For more than 5,000 years, the Fraser was perhaps the world's foremost producer of these remarkable fish. Entire cultures were built around the coming of the salmon; the enormous and dependable runs allowed the evolution of sophisticated societies and the creation of art that has come to symbolize not only British Columbia, but Canada.

Sockeye once also spawned in virtually every river from the Klamath in northern California north and east to Bathurst Inlet in the Canadian Arctic. Along with the larger chum salmon and the smaller pinks, they encouraged coastal bear populations that were larger in size, as well as in number, than anywhere else in North America; sustained orca populations in Juan de Fuca Strait and Puget Sound and, as scientists are only beginning to learn, fertilized the primordial rainforest itself.

Today, however, sockeye, as well as almost every other salmon species, are greatly diminished in number in most places, and the bears and the whales that depended on them are declining markedly.

Though paleontologists and archaeologists believe the enormous spawning runs – salmon returning in numbers large enough to reliably support societies and species for months at a time – go back five or even six millennia, salmon as a species are far older than the Fraser River itself. The ancestors of today's salmon date back to the Early Cenozoic, between 50 and 60 million years ago. Over time, the species evolved and changed, becoming ever more dependent on the Pacific Ocean.

Sockeye and chum salmon are believed to have diverged physically more than five million years ago and perhaps a million years ago, evolutionary changes caused sockeye and pink salmon to become separate species. Fossils found along the South Fork of the Skokomish River in Washington's Olympic Peninsula, and dated to about 900,000 BP, were remarkably similar in size and characteristics to today's sockeye, but with small details reminiscent of pinks.

Not all sockeye are anadromous; some wild salmon live their entire lives in fresh water, not only in British Columbia, but in Alaska, Siberia and Japan. (These smaller kokanee salmon, as the freshwater variety is called, have also been installed in lakes and rivers in many Western states.) And as recent discoveries in the BC Interior seem to indicate, these freshwater stocks go back at least 18,000 years.

Nevertheless, according to years of painstaking work by archaeologists, it seems clear that in most of British Columbia and at least part of the Pacific Northwest, it was nearly 4,000 years after the ice sheets disappeared about 11,000 years ago before sockeye salmon stocks had recovered enough to constitute a significant part of the diets of early

COURTESY OF TIMOTHY KNEPP / US FISH AND WILDLIFE SERVICE

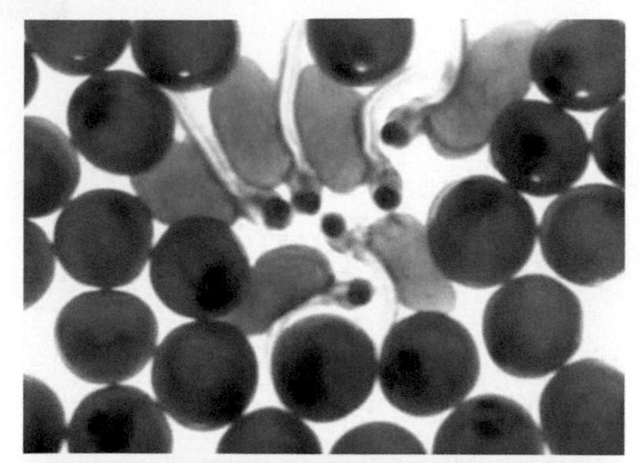

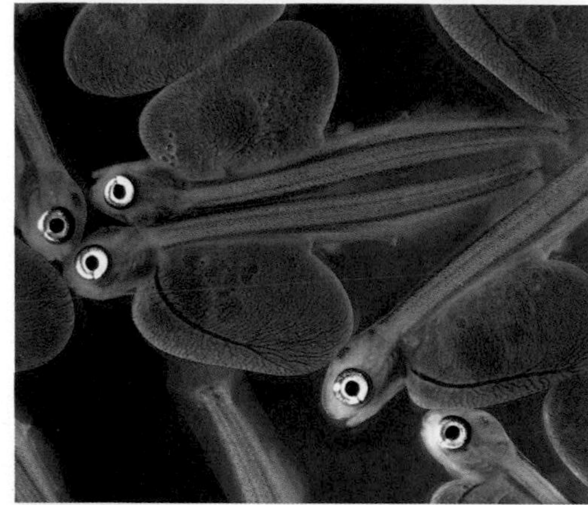

Salmon develop rapidly before headng for the sea, alevins hatch from eggs (top), to become fry with a pendulous egg sac (centre) and then finger-length smolts. However, it is as spawning adults (bottom) that we know sockeye salmon best.

IMAGES COURTESY OF US FISH AND WILDLIFE SERVICE

North Americans. And it was even longer before salmon could sustain populations of any size. Early communities along the Fraser River may have caught salmon when and where they were available, but for millennia they were not a staple, but rather an alternative to a diet that was mainly meat, plants and fruit.

Like its larger chinook relatives, the tale of the sockeye is nothing short of remarkable. Hatched in the early spring from gravel beds in clean, oxygen-rich tributary streams, small rivers or lakeside beaches, each tiny fry has a pendulous yolk sac that sustains it for the first two or three weeks. Then it's on its own. Sockeye young, called smolts, spend up to three years in freshwater lakes and rivers before the travelling to the ocean in May and June. The presence of lakes seems crucial to the timing. Those that are born in river systems without lakes usually migrate in their first year, while those that can escape to lakes, generally head for the ocean in the spring of their second (or sometimes even their third) year.

Sockeye have long been the second-most abundant species in British Columbia waters (after the smaller pinks) and the most economically valuable. The young adults – dark blue in colour, with distinctive scales and dark tails – move northward along the coast and west at the Gulf of Alaska into the Pacific. Unlike other Pacific salmon but like some whales, sockeye are filter feeders, feeding on plankton strained from large volumes of water.

For the next two years, they cruise the ocean, apparently travelling enormous distances and consuming a diet that, according to US biologists, "turns mainly to amphipods, copepods [both tiny crustaceans that look like little white bugs], squid and some fishes." Scientists are not yet sure where each population goes. Increasingly however, thanks to genetic testing, scale growth rings, parasites and other identifying markers, they do know where each salmon caught in the open ocean originates.

Then, at four or five or occasionally even

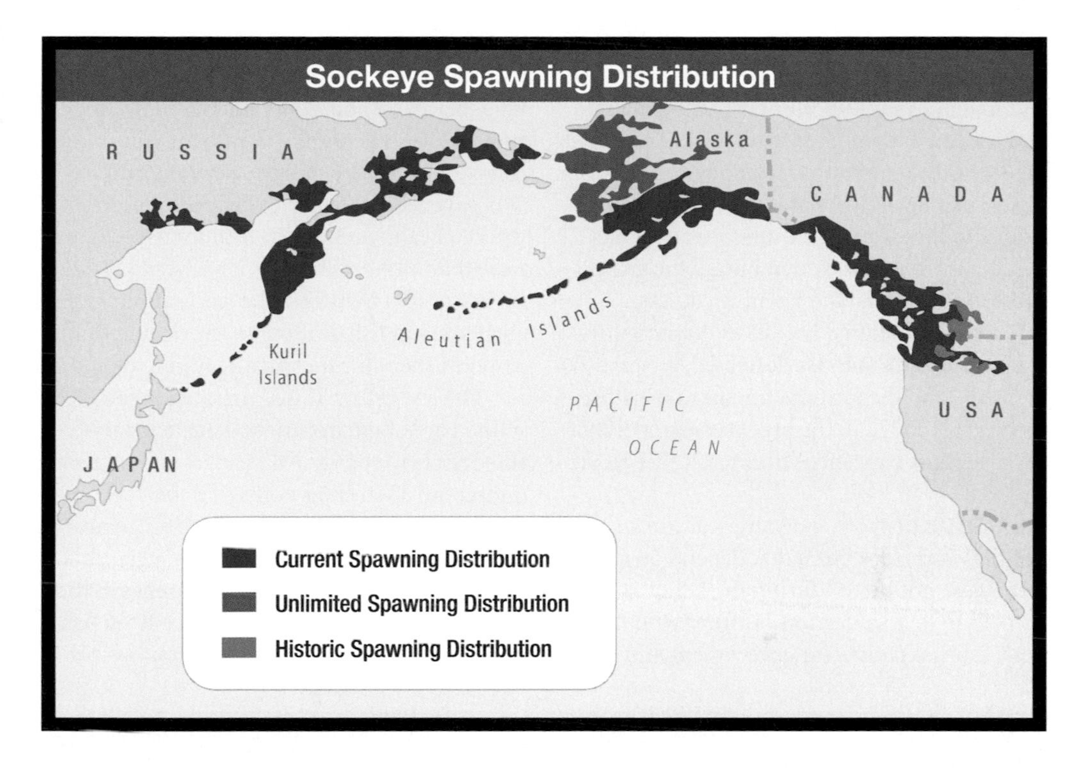

six years of age, the urge to spawn overcomes each sockeye and it heads for the rivers or stream in which it was born. Scientists are still trying to understand the spawning impulse, for it has obvious implications, both for the fish and the populations who have long depended on it. Investigations conducted during the past 50 years have shown that salmon are able to use the sun's position and Earth's magnetic field, among other positioning systems, to orient themselves at sea and, once in fresh water, each fish can "smell" the distinctive chemical odor of its home stream. As they head upriver in the late summer or early fall (or, in the case of some populations such as BC's endangered Cutlus Lake sockeye, in November or December), the bodies of sockeye turn bright red, their heads green and the males develop large hooked jaws.

Once the urge to spawn takes over, each fish will either overcome every obstacle to reach its final destination, or die trying. Surprisingly, each year's run is different, with dominant runs occurring every fourth year. Before the coming of Europeans, so many fish choked the smaller tributaries during these bountiful years that some fish were actually squeezed out of the water onto the banks. And every year, the narrows and rapids of the Fraser and its hundreds of tributary rivers and streams were crowded, not only with people, but with bears, eagles, raccoons, otters and many other species that had come to depend, at least in part, on these remarkable fish.

Overfishing – mainly by commercial fisheries that use huge nets to catch spawning salmon in great numbers as they stream toward the estuaries prior to heading upriver – pollution and degradation of the spawning streams are the main threats to sockeye populations. However, hydro manipulation of rivers, including water storage, withdrawal and diversions for agriculture, has also reduced or eliminated spawning streams, and caused high mortality to juvenile salmon. Dam turbines and sluiceways have also killed both adults and juveniles in large numbers and logging and mining have altered or destroyed spawning streams; removing trees along the edges of streams have eliminated places for young smolts to hide, caused eroding stream banks, overheated water and often buried the gravel substrate so crucial to spawning success.

Even in National Forests in Washington State, stream habitat has decreased by nearly 60 per cent. And the introduction of non-native species has added to the predator gauntlet that sockeye must run before they reach their goal.

In Canada, both spawning salmon and the grizzlies and black bears that depend on them were "just not there" during the fall of 2009, according to a wilderness outfitter who has done salmon counts for government and First Nation fisheries for 20 years in British Columbia. Others were reporting the same thing in many places up and down the BC coast. And though 2009 was not the dominant year in the four-year cycle, the strong spawning run in 2005 gave federal fisheries officials reason to expect a healthy return. In fact, some predicted more than eight million sockeye would tackle the Fraser River. Instead, the run was just over a half-million fish, leading to the closure of the sockeye fishery for the third year in a row.

The low return raised many questions on the part of environmental agencies and Aboriginal groups, as well as anger and despair among the dwindling ranks of fishermen. Some blame warmer ocean and river temperatures associated with climate change, while others feel that dwindling food supplies in the open ocean or the impact of sea lice from fish farming on tiny juveniles may be responsible.

PAUL SMITH

Kermode bears, such as this one, depend on salmon for the fat and protein they need to get them through the winter months.

Chief Houdikoff of Attu village, Alaska, and his son net sockeye salmon, c. 1938.

COURTESY OF VICTOR B. SCHEFFER / US FISH AND WILDLIFE SERVICE

The one thing on which all agree is that it is time that the provincial and federal governments take the state of the Fraser River salmon fishery seriously, for something that once defined British Columbia is on the verge of collapse … or worse.

COSEWIC, the Committee on the Status of Endangered Wildlife in Canada, lists just two sockeye salmon populations as endangered – the all-but-extirpated late run that spawns in Cultus Lake on the Washington State border and the Saginaw Lake population, located on the Sunshine Coast north of Vancouver.

Yet many other populations deserve the listing, Vicky Husband, conservation chair of the Sierra Club of Canada's BC Chapter, believes. "It's not just the Fraser River where sockeye have been going 'missing' all these years. Sockeye are going missing from the entire BC coast. Most of these sockeye stocks are from the coast's most pristine areas, and we're losing them because of mixed stock fishing, global warming, and because the Department of Fisheries and Oceans hasn't been protecting them against overfishing."

There are some bright spots in this dismal picture. On Vancouver Island's West Coast, the number of spawning coho has jumped dramatically, thanks to the recovery efforts of the Central Westcoast Forest Society. In 1995, just 562 coho salmon were counted along log-jammed Kootowis Creek on Kennedy Flats between Ucluelet and Tofino. Society members went to work cleaning up the creek, which was choked with timber and debris. Six years later, the number of spawning coho had increased by more than 10 times, to 6,789.

Perhaps sockeye all over the West Coast need that kind of attention, the kind that has been paid to Idaho's Snake River population since 1991. Admittedly, the Snake River sockeye are considered unique. They travel a greater distance (about 900 miles or 1440 kilometres) and climb to a higher elevation (6,500 feet or 2030 metres) than any other sockeye population, and they are the world's most southerly extant sockeye population. When it was listed, the only wild sockeye left was that in Redfish Lake in the Stanley Basin or upper Salmon River drainage in the Sawtooth Mountains. Other lakes in the basin had historically supported anadramous sockeye populations, but they were gone, and though kokanee or freshwater salmon were also found, only the Redfish Lake sockeye made the amazing, four-month spawning run up the Columbia and Snake Rivers to the Salmon River and through what is now the Sawtooth National Recreation Area to the lake where, for thousands of years, they had spawned.

Sockeye Salmon Fast Facts

Onchorhynchus nerka

LENGTH & WEIGHT

Perhaps the most distinctive of the salmon species, sockeye are up to 86 centimetres or 34 inches in length, can weigh up to 12 pounds or 5.5 kilograms and are four to six years of age when they return to spawn. During most of their lives they are blue or silver in color, with distinctive scales and dark tails. When they spawn, however, their bodies turn bright red, their heads become green, and the males develop large hooked jaws.

DISTRIBUTION

Sockeye can be found in rivers and lakes from the Columbia River and its tributaries north and west to the Kuskokwim River in western Alaska (as well as along the Russian and Chinese coasts and as far south as northern Japan). When at sea, sockeye range farther than any other salmon species; though biologists are not certain where they roam, many seem to spend most of their lives in a vast area of open water south of the Aleutian Islands

HABITAT

Clean, cold, well-oxygenated water is crucial to the development of sockeye fry, as are deep, still pools where young fish can grow before heading to the sea. The returning fish require rivers or streams with a gravel substrate for spawning.

DIET

Unlike other Pacific salmon species, sockeye are filter feeders. They take in water filled with plankton or other small aquatic organisms such as shrimp and, with their gill rakers, hold in the plankton as the water exits their mouths. Feeding on plankton is believed to be responsible for the sockeyes' bright red flesh, as well as their low mercury concentrations.

MATING & BREEDING

Famously, sockeye will struggle back upstream to spawn in the fall, sometimes travelling for weeks against strong currents and leaping up thundering falls. During this time, the males' coloration changes to bright red, with a pale green head, while females may develop green and yellow marks. Finally, arriving bruised and exhausted, the females create three to five redds (or nests) over two or three days, laying their eggs in the gravel of the fast-flowing stream where they were born. The males then fertilize them. The dying parents will defend their nests to the end, and then the eggs lie hidden until the fry emerge in the spring. The newly hatched salmon are tiny, with pendulous yolk sacs. Each little fry spends between two and three weeks in this alevin stage, growing and absorbing its yolk sac.

Most sockeye spawn in rivers that feed into lakes, though some stocks or individual populations spawning along sandy lakeshore beaches. Unlike Atlantic salmon, which often return to the sea after spawning, most Pacific salmon die within days of spawning (the related cutthroat and steelhead are exceptions).

Sockeye young spend up to three years in freshwater lakes before the young fish, called smolts, travel to the ocean in May or June. Sockeye salmon migrate thousands of kilometres, far out into the Pacific and Gulf of Alaska – in fact, scientists are not yet sure precisely where they go – before returning to their home streams.

Sockeye runs are cyclical and much larger in some years than others; scientists are still unsure why this is the case.

In the Pacific Northwest, there is a type of sockeye, known as kokanee, which spends its entire life cycle in fresh water. Kokanee salmon have been widely used to stock inland rivers and lakes.

NAMES

The name "sockeye" is thought to have come from sθ ´ əqəy, pronounced "sukkai", the name for the species in Halkomelem, the language of the people of the lower Fraser River. These magnificent fish are also called red or blueback salmon.

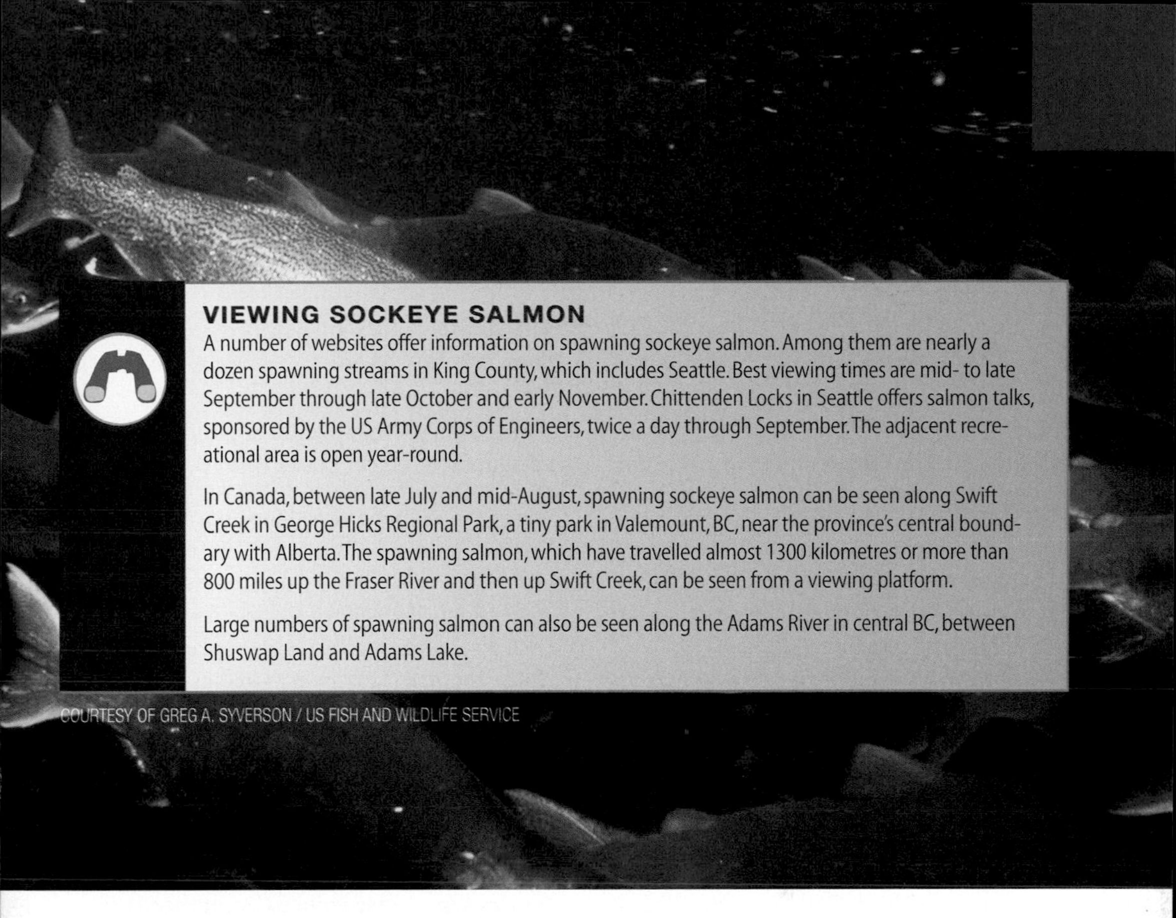

VIEWING SOCKEYE SALMON

A number of websites offer information on spawning sockeye salmon. Among them are nearly a dozen spawning streams in King County, which includes Seattle. Best viewing times are mid- to late September through late October and early November. Chittenden Locks in Seattle offers salmon talks, sponsored by the US Army Corps of Engineers, twice a day through September. The adjacent recreational area is open year-round.

In Canada, between late July and mid-August, spawning sockeye salmon can be seen along Swift Creek in George Hicks Regional Park, a tiny park in Valemount, BC, near the province's central boundary with Alberta. The spawning salmon, which have travelled almost 1300 kilometres or more than 800 miles up the Fraser River and then up Swift Creek, can be seen from a viewing platform.

Large numbers of spawning salmon can also be seen along the Adams River in central BC, between Shuswap Land and Adams Lake.

COURTESY OF GREG A. SYVERSON / US FISH AND WILDLIFE SERVICE

The Redfish Lake population was never huge, but the creation of Sunfish Dam in 1910 across the Salmon River about 20 miles below Redfish Lake, even with an inadequate fish ladder, all but prevented the spawning salmon from reaching their destination. Realizing that the salmon population was dropping like a stone, part of the dam was removed in 1934 and between 1954 and 1966, between 11 and 4,361 fish returned to Redfish Lake each year (an average of more than 700 a year). However, commercial fishing continued to take a huge toll on sockeye generally and introduced and native species in the Columbia River consume up to 20 per cent of juvenile salmon on their way to the sea. Even before they begin their long trip to the Pacific, it's estimated that introduced rainbow trout devour up to 60 per cent of sockeye eggs, fry and smolts.

Little wonder then, that in 1985, '86 and '87, just 11, 29 and 16 Redfish Lake sockeye made it home to spawn. And since 1987, only 18 naturally occurring sockeye have returned to the Stanley Basin.

Beginning in the 1990s, therefore, a captive breeding and release program was begun and in 1999, the first adults from this stock program returned to Redfish Lake. Between 1999 and 2005, a total of 345 captive brood adults migrated to the Stanley Basin from the Pacific. Though catch rates for spawning salmon have been reduced from 40 per cent to just eight per cent in 2005 and to zero in 2008, other threats continue to diminish the brood stock, including water withdrawals in the upper Salmon River hydrosystem, which are believed to kill up to 20 per cent of migrating juvenile salmon.

Sockeye in Lake Ozette, which lies in Washington's Olympic National Park, face quite different dangers. Located very close to the Pacific, on the northwest edge of the Olympic Peninsula, the population was listed as threatened in 1999, a result of declining numbers that had largely to do with degraded spawning habitat and predation by non-native species. The plan for the lake's sockeye will address these problems and continue to prohibit the harvest of salmon from the lake and its feeding rivers and streams.

White Sturgeon

Acipenser transmontanus • Endangered (Kootenay and Kootenai River populations; Nechako River population; Upper Columbia River population; Upper Fraser River population)

VIRTUALLY unchanged over the past 70 million years, sturgeon are among the oldest fish in existence. Even more remarkable, their ancestors stretch back almost 400 million years, to the dawn of vertebrate life.

Fish were the first vertebrates – animals with backbones – on Earth, appearing about 500 million years ago. Prior to the evolution of true vertebrates, marine creatures developed that lacked jaws and had a cartilage rod – a notochord – rather than a backbone. Two descendents of these primitive creatures, lampreys and hagfish, are with us today; though often described as fish, the argument continues over whether they are true fish. Sturgeon are the only other freshwater fish still living that lack vertebrae or a backbone.

The earliest fish were "shell-skinned", armored with scales and bony plates, but about 400 million years ago, the first of what have been called the "bony fishes" appeared. Though descendants of the earliest of these, lung fish – an ancient group with functional lungs that ultimately led to terrestrial creatures, including mammals – and their lobe-finned successors are rare, they can still be found today. For example, the coelacanth, which was believed to be extinct until it was discovered in the tropical waters off India in 1939, is a lobe-finned fish.

However, it was their successors, the ray-finned fish, that ultimately became the Earth's most successful vertebrates. Today, there are an estimated 30,000 species of bony fish in existence. Among these, sturgeon were the first to appear and were the dominant species for about 175 million years. Then, during the Triassic period about 225 million years ago – as Pangaea was breaking up and the seas were increasingly dominated by marine carnivors – they were almost completely replaced by sharp-toothed, heavily-armored garfish, and bowfins; both have rudimentary lungs. These, in turn, largely disappeared as ray-finned fish – known as teleosts – became more and more common. Today, there are more than 20,000 species of ray-finned fishes.

The sturgeon that survived these evolutionary waves were quite different than their ancient ancestors, which had been covered with hard, shiny, interlocking plates. Beneath this armored exterior, prehistoric sturgeon had strong jaws and a bony skeleton. By contrast, though they have been called living fossils, their modern descendants have lost most of the heavy scales.

DENNIS FAST www.dennisfast.smugmug.com

They retain only a bony cap over their skull, which is made of cartilage, and five rows of bony plates, called scutes, running the length of their body. Unlike early sturgeon, they have a weak jaw and feed through a round, toothless tube-like mouth (see Fast Facts on page 249).

Most biologists consider the surviving sturgeon species to be degenerative forms of their ancient ancestors and they differ from modern fish in many ways. In fact, with their cartilaginous skeleton and their shark-like tail

The Snake River twists through Hells Canyon on the Idaho-Oregon border.

COURTESY OF X-WEINZAR / en.wikipedia.org / creativecommons.org/licenses/by/3.0

fin, some have suggested that sturgeon are more like sharks than modern fish.

In optimum circumstances, anadromous white sturgeon (those that migrate between freshwater and salt water) can reach six metres in length and weigh more than 1,750 pounds or 800 kilograms. Members of the species that are confined to fresh water are considerably smaller.

Found only in the Northern Hemisphere, but once widely distributed, almost all the world's 26 and North America's eight species of sturgeon are threatened or endangered. In large part, this is due to two conflicting realities: the very low reproductive rate of these ancient species and the highly destructive capacity of human technology.

Depending on the species, female sturgeon may not mate and begin to spawn until they are between 22 and 30 years of age. And spawning, which may only occur every four to nine years, requires very specific conditions to produce viable young. White sturgeon, for example, require swift-flowing streams with pebbly or cobbled beds for successful spawning.

Yet these ancient creatures have proved able to adapt to changing conditions in the past. When Bonnington Falls, just upstream of the confluence of the Kootenay and Columbia Rivers, became impassable as a result of isostatic rebound – the rising of the earth's surface after the enormous sheets of ice melted – white sturgeon both upstream and downstream of the falls were able to adjust to the new reality. Those in the Columbia River system downstream of the falls found new spawning channels and continued their anadromous existence that took them to the Pacific and back, while those in the Kootenay River system above the falls gave up the sea forever and adapted to a purely freshwater existence.

Both sturgeon populations had very long life spans, with individuals living 100 years or more. And for thousands of years, coastal peoples lived in harmony with these ancient creatures, taking no more than the species' low

STEFAN CLAESSON / NOAA PHOTO LIBRARY / DEPARTMENT OF COMMERCE GULF OF MAINE COD PROJECT, NOAA NATIONAL MARINE SANCTUARIES; COURTESY OF NATIONAL ARCHIVES

reproduction rates could sustain. But the human-engineered changes of the past century have proven deadly. Overfishing, dam building that has altered the flow and depth of nearly every major river in Western North America, destruction of marshes and sloughs that have long served as nurseries for young fish, as well as pollution have combined to put these relics of the dinosaur age on the path to extinction.

White sturgeon appeared for the first time in scientific literature in 1836, with the publication of Sir John Richardson's *Fauna Boreali-Americana*; Richardson named them *Acipenser transmontanus,* the "sturgeon from across the mountains". The only other West Coast species is the endangered green sturgeon.

Just 50 years after Richardson's publication, commercial fishing began and within 15 years, overfishing was threatening the species. In 1892, for example, more than 25,000 tons of white sturgeon were taken from the Columbia River and by 1899, what had once seemed an endless supply of the huge fish was nearly gone. By the late 1940s, however, the population had recovered somewhat and, though restrictions were instituted on the maximum size of fish that could be legally taken, the commercial fishery increased steadily beginning in the 1970s.

Though sturgeon fishing has been illegal in the Kootenay River system in Montana, Idaho and BC for more than 20 years, the landlocked white sturgeon population is also in steep decline. Confined to Kootenay Lake and about 270 kilometres of the Kootenay (or Kootenai, as it's spelled in the US) River between the lake and Montana's Kootenai Falls, these remarkable fish have been dramatically affected by Libby Dam, which began operation in 1974. Disrupting the natural flow, altering daily and seasonal water temperatures and burying the crucial pebbled spawning sites in

Left: A turn of the century trophy photo on the Columbia River in Northwest Oregon. Below: Anglers display their catch of white sturgeon in San Francisco Bay.

NMFS FILE PHOTO / NOAA PHOTO LIBRARY/ DEPARTMENT OF COMMERCE

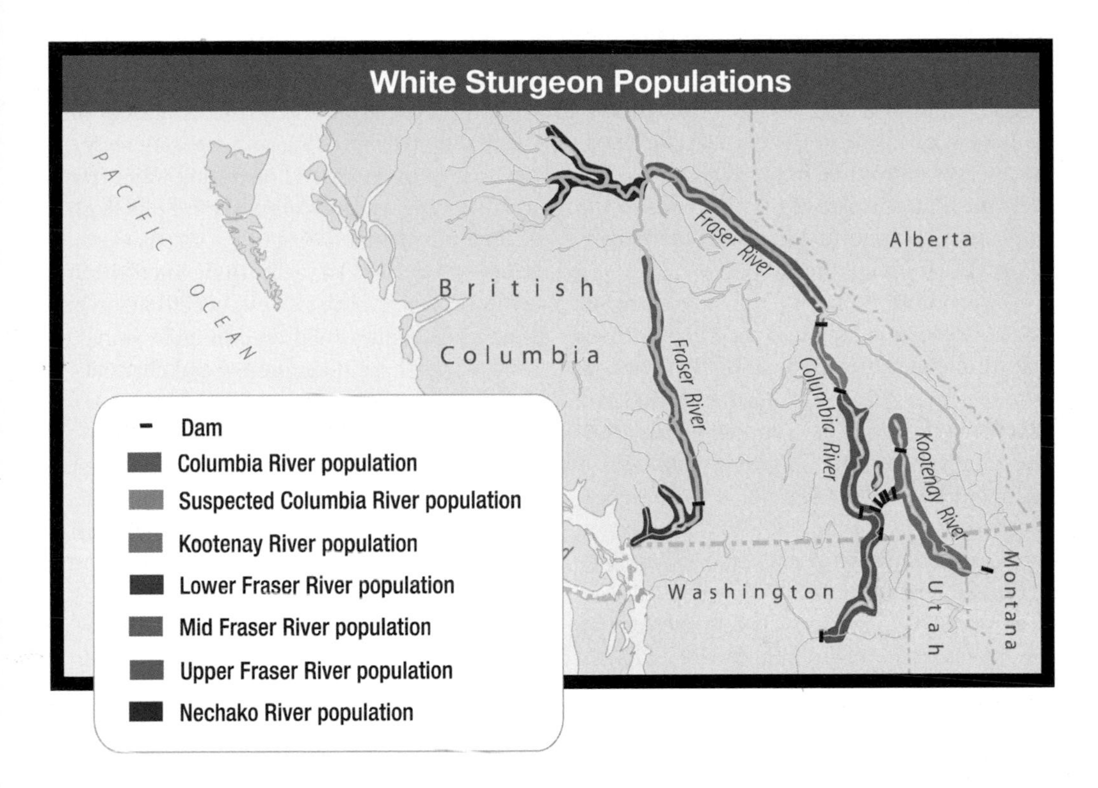

silt and sand, the dam has made reproduction in the river all but impossible. As a result, the total population had declined to approximately 1,468 wild fish by 1997, and was estimated at 1,000 in 2009, according to the U.S. Fish and Wildlife Service. And of the remaining wild sturgeon, almost none were under the age of 35. At an attrition rate of approximately nine per cent, researchers predict that all remaining wild fish will be gone by 2065 and that the wild population will be functionally extinct by 2035.

In Canada, in 2007 the Fraser River Sturgeon Conservation Society estimated that there were just 84 fish over 279 centimetres or 108 inches in length, and between 7,000 and 11,000 over 160 centimetres or 62 inches.

Toxins are another threat to these magnificent fish, for the accumulation of dioxin, PCBs and mercury is particularly high in sturgeon meat, in part thanks to their eating habits. As scavengers, sturgeon eat raw sewage, dead fish and paper mill wastes, which include dioxin. Thanks to their longevity, PCBs – which were outlawed in 1977 – are also found and along with other fish, sturgeon in the Columbia and Willamette Rivers have high levels of mercury. Though these toxins accumulate mainly in a sturgeon's liver and pancreas, the Oregon Fish and Wildlife Department has issued warnings on all fish caught in the Willamette and Lower Columbia for women of childbearing age, children under six and anyone with liver or kidney damage. Even healthy adults are warned to eat no more than one meal per month and to reduce exposure to toxins by removing the skin, fat, eggs and internal organs.

In the Fraser River, in 1993 and 1994 there was an unexplained die-off of more than 30 large, mature sturgeon. In response, Fraser River First Nations instituted a ban on sturgeon fishing and called on the government to eliminate the harvesting of sturgeon in British

Columbia. The province responded by changing recreational fishing to only catch-and-release. Despite these measures, after an initial increase from a low of 47,431 in 1999 to 62,611 in 2003, the number of juvenile fish began to drop. By 2007, the total estimate of numbers was 46,108 sturgeon, a population decrease of more than 26 per cent from four years earlier.

In an effort to save the Kootenai sturgeon that have meant so much to their people for so many millennia, the Kootenai of Idaho began to raise young sturgeon in a hatchery in Bonners Ferry, releasing them annually since 1999 into the river. Working in partnership with them, in 1998 British Columbia Environment approved the use of the Kootenay Trout Hatchery near Fort Steele, as a backup or fail-safe facility for fertilized eggs and juvenile sturgeon, to ensure that at least some fish would survive should a catastrophe occur at the Idaho hatchery. More recently, the hatchery has begun raising sturgeon young to be released into the upper Columbia River.

Close observation at the hatcheries, as well as trawling for young of the year in the lower Columbia, has led to many discoveries. Among them, that hydropeaking (the rapid increase or decrease in the release of operating water from hydroelectric reservoirs in response to a fluctuation in the demand for power) displaces white sturgeon eggs and larvae from the incubation areas. Biologists have also discovered that white sturgeon juveniles are preyed upon by minnows of northern pikeminnow and channel catfish and walleye.

Perhaps to outwit those predators, in early 2008, observers witnessed up to 60,000 sturgeon hatchlings massed in a dense "sturgeon ball" at the base of the Bonneville Dam in the Columbia River Gorge east of Portland, Oregon.

Meanwhile, the engineers at Libby Dam continue to experiment with flow capacity, in the hope that a solution to the spawning problem can be found without disrupting power production or flood control.

This aerial view looks east of the Columbia River's Bonneville Dam (near right). The dam features fish ladders that assist salmon and steelhead in their upstream spawning migrations, and these, in turn, draw California sea lions to the base of the dam.

COURTESY OF THE US ARMY CORPS OF ENGINEERS

White Sturgeon Fast Facts

Acipenser transmontanus

LENGTH & WEIGHT

White sturgeon are the third-largest sturgeon species, after Russia's beluga and kaluga. The largest anadromous individual on record weighed 816 kilograms or 1,798 pounds and was 6.1 metres or 20 feet long. The largest fish caught in Canada weighed 629 kilograms or almost 1,400 pounds, while the biggest white sturgeon in the landlocked Kootenay River population was a 159-kilogram (or 350-pound) fish captured in Kootenay Lake in 1995.

LIFESPAN

White sturgeon (which are actually gray in color; it is the flesh that is white) can live to be 100 or more.

HABITAT

Naturally anadromous, white sturgeon spend much of their lives at sea, feeding at the bottom of slow-moving bays and estuaries. However, they spawn only in fresh water. Though they have been found in coastal estuaries and rivers from central California to the Gulf of Alaska, the three main watersheds where they spawn are the Sacramento, Columbia and Fraser Rivers and their tributaries. Landlocked or freshwater populations exist in the Upper Columbia and Upper Kootenay Rivers.

MATING & BREEDING

Females do not breed until they are about 22 years of age and then spawn increasingly infrequently. Younger females spawn about every four years, while older females may only spawn every eight or nine years. However, the number of eggs they lay increases dramatically over time. In late spring and early summer, they congregate in rivers with fast-moving water, gravely or rocky bottoms and temperatures between 14 and 19 ° C or 58 to 66 ° F. The males (which mature at about 16 years of age) broadcast their sperm into the water as the females release anywhere between 100,000 and a million sticky eggs over the gravel. The brown fertilized eggs sink and adhere to the gravel. In a week or so, the tiny larvae hatch, resembling a tadpole, and drift downstream. Over the next month, they live off the yoke sac as they develop a full set of fins, rays and scutes. Then, as small juveniles, they begin to feed on insects, tiny fish and small crustaceans.

DIET

Sturgeon eat a variety of fish and crustaceans, but generally feed on dead fish, crustaceans and molluscs.

NAMES

Its Latin name, *Acipenser transmontanus*, means "sturgeon beyond the mountains"; white sturgeon are also known as Pacific sturgeon, Oregon sturgeon, Columbia sturgeon, Sacramento sturgeon and California white sturgeon. And because of the diamond-shaped plates along the sides of their bodies, white sturgeon are also called "diamond-sides".

VIEWING WHITE STURGEON

Though they are found along the West Coast from central California to the Aleutian Islands, the chances of seeing a large sturgeon in the wild are relatively remote. However, the Kootenay Trout Hatchery outside Fort Steele, BC has excellent visitor facilities, including aquariums, and offers self-guided tours, and beautifully groomed hatchery grounds where visitors are encouraged to picnic. The accessible facility is open year round from 8 a.m. to 4 p.m. Admission is free. From Cranbrook, travel 32 kilometres southeast on Highway 3 to the Wardner/Fort Steele Road and follow the signs.

Catch-and-release fishing is offered on the Fraser River between March and November.

The Grasslands

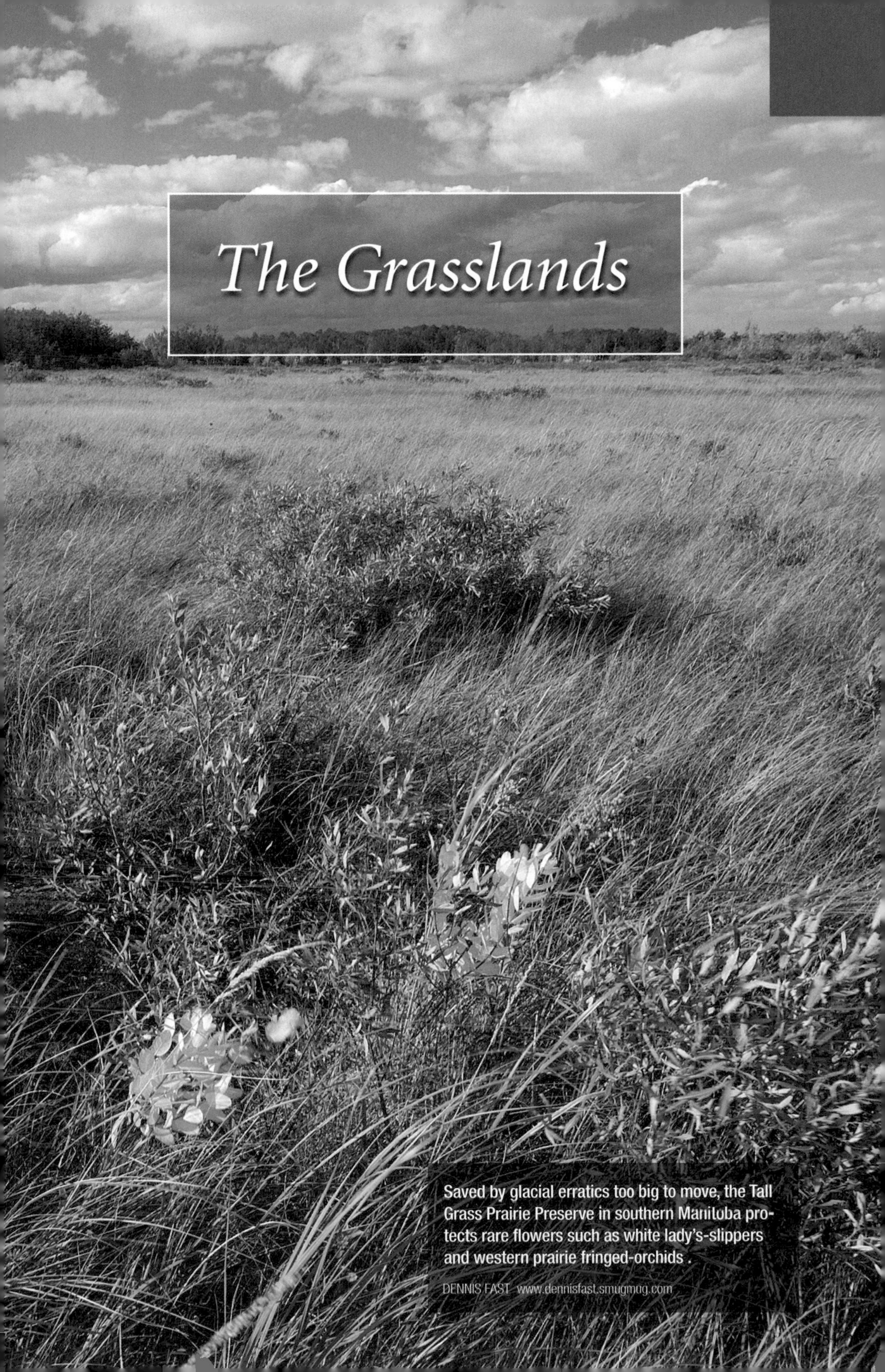

Saved by glacial erratics too big to move, the Tall Grass Prairie Preserve in southern Manitoba protects rare flowers such as white lady's-slippers and western prairie fringed-orchids .

DENNIS FAST www.dennisfast.smugmug.com

Deep-rooted and soaring skyward, big bluestem is the dominant grass of the tall grass prairie.

DENNIS FAST www.dennisfast.smugmug.com

The Grasslands

Towering Douglas-firs may be more spectacular and great swaths of fireweed more colorful, but grasses are, without any qualification, the most important plants in the world.

Humankind would not have evolved the way it did without grasses, for they provide our cereal grains – wheat, barley, oats, rice, maize, sorghum and millet – and these, in turn, supply 75 per cent of our food energy and half our protein.

Grasses also feed the animals we rely on for much of the rest of our sustenance – our meat, milk and eggs. In fact, it appears they may even be responsible for the very evolution of those animals. Recent research indicates that the origin of grasses – and the grazing animals that eat them – may be much older than previously believed, and it seems increasingly clear that the plants of the grasslands and the grazing mammals they sustained evolved together, each spurring changes in the other over millions of years. It's a perfect example of symmetrical coevolution, which scientists now believe began long before the demise of the dinosaurs 65 million years ago.

That timeline was revamped recently, when analyses of fossilized dinosaur dung in India revealed the remains of at least five types of grasses. Suddenly, it was clear not only that dinosaurs had dined on grass, but that grasses had evolved much, much earlier than previously believed. Swedish paleobotanist Caroline Strömberg, who identified the plants – early ancestors of rice and bamboo – said the fossilized faeces likely belonged to titanosaurs, which were common in the area in the late Cretaceous period. Her research pushes the oldest known grass fossils back more than 30 million years, for Strömberg believes the origins of grasses may go back as much as 83 million years.

American evolutionary biologist Elizabeth Kellogg said Strömberg's find "completely revises what we've thought about the origin of grasses", and also gives clues to a mystery regarding an ancient class of mammals called gondwanatheres. These groundhog-sized creatures, which appeared in the waning days of the dinosaurs, had spread to many parts of Gondwana, the ancient southern supercontinent. Fossilized remains of gondwanatheres have been found in India, Madagascar and South America, all pieces of what was once Gondwana, or Gondwanaland. (Its northern counterpart was Laurasia, which included ancestral North America and Asia.)

Gondwanatheres had long front teeth with a flat chewing surface, characteristics found in modern grazers such as horses and bison. Prior to Strömberg's discovery, paleontologists

MELANIE FROESE

Based on skeletal remains, paleontologists believe gondwanatheres may have looked like this, with teeth adapted to eating grasses.

couldn't understand why gondwanatheres had teeth capable of enduring constant abrasion when there didn't seem to have been grasses for them to eat.

Despite these Cretaceous beginnings, grasses evolved slowly, likely because of global climatic conditions and the positioning of the Earth's continents. During the Jurassic period (between 208 and 146 million years ago) Gondwana and Laurasia had begun to break apart. Once clustered around the equator, they drifted apart, moving on the Earth's tectonic plates. Still, even by the Cretaceous period, between 146 and 65 mya, the global climate was still considerably warmer than it is today. Tropical conditions extended into the central United States and it was so warm in the polar regions that trees grew in both the Arctic and Antarctic, and alligators lived on Ellesmere Island in Canada's far north.

As the continents moved apart, creating the primeval Atlantic Ocean, the continental interiors became drier and, beginning about 50 million years ago, cooler as well. In the Americas, cooler, drier climates and the accompanying increase in naturally-occurring fires, forced grasses to develop survival mechanisms as they moved north and south from the tropics. These techniques included deeper roots, tougher tissues, hormonal growth regulation and symbiotic relationships with soil organisms.

In North America, grazing mammals were forced to simultaneously evolve in order to deal with the changes in the grasses. Early grazers – including such remarkable beasts as rhinoceroses, tapirs, brontotheres, small horses, camels and oreodonts – developed hard teeth, just as the gondwanatheres had millions of years before, with enamel ridges on their crowns – as well as complex digestive systems. Longer legs allowed them to outdistance the great carnivores of the northern plains – the lions, sabre-toothed tigers, short-faced bears and dire wolves. These evolutionary changes continued over time; grasses became ever hardier and herbivores developed more complex grazing systems. By the time *Bison priscus*, which migrated from Asia sometime around 350,000 BP, settled in the Mammoth Steppes of Beringia and Alaska, grazing mammals were wonderfully advanced, with high-crowned, enamelled teeth that continued to grow at the roots, long, broad faces to accommodate the teeth, four-chambered stomachs, long legs and hard hooves.

Earth's continents had long since reached the positions they occupy today and, with a global climate that was much cooler, grasslands accounted for nearly a quarter of the globe's terrestrial vegetation. In the northern hemisphere, great glaciers waxed and waned, and in North America's heartland, from the western mountains to the forests of the east and from

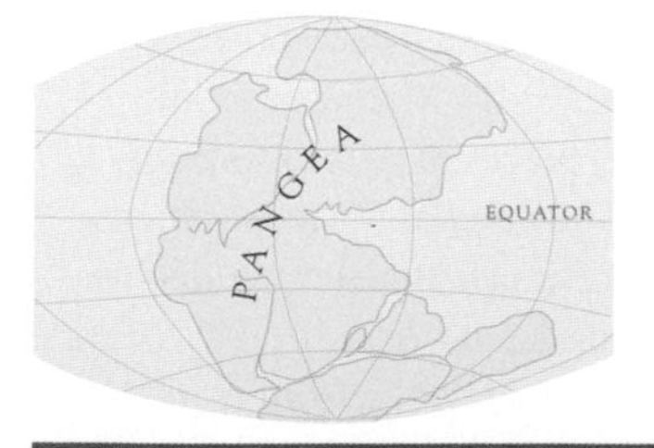

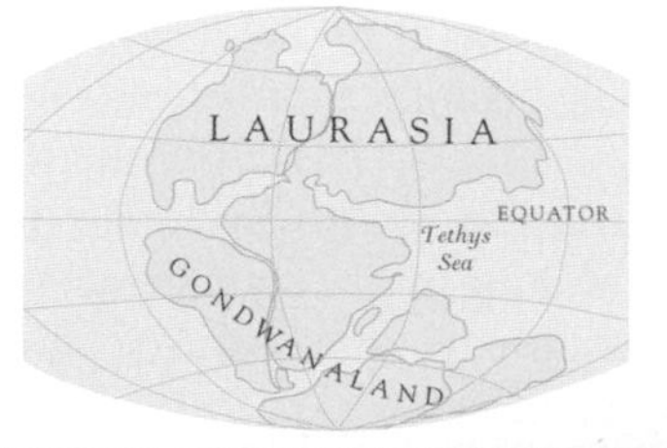

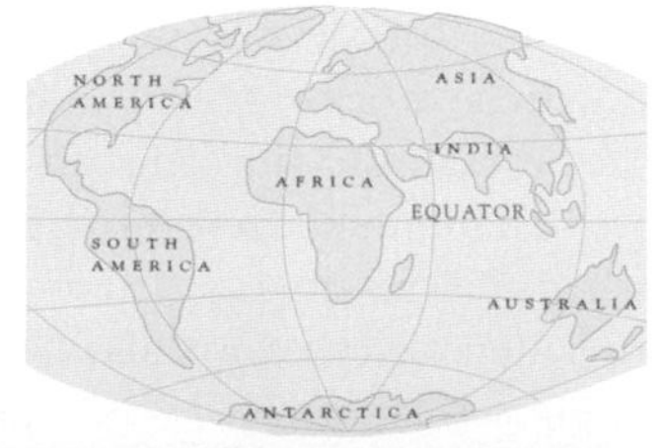

Pangaea was not the first of Earth's supercontinents, but it is the one we know best, for as it rifted apart, it created the continents we know today.

Grasses of burnished gold stretch to the horizon in Badlands National Park, South Dakota.

COURTESY OF WING-CHI POON / CREATIVE COMMONS ATTRIBUTION SHAREALIKE 2.5 / WWW.CREATIVECOMMONS.ORG

the aspen parklands of the north deep into what is now Texas, grasses and prairie flowers, or forbs, ruled.

This is not to say that grasslands were everywhere the same. Far from it. In the rain shadow of the Rockies, drought-hardy foot-high buffalo grass, needle grasses, sagebrush and several species of small cacti survive an average annual rainfall of as little as 12 inches or about 30 centimetres. This short grass prairie, long the domain of the huge plains grizzly, also harbored large populations of swift foxes, black-tailed prairie dogs and greater sage-grouse. Today, all are greatly diminished in the US and listed as endangered in Canada.

To the east, beginning about the 100th meridian, where increased precipitation supports a greater abundance of species and more vigorous growth, is North America's mixed grass prairie, the land of the incomparable pronghorn, which, like the plains bison, was nearly extinguished as the Great Plains were settled. Here, too, is the burrowing owl, listed as endangered in Canada and declining in many of its former habitats in the central US. Among more than 150 species of grasses and flowers found here are blue grama grass, June grass, western wheatgrass and the magnificent and endangered western prairie-fringed orchid. In well-watered swales or draws, little bluestem stretches waist high, often among clumps of aspen, bur oak and birch and even white spruce or pine (found along Nebraska's Niobrara River and in South Dakota's Black Hills).

Both the short grass and mixed grass prairies, as well as many of their dependent species, have been enormously impacted by agriculture, ranching, industrialization and urban sprawl. Yet neither is as imperilled as the grasslands of the eastern Great Plains. Here, where a north-south swath of exuberant growth once stretched from southern Manitoba to San Antonio, Texas, is one of the most endangered ecosystems on Earth. Vast tracts of big bluestem once stretched higher than a man on horseback, turning the fertile plains into a sea of grass and rendering the first Europeans to see them all but speechless. "For miles we saw nothing but a vast prairie of what can compare to nothing but the ocean itself," wrote one. "The tall grass looked like the deep sea; it seemed as if we were out of sight of land . . ."

This was one of the most bountiful ecosystems on Earth, richer by far than the towering coastal forests and rivalling the plains of Africa with their huge herds of wildebeast, springbok and elephants. By the time Europeans arrived, an estimated 30 million plains bison roamed the grasslands, while perhaps 40 million pronghorn populated the short grass and mixed grass prairies. These huge herds were just the most obvious of the thousands of species that thrived on the Great Plains. Here, too, were birds by the million; greater prairie-chickens danced to herald spring; meadowlarks filled May mornings with their flutelike mating songs and tiny songbirds flashed among the greening thickets or created hidden symphonies in the grasses. In the summer months,

First on foot and later on horseback, Native North Amerians drove herds of bison over steep embankments called buffalo jumps.

NATIONAL ARCHIVES CANADA / ACCOUNT NO. 1946-110-1 / C-000403 / DON ET MADAME J.B. JARDINE

long-legged, big-eyed burrowing owlets lined the entrances to their underground nests, and whooping cranes, with their primeval migration calls, pointed the way south as autumn approached.

Other creatures thrived on the bounty of the grasslands. Plains grizzlies stalked the enormous herds of bison, often using riverine corridors to travel from the Rockies as far east as the lower Missouri and Manitoba's Red River, while prairie wolves roamed the peripheries of the herds, searching for stragglers.

Beneath the earth's surface, billions of prairie dogs and ground squirrels built their tunnels and nests, in the process aerating the densely rooted soil and providing homes for other creatures, including black-footed ferrets and burrowing owls.

The coming of Europeans changed all this. In less than two decades between the mid-1860s and the mid-1880s, bison were slaughtered by the millions. Losing their main prey, grizzlies and wolves, on the verge of starvation, withdrew to the western mountains or turned in desperation to the herds of cattle that had begun to replace them. By 1920, the great bears occupied less than five per cent of their original grasslands range.

Prairie dogs suffered the same fate. Driven from the landscape by campaigns of poisoning, the intensive agriculture that often followed, and urbanization, during the 20th century an estimated 98 per cent were destroyed. And as they disappeared, so did the creatures that depended on them for nests and shelters from prey.

North America's grasslands once covered nearly 40 per cent of the continent; today, less than two per cent of the tall grass prairie and seven per cent of the mixed grass prairie remain, while the short grass prairie has been declared endangered in a number of states and provinces, including Colorado and Alberta. Realizing at last how many species depend on these remarkable ecosystems, many organizations, as well as national, state and provincial bodies are now setting aside some of the existing remnants of native grasslands and re-establishing native grasses and forbs in areas that have been farmed or ranched in the past. Recreating North America's magnificent grasslands in all their diversity and complexity may well be impossible, however, for even where native plants have been carefully introduced, there are generally not more than 30 or 40 species where once 200 or more grew. Nevertheless, in reserves and parks from Manitoba in the northeast to Texas in the southwest and in hundreds of places in between, the progress made to date gives one hope that someday our grandchildren may gaze out over a blanket of wild flowers in the spring or waving grasses in the fall and understand what we almost lost.

Big Bluestem

Andropogon gerardii

GLEAMING purple-blue in the late summer sun, sustaining vast herds of bison and creating such dense sod that early settlers built their homes with it, big bluestem was once synonymous with the tall grass prairie. With roots extending nearly nine feet or three metres into the earth, and seed heads soaring an equal distance skyward, it is the dominant grass of the tall grass prairie. Prior to European settlement, it often grew in such dense stands that it shaded out other plants, creating large swaths of pure bluestem. For the great herds of bison, as well as elk and deer, nothing could have been better, for big bluestem has been called the "ice cream" of North America's grasses, providing top-notch forage and yielding between three and five tons of hay per acre.

Big bluestem is a warm-season grass, beginning its annual growth in June when temperatures rise and rapidly stretching to full height. The tall, slender stems turn from green to blue in mid-summer and are topped by a tri-pronged seed head that turns first reddish-orange, then bronze in the fall. The shape of the seed heads have earned big bluestem many nicknames, including "turkey feet" and "beard grass"; its scientific name – *Andropogon* – comes from the Greek *andro,* meaning "man" and *pogon*, meaning "beard".

Though its seeds drop to the ground, the plant also grows from its extensive root system, forming large clumps and earning yet another name, "bunch grass". Because of its excellent forage qualities, and unparalleled ability to hold down soil against relentless prairie winds, big bluestem is once again being cultivated as fodder in both the US and Canada.

Big bluestem once stretched across much of North America, from the moist valleys and well-watered draws of southern Saskatchewan east to Quebec, south through Montana and Wyoming to Arizona and northern Mexico, and eastward to Florida and Maine.

Today, only remnants of wild big bluestem remain, largely in reserves. But a growing demand has prompted plant breeders to invest in bluestem cultivars, and governments to encourage the planting of native species along highways and pipelines.

To see big bluestem at its glorious best, visit preserves from mid-summer to early fall.

DENNIS FAST www.dennisfast.smugmug.com

The Aspen Parkland

YELLOW-HEADED and red-winged blackbirds flash among the burgeoning bulrushes and a blue-winged teal skirts the edge of the marsh. The clear rising notes of a western meadowlark fill the air. Across a nearby meadow a moose, followed closely by two gangly calves, emerges from a thicket of aspen and birch. Like shadows in a rising mist, they drift toward the pond.

This is spring in North America's aspen parkland, a place of rolling hills, verdant vales and "prairie potholes", as the thousands of small marshes that dot the region are often called. Arcing nearly 1500 kilometres or about 950 miles from northeastern British Columbia across the prairies to northwestern Minnesota, and south along the mountain valleys to Colorado, the parkland is known by many names. Environment Canada calls it a "transition zone" between two ecosystems, a confluence between the central grasslands of the Great Plains and the vast northern boreal forest. Yet this is more than just a meeting place; it's a distinct ecosystem, and one that's so crucial to the continent's waterfowl and wetlands species that many call it North America's "duck factory".

All this is a legacy of the last glaciation. Scraped to the underlying bedrock by the advancing ice sheets, the parklands were then molded and sculpted by meltwater laden with soil and sand as the world warmed again.

The aspen parkland is markedly different in different places, and this variety is a good part of the reason for its prodigious fecundity. In Minnesota and southeastern Manitoba, the parkland borders an expanse of tall grass prairie. Many believe the two are simply different faces, or phases, of the same environment, for both thrive on the deeply fertile soil left by Glacial Lake Agassiz and on an annual rainfall that averages 18 inches or almost 50 centimetres. Left untrampled by what were once vast herds of bison, fired only for agricultural purposes and given enough time, the riparian forest that lines the Red River and its many small tributaries will inexorably turn the tall grass prairie into parkland. Recognizing this, Minnesota refers to its share of the region as the northern tallgrass aspen parkland. And many Environment Canada maps show the two ecoregions as occupying the same territory.

Creating a third layer over much of both, are the region's wetlands, which are perhaps even more important than the grasslands and dappled woodlands. They include the oxbow lakes of the Red and Assiniboine River valleys, and the lush marshes that rim Lake Manitoba's southern shore to the thousands of prairie potholes that dot the rolling landscape from Iowa northwest to British Columbia.

These small marshes and hilly terrain – which geographers know as "knob and kettle topography" – owe their existence to the Laurentian ice sheet. Though the ice melted rapidly, relatively speaking, at the end of the last glaciation, its retreating edges regularly

A blue-winged teal rests on one leg.

DENNIS FAST www.dennisfast.smugmug.com

stalled. When that happened, the resulting meltwater, which was filled with clay, silt, sand, gravel and even huge boulders, created mounds and ridges along the stationary edges of the ice.

Large blocks of ice also calved off the thinning edges and were often buried under sediment. Insulated by thick layers of soil, these chunks of ice took years or even decades to melt. When at last they did, the overlying sediments collapsed, creating depressions in the hilly terrain. Surrounded by the moraine remnants or "knobs", the depressions usually filled with water. Sometimes these were seasonal wetlands, formed by spring run-off and disappearing through evaporation during the summer. But thousands became permanent wetlands. In either case, they served as perfect nurseries for dozens of species of dabbling and diving ducks, as well as swans, geese, teal and other water birds, an estimated 50 to 80 per cent of North America's waterfowl.

The parkland is named for its trembling aspen, but at its southeastern extent, it once boasted a forest that might have been more at home in Eastern North America. Towering American elms, huge bur oaks, giant cottonwoods and arching Manitoba maples once grew along the banks of the Red River and its many tributaries, including the Pembina, Morris, Rat and Assiniboine Rivers. An understorey of shrubs might have made these mature riparian forests all but impenetrable, were it not for the millions of bison that once called the prairies home. Just as grazing cattle keep undergrowth at bay and prune lower branches in wooded areas today, browsing bison turned the forests of the past into open woodlands, allowing warm-season grasses to flourish along the rivers. The parkland landscape was shaped by drought, wind and fires, which prevented trees from growing in areas, or periodically burned them, allowing fields of grasses to grow.

Arching northwest from the Red River Valley, the aspen parkland rises slowly in elevation. Meadows are interspersed with bluffs of aspen and bur oak and willows line the streams. Much of most fertile land is now cultivated, but glimpses of the parkland's many faces can still be seen (see Viewing the aspen parkland on page 263). Throughout the region, rivers and streams have created steep valleys that are protected from some of the harsher elements and support forests of poplar, spruce, birch, and willow along their shady banks and dry, grass-covered slopes facing the sun. Among the most dramatic is Alberta's North Saskatchewan River Valley, which stretches 1287 kilometres or 800 miles from the Rockies, through the province's capital, Edmonton, before joining the South Saskatchewan, It then continues for another 550 kilometres or 342 miles across the Prairie Provinces to Lake Winnipeg.

Summers in the parklands are short and warm, and winters long and cold; for plant life to survive, it must be hardy and able to tolerate occasional droughts and floods. The meadow areas are covered in a variety of grasses, sedges and wildflowers; snowberry and wild rose bushes are common. These fields have rich soils from years of plant materials building up and breaking down. With woodlands

DENNIS FAST www.dennisfast.smugmug.com

COURTESY OF JEAN IRON

Above: A least bittern strikes a characteristic pose as it hides in wetland vegetation.

Left: Plains bison at Riding Mountain National Park.

dominated by aspen or spruce, with jack pine and lodgepole pine growing in sandier areas and willow preferring areas with wetter soil, a range of species were able to thrive. An understorey of shrubs, herbs and other groundcover grows in aspen groves, where plenty of sunlight reaches the ground. Spruce groves do not support as wide a variety of plant life, because their thick boughs restrict light and their decaying needles create an acidic soil that many plants cannot tolerate.

The result, for millennia, was a region that was home to a huge range of species: grizzly and black bears, wolves, lynx, wolverines and cougars, moose, deer, elk and bison, beavers, muskrats, martens and fishers, as well as hundreds of bird species from tiny ruby-throated hummingbirds to majestic bald eagles. Many birds, such as grebes, yellow rails, and sharp-tailed sparrows, have their northern breeding habitat in the parklands. Some, like the threatened least bittern, are very rare. Apart from the California Coast, Manitoba's southern wetlands have long been among the few places in the West where these small herons nest.

Rabbits and hares dined on saplings, while burrowing animals such as ground squirrels, pocket gophers, badgers, foxes and coyotes created mounds of exposed soil where new seeds could become established. Sharp-tailed grouse, pine grosbeak and others helped to distribute the seeds of shrubs and trees, allowing them to spread.

Everything changed with the arrival of Europeans. Fur traders emptied the rivers and streams of beavers, muskrats, martens and fishers by about 1820, plains bison were all but gone by 1870 and wolves and lynx were hunted to extirpation by the mid-20th century. Beginning in the 1870s, settlers snapped up land in the fertile parklands, and logging became big business. Logging and milling operations targeted huge white spruce, bur oak, jack pine, white birch and balsam poplar. And even today, despite growing concern about the little that remains in a natural state, trembling aspen is increasingly under attack on many sides. Though it's the most widely distributed tree species in North America and the most abundant deciduous tree in Canada's boreal forest, natural aspen parklands are greatly diminished. Just 10 per cent of virgin parkland remains in Manitoba and only half that in Alberta.

Elsewhere, aspen trees are swiftly dying of a sharp spike in insect predation and disease. In Colorado, for example, between 2002 and 2008, the number of dying aspen was triple the norm. And by 2006, concentrated patches of dead trees covered an area of 200 square miles or 56,000 hectares.

The trees in Colorado may be suffering from the same problems now seen in many stands in Canada. Though the suppression of naturally occurring fires following the settlement period initially encouraged the growth of trees, for several reasons, things have dramatically changed in the decades since. While a combination of grasslands fires and grazing bison once controlled aspen encroachment onto prairie grasslands, the elimination of these natural disturbances has led to overmature aspen stands that rapidly deteriorate. This decay leads to an inability to reproduce by suckering, which allows stands of aspen to break up, resulting in a prolonged transition to shrubs or softwoods.

Since technological advances opened the door for aspen to be used for high quality paper production in the mid-1950s, aspen forests have also been widely harvested across Western Canada. Aspen is also used for a variety of wood products, including oriented strand board – OSB, as it's known, has largely replaced plywood and waferboard in new residential construction across North America – as well as lumber. Today aspen accounts for almost 40 per cent of the gross volume of forest products sold each year. In one region of Alberta, the harvest and production of roundwood hardwood, which principally uses aspen, increased from under 800,000 cubic metres in 1970 to more than eight million cubic metres in 1995.

Moreover, due to a combination of warmer, drier weather, which has led to increased forest fires, along with forest tent caterpillar infestations, studies have shown that average stands of aspen decreased by fifty per cent in the late 1970s and by another thirty-four per cent in the late 1990s. In short, right across the parklands, aspen forests have been noticeably dying back over the past 40 years. These changes have raised alarms in many constituencies about the future of the parkland forests.

Infestations of tent caterpillars, pictured here, have severely impacted aspen forests.

IMAGES BOTH PAGES: DENNIS FAST www.dennisfast.smugmug.com

In the western parklands, oil and natural gas have been a lucrative source of income for the provinces and states, and exploration and drilling have significantly altered the natural landscape. Large cities and their requisite infrastructure have also broken up the parkland region.

Aside from direct changes made to the environment, other actions, such as the decimation of the bison herds and suppression of natural fires, have upset the region's equilibrium. In some areas, invasive introduced plant species have become a problem. In others, aspen and spruce groves are encroaching on grasslands because the natural forces that kept the climax vegetation from taking over have been dramatically reduced. In protected natural areas, proscribed burns are now used to take the place of lightning to keep the trees in check.

Because of its diversity, the aspen parkland, and many of its dependent species, require large, undisturbed areas to remain healthy and balanced. Instead, the remaining natural parkland is spread throughout the region, creating islands of wilderness, placing stress on animal populations and destroying natural migration corridors.

VIEWING THE ASPEN PARKLAND

Among the best ways to view the diversity of the aspen parkland is to follow Canada's Yellowhead Highway (No. 16) from Portage la Prairie in southern Manitoba northwest to the Rockies. Along (or just off) its 1500 kilometre or 980-mile route, travellers will find dozens of places to explore wild aspen parkland, including:

Delta Marsh: Rimming the south shore of Lake Manitoba, north of Portage la Prairie, this is one of North America's largest lacustrine marshes, and recognized as a wetland of international significance. Separated from the lake by a beach ridge more than 40 kilometres or 25 miles long, it serves as a bottleneck in one of the hemisphere's busiest migration corridors, attracting hundreds of thousands of birds every spring and fall, from warblers, finches and song sparrows to endangered piping plovers (see page 298) and throngs of migrating geese, swans and ducks. It's also home to one of the busiest bird banding stations in North America. There is a trail with a boardwalk and viewing towers.

Riding Mountain National Park: Part of the Manitoba Escarpment, this "island" reserve covers 2973 square kilometres or 1,150 square miles of rolling hills and valleys, stretching west of the dramatic Manitoba Escarpment. Though surrounded by agriculture, the park includes expanses of boreal forest, huge meadows of grasslands, and significant wetlands and is home to wolves, moose, elk, black bears, hundreds of bird species and a captive bison herd. It has numerous hiking trails, and a full range of visitor services including swimming and boating, accommodation, restaurants and shopping. From the Yellowhead Hwy travel north about 30 km or 19 miles on Hwy 10 (which continues north through the park).

Last Mountain Lake National Wildlife Area: This "Wetland of International Importance", located just east of Lanigan, Saskatchewan, is in the heart of the central flyway. It serves as an important migratory stopover for more than 280 bird species during migration, including cranes, geese and ducks, as well as birds of prey, songbirds and shorebirds that spend from a few days to a few weeks during spring and fall migrations. More than 100 species breed here and the bird sanctuary offers habitat for nine of Canada's 36 species of vulnerable, threatened and endangered birds, including Peregrine falcons, piping plovers, whooping cranes, burrowing owls, ferruginous and Cooper's hawks, loggerhead shrikes, Baird's sparrows and Caspian terns. Turn off the Yellowhead onto Hwy 20 and go south for 47 km to the local road to the NWA, which has a self-guided wetlands trail with a boardwalk, a grasslands trail, a 16-km driving route, an interpretive kiosk and a picnic site.

Waneskewin, on the outskirts of Saskatoon: This is an excellent view of the aspen parkland and its people prior to the coming of Europeans. Located three km or two miles north of Saskatoon on the North Saskatchewan River, it shows how North America's First Nations lived with the bounty of the aspen parkland for more than 6,000 years. Waneskewin has an excellent interpretive centre and a gift shop, as well as many trails and a picnicking area.

Elk Island National Park, in central Alberta: Less than an hour east of Edmonton, this oasis of lakes, wetlands and rolling hills is fenced on all sides. Within its boundaries, are herds of free roaming plains bison, wood bison, moose, deer, and elk and more than 250 species of birds. Elk Island offers camping and picnicking facilities, an interpretive centre and many trails.

Moose, relatively recent arrivals in North America, are among the continent's largest ungulates.

DENNIS FAST www.dennisfast.smugmug.com

Though naturalists have been aware of the importance of wildlife corridors for decades, it took the construction of British Columbia's Coquihalla Highway, and the enormous number of collisions with migrating deer and elk that immediately followed its opening, to make most Canadians realize that humans and waterfowl are not the only animals that create regular routes between places for hunting, seasonal migrations or other aspects of life. In *The Hidden Life of Dogs*, her book on the migratory habits of several species of dogs, author Elizabeth Marshall Thomas wrote about wolves on Baffin Island; so regular are their hunting circuits that over generations they had worn paths into solid rock.

To try to restore some of the balance, groups such as the Nature Conservancy and the Canadian Parks and Wilderness Society are trying to protect large pieces of the remaining natural landscape, and are advocating for the creation of corridors of undeveloped or re-naturalized land between these protected areas to ensure the long-term viability of animal populations. To this end, they are attempting to work with government, rural communities, and private landowners, impressing on them the benefits of conserving biodiversity and ecological integrity in the parklands region.

The Tall Grass Prairie

THE MAGNIFICENT, and magnificently fertile, tall grass prairie owes its being to what classical Greek philosophers believed were the four fundamental elements: earth and air, water and fire. Today's scientists have identified 92 naturally occurring elements on Earth and synthesized another 20, but intriguingly, the fundamental four of classical Greece were the key ingredients for the development of North America's lush, bountiful tall grass prairies.

When one thinks of the ancestral Great Plains, one imagines grasslands as far as the eye can see, capped by a great dome of brilliant blue. The land is an ever-changing palette of color as the seasons turn, its rolling hillsides covered with seemingly endless herds of bison, its riverside copses home to elk, deer and those great hunters, the plains grizzly and prairie wolf. In the spring, the air rings with the music of mating songbirds and in the autumn the skies are filled with migrating flocks of waterfowl so large they threaten to blot out the sun. In short, this was Nature at its finest, throbbing with life, and unquestionably, elementally, earth and air.

But this vibrant ecosystem would not have developed without the other elements of ancient Greece – water and fire.

Water was crucial in three forms: ice, an enormous lake, and rain. Between about 25,000 and 10,000 years ago, enormous sheets of ice, created by global cooling, covered almost all of Canada and much of the northern United States. This was not North America's first glaciation; far from it. Scientists believe continental ice sheets formed this way many times over the past two million years. Some, such as the Illinoian glaciation about 160,000 years ago, virtually inundated the northern Great Plains, covering most of Illinois and stretching south to southern Missouri.

However, the most recent glaciation, the

DENNIS FAST www.dennisfast.smugmug.com

At Manitoba's Tall Grass Prairie Preserve, lush grasses are surrounded by aspen woodlands.

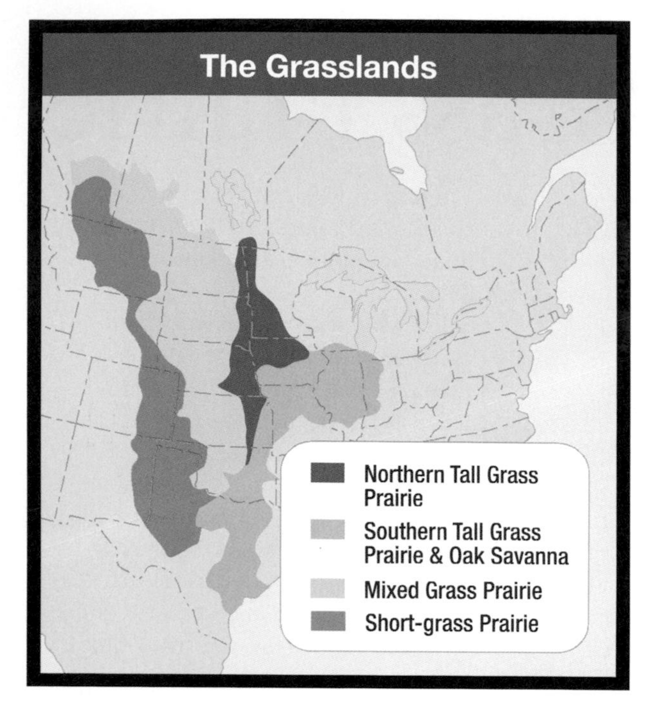

Wisconsin glaciation, is the one we know best, for its indisputable imprints are everywhere on the land. Among these are many of the features that distinguish the northern tall grass prairie, as well as the northern extents of the mixed grass and short grass prairies to the west. These glacial features include end moraines, alluvial fans, drumlins, prairie potholes, beach ridges, glacial spillways and erratics. For geologists, these features are like chapters in a glacial textbook and over the past century, they have been examined, studied and increasingly understood.

Though glacial features can be also seen in the Rockies and along the Pacific Coast, in few places can they be analysed as easily as on the Great Plains. From a hub west of Hudson Bay (as well as from other centres east and north) the enormous Laurentide ice sheet grew. By 18,000 BP, it covered virtually all of Western Canada, much of Minnesota and North Dakota, about half of South Dakota (as well as the entire Great Lakes and many of the northern states to the east). In the Western US, the ice margin curved north, outlining what would eventually be the modern Missouri River Valley and then, in western Montana, turned sharply north into Alberta, before heading west again toward the Rockies, where it collided and coalesced with the Cordilleran ice sheet.

For about three millennia, temperatures stabilized and the glacier margins were relatively stationary; the rate of growth at the centre of the ice sheet (provided by new snow each winter) was approximately equalled by its rate of summer melting along the edges. These margins are quite obvious today, for they are marked by moraines – hills or high ridges of clay or sand that formed from the sediments flowing off the ice in the glacial meltwater. Even more important to the plants that would eventually grow on them, these moraines not only mark the extent of the ice at a particular time, their makeup also indicates the direction of the flow of each lobe of ice, which in turn determines the type of soil each contains. And these things helped to influence whether farmers would one day find them suitable for agriculture, or whether tall grass prairie would find a small haven against development.

Sandy, gravelly soil and steep slopes – a legacy of ice that had travelled over Cretaceous limestone – often combined to make the moraines unattractive to farmers; some were overlooked or used only for grazing cattle. And in southwestern Minnesota and northwestern Iowa, the gray till of the Des Moines lobe, which had crossed Cretaceous shale in Saskatchewan and North Dakota, combined with the significant slopes of the Bemis Moraine to make at least part of the region both perfect for tall grass prairie and perfectly unsuitable for farming, thus saving a section of this remarkable ecosystem.

Glacial erratics – boulders transported by the ice and dropped when it melted – also helped to preserve small parcels of grassland. Not only can they often be traced back to their northern points of origin, but they were sometimes so large they proved virtually impossible to clear. In Manitoba's southeast corner, near the tall grass prairie's northern limit, the glaciers stalled for a time and spat out huge granite boulders from the Laurentian Shield far to the north. These glacial erratics (or "buffalo rubbing stones", as some would later call them) made farming impossible when the region was settled in the late 19th century. And thus, another small piece of northern tall grass prairie survived.

When global temperatures began to rise about 15,000 years ago, the huge Laurentide ice sheet began to melt. Though its disappearance took thousands of years, the immediate result was water – and lots of it. For the first 3,000 years, meltwater from the Keewatin dome flowed east and south, along the Missouri and Mississippi River Valleys and into the Gulf of Mexico. But by about 12,000 years ago, the ice margin had retreated north of the continental divide. From this point, which here is just south of Lake Traverse on the Minnesota-North Dakota border, the land sloped north. Dammed on the north by the enormous ice sheet, water pooled along its southern edge. As the ice melted, the great lake grew, spilling south whenever its level grew higher than the height of land (or later east, into the Great Lakes, and even north into the Arctic Ocean about 10,000 years ago), but always growing again. Eventually, Lake Agassiz covered an area greater than the combined area of today's Great Lakes.

Lake Agassiz was not the only glacial lake on the northern plains. Several other, smaller lakes formed at the edge of the ice to the west and north, but since all were upstream from Lake Agassiz, which occupied the low centre of the continent, when the ice dams that contained them gave way, the water came thundering down, adding to the huge lake. The torrents

Okotoks Rock, located just outside the southern Alberta town of the same name, is purported to be the world's largest glacial erratic.

JERRY KAUTZ

Konza Prairie Preserve, with its lovely mosaic of woodlands and grasslands, lies in the Flint Hills of northeastern Kansas.

COURTESY OF EDWIN OLSON / CREATIVE COMMONS ATTRIBUTION SHARE-ALIKE 3.0 / en.wikipedia.org creativecommons.org/licenses/by/3.0

carried huge loads of sediment and, where they spilled into Lake Agassiz, built large deltas at the edge of the enormous lake.

Geologists have mapped thirty-three such deltas around the shifting edges of Lake Agassiz; the largest of these are western Manitoba's Assiniboine Delta, with the magnificent, shifting dunes of its Spirit Sands region, home to the endangered northern prairie skink (see page 304) and, farther south, the Sheyenne Delta, which includes the Sheyenne National Grassland area of North Dakota. Today this rolling prairie, dotted with groves of bur oak, resounds in April with a sound rarely heard in the twenty-first century – the drum rolls of mating greater prairie-chickens.

Lake Agassiz left many other clues to its existence, including its watery remnants, Lake Winnipeg and Lake Manitoba. The former is the eleventh-largest lake in the world and both have supported magnificent fisheries (including endangered lake sturgeon, see page 308) for millennia. But just as important was the thick soupy layer of clay the lake laid down along the broad Red and Bois de Sioux Rivers, which mark the border between Minnesota and North Dakota before flowing north into Lake Winnipeg. In time, this would become one of the most fertile areas in the world, a veritable cornucopia, North America's Fertile Crescent.

Ice and water had ruled for more than 15,000 years. Now the warming air and, ultimately, fire were poised to take over. In succession, spruce and tamarack gave way to poplar and black ash, then to balsam fir, birch, elm, oak and hazel and finally to grasslands. To the south, beyond the reach of the ice, a mosaic of oak woodlands, grasslands and wetlands grew up between the Flint Hills of Kansas and San Antonio, Texas. Today, this oak savanna is known as the "crosstimbers and southern tallgrass prairie" or the "blackland prairie".

Now the stage was truly set for the last of the four fundamental elements – fire – and the creation of what has been called the most magnificent grassland in the world. North Americans know all about climate change and global warming; they're in the news every week. But though our warm springs and hot summers seem dramatic, they're just a hint of the hot, dry climate that was the norm between 8,500 and 4,000 years ago. This was the peak of the Hypsithermal, the long global warming that released North America from the last of the glaciers and then turned much of the continent's interior into a baking inferno. The woodlands and tall grasses that had been established across most of the northern Great Plains gave way to sage, short grasses and cacti, with wolf willow and cottonwood along the draws and ravines, where water would briefly collect in the spring or following a rain. For thousands of years, much of central North Dakota would have looked like the short grass prairies of southern Wyoming.

But just as it does now, even during the height of the Hypsithermal, more snow and rain fell in the eastern Great Plains than it did farther west. Here, the tall grasses hung on, sending their roots deep into the fertile soils to capture the moisture below.

For millennia, grassland fires have been started naturally by lightning. Surprisingly perhaps, these fires seem to have occurred more often during the Hypsithermal, for long periods of drought are often punctuated by powerful summer storms, complete with lightning that easily ignited the tinder-dry grasslands. It took the people of the plains little time to realize that in the wake of such storms, the dry grasslands, their thick thatch burned off and nourished by the rain that often followed, turned green almost overnight. And on the heels of that new growth came the bison – in the hundreds or even the thousands.

Soon, firing the plains became a rite of spring. As soon as the grasses on the open plains had dried following the spring melt, and often before the snow was gone from the woodland verges, people from the tall grass prairies to the high plains set the grasslands alight, drawing in the bison.

But on the tall grass prairies, fire did more than attract the huge herds. It also kept a remarkable ecosystem in perfect balance, ridding the land of its thick annual layer of thatch and keeping encroaching aspen (in the north) and eastern red cedars (in the south) at bay. The result was an undulating ocean of grass, one of the most bountiful ecosystems on Earth.

Today, little remains. Of the magnificent northern tall grass prairie, less than one per cent survives in an unaltered state, though in many places restoration efforts have begun. Farther south, the devastation has been almost as complete. Texas, for example, once had more than 12 million acres of blackland prairie; today, less than 5,000 acres survive. Unfortunately, the remaining areas are highly fragmented, creating concerns about the many species that migrate over considerable distances. Today, the Nature Conservancy and other organizations, as well as government departments in both Canada and the US are working to create wildlife corridors to sustain populations of elk, prairie wolves and other species.

Today, controlled burns, such as this one in South Dakota, are used to renew native grasslands.

IMAGES COURTESY OF US FISH AND WILDLIFE SERVICE

VIEWING THE TALL GRASS PRAIRIE

The Manitoba Tall Grass Prairie Preserve, southeast of Winnipeg: This remnant of tall grass prairie was saved by the huge boulders dropped here by the retreating Laurentide ice sheet. Dubbed "sleeping sheep" by Ukrainian settlers, the glacial erratics proved too difficult to clear, thus saving some of the rarest plants on the continent, as well as territory used by elk and wolves that migrate from Minnesota, often during the winter months. Plants include small white lady's-slippers and western prairie fringed-orchids (see page 272) . Birders may see sandhill cranes, bobolinks and many of the grassland sparrows, as well as more than 20 species of butterflies, including monarchs and the rare Powesheik skippers. From Winnipeg, go south on Hwy 59 for about 90 km and turn east on 209 at Tolstoi for 3.2 km. To access the interpretive trail, enter the parking area on the south side of the highway. Open year-round.

Bluestem Prairie Scientific and Natural Area, northwestern Minnesota: This area once lay beneath the southern edge of glacial Lake Agassiz. As the lake level rose and fell, its waves created a series of sandy, gravelly ridges. Thousands of years later, farmers found these ridges and the wetlands that formed in the swales between them difficult to plow, saving their array of grasses and flowers for posterity. The lush growth in the swales includes head-high stands of big bluestem, tall prairie cordgrass, lovely blue-eyed grass and threatened western prairie fringed-orchids. In late April and early May, prairie crocuses can be seen and the booming mating calls of greater prairie chickens can be heard. Owned and managed by the Nature Conservancy, the 4,658-acre Bluestem Prairie Preserve is bordered on its north edge by Buffalo River State Park, which offers campsites and swimming. Travel 14 miles east from Moorhead, North Dakota on US 10. Turn south on SR 9 for 1.5 miles, then east on 17th Avenue South, a gravel road, for another 1.5 miles to the parking area on the left side of the road. A nature trail follows a prominent beach ridge, and connects with trails in the state park

Sheyenne National Grassland, southeastern North Dakota: Like Minnesota's Bluestem Prairies, this huge area of protected tall grass prairie (interspersed with private land) once lay at the edge of Lake Agassiz. Within the protected area are 850 species of flowers, including more than 35 that are rare or sensitive. This bounty of flowering plants attracts a multitude of butterflies, including rare Dakota skippers and regal fritillaries. This is also the southwestern extent of American elm. Along the Sheyenne River, designated a state wild and scenic river, are many small springs and streams.

VIEWING THE TALL GRASS PRAIRIE

Moose and elk can be found here, as well as beavers, coyotes and deer. Greater prairie chickens abound, as well as sharp-tailed grouse. The Nature Conservancy owns two preserves within the national grassland: Brown Ranch and Pigeon Point. The latter, "a unique ecoregion", has the "highest species diversity of any place in the state". For information on and maps for the grassland, contact the Sheyenne National Grassland, 701 Main Street, PO Box 946, Lisbon, ND 58054. To reach Pigeon Point on the Sheyenne River, take SR 27 east from Lisbon, ND, for 16 miles. Turn north on CR 53 and continue four miles north and two miles west to the gate.

Pipestone National Monument, Pipestone, Minnesota: At least part of the beauty of Pipestone is found, not surprisingly, in the rock. A low escarpment of Sioux quartzite blushes in the setting sun, its magic enhanced by Pipestone Creek, which plunges over the cliff face and meanders west. The cliff has three layers of rock: mud (the pipestone or catlinite), sand (which became quartzite) and gravel (which turned to conglomerate), all deposited between 1.6 and 1.75 billion years ago. Time and pressure turned the mud to pipestone; traces of hematite gave it its red colour. Unlike most catlinite, this material contains almost no quartz and is therefore dense, but easy to carve, rather like a human fingernail. Quarried here for at least 3,000 years, pipestone was traded as far east as Georgia and west to the Pacific. Aboriginal Americans still quarry the stone and sell magnificent pipes.

Running north south through the national monument, the escarpment is surrounded by tall grass prairie and sits on the Coteau des Prairies (or Prairie Highlands). In addition to the pipestone, visitors can see stone of another kind – glacial erratics. One of the largest boulders, now split by time and frost into three, sits at the entrance to the national monument. It is called the Three Maidens. An excellent interpretive centre and the wheelchair accessible Circle Trail assist in appreciating both the rock and the rare tall grass prairie. Continuing south on Hwy 75, the escarpment and an expanse of tall grass prairie can also be seen at Blue Mound State Park. Here, a bison herd has been reestablished, and an interpretive centre gives information about the region's geology, wildlife and history. Travel south on I-29 to South Dakota Hwy 34. Head east toward the Minnesota border, where the highway becomes SR 30. Continue east to Pipestone. Follow the signs to the national monument. Camping and accommodation are available at Split Rock Creek State Recreation Area, seven miles southwest of Pipestone off SR 23 and at Blue Mounds State Park just north of Luverne off US 75.

DENNIS FAST www.dennisfast.smugmug.com

Western Prairie Fringed-orchid

Platanthera praeclara

Endangered (IUCN and Canada); threatened (US)

IN CANADA, the beautiful, elusive western prairie fringed-orchid is found only in Manitoba, and then only in and around the Tall Grass Prairie Preserve near Vita in the province's southeastern corner. Yet in this small area is the largest single population in the world, about half of all known plants. In the US, populations of this magnificent orchid are found in North Dakota, Iowa, Kansas, Minnesota, Missouri and Nebraska.

Flowering between the third week of June and mid-July, each tall flower stalk bears as many as 30 creamy white flowers. Each is about 2.5 centimetres or just over an inch wide, with three deeply fringed lobes capped by a hood, and lasts for approximately 10 days. Opening from bottom to top, the plant blooms over a period of about three weeks.

Growing from a tuber, one slender shoot is sent up each spring, with the stem embraced by a series of elongated leaves that provide food for the tuber and allow the growth of the following year's bud. Very rarely, a tuber produces two buds (and eventually two flower shoots); even then, the number of flowers depends on ample spring rains, as well as the amount of snow cover or snow depth.

Though each plant may live for several years, the survival of the species is dependent on several factors. Pollination depends on nocturnal sphinx moths (or hawkmoths), which are drawn to the flowers by their night-time fragrance. Gathering the nectar, the moths transport the pollen from flower to flower, allowing each plant to produce thousands of tiny seeds that are then dispersed by the wind. However, in order to grow, the seeds must establish a root connection with a particular soil fungus, which feeds the developing orchid for two or more years before the first stem and leaves appear.

Despite its complex life cycle, this tall, delicate orchid once grew throughout the western central lowlands and eastern Great Plains, from North Dakota south to Oklahoma and from Iowa west to Nebraska and Kansas, sending its spike of exquisite flowers a metre (three feet) or more into the sunshine, wherever calcium-rich prairies and wet meadows could be found. Today, outside of Manitoba, 90 per cent of known orchids are located within the Sheyenne National Grassland, which is managed by the USDA Forest Service.

DENNIS FAST www.dennisfast.smugmug.com

The destruction of the tall grass prairie for agriculture, along with the application of insecticides, fungicides and herbicides, which kill the orchid's pollinators and fungal hosts as well as the plants themselves, are the main reasons this once abundant orchid is on the brink of extinction. Other destructive factors include the draining of wetlands, trampling by livestock and even the collection of flowers. Surprisingly perhaps, it seems that occasional grassfires, which eliminate shrubs that might shade out the plants, are beneficial to the survival of this beautiful wild orchid. The best time to view the flowers is from late June through early July.

Swift Fox

Vulpes velox • Endangered (Canada)

QUICK, curious, smart and shy, the swift fox once thrived throughout the Great Plains, from the foothills of the Rocky Mountains in the west to southern Manitoba in the east and as far south as Texas. However, a combination of fur trapping, incidental poisoning and habitat loss dramatically reduced both the numbers and range of swift foxes.

About the size of a house cat or a jackrabbit, the swift fox is among the smallest North American canids. During the summer months, its short, reddish-gray fur can be distinguished from that of red foxes by black patches on each side of its nose and by the black tip on the end of its bushy tail. In the winter, however, its fur grows long and thick, changing to buff-gray on its head, back and upper tail, with white through buff fur on the throat, chest and belly. Little wonder swift foxes held such appeal for fur trappers and traders. And appealing they were; between 1835 and 1839, 10,614 swift fox pelts were traded at the posts of the American Fur Company in Montana and the Dakotas, while between 1853 and 1877, the records of the Hudson's Bay Company show that it sold 117,025 pelts.

The swift fox's magnificent coat was part of the reason for its decline during the 19th century.

PARKS CANADA

Things went from bad to worse after the 1870s, as settlers and ranchers poured onto the prairies and embarked on poisoning, shooting and trapping campaigns aimed at wolves and coyotes. Though swift foxes are actually aids to rural landowners, since they feed on rabbits, mice and other small rodents, they were killed along with their larger relatives.

By 1938, they were gone from the Canadian prairies – actually the last specimen known to have been killed for a museum collection was trapped a decade earlier – and, based on a complete lack of reports for more than a half-century from Montana, Wyoming, the Dakotas and Nebraska, appear to have been extirpated there as well by the 1960s. Though recommended for a listing of "threatened" by the U.S. Fish and Wildlife Service in 1995, it was decided that other species were of higher priority. This prompted state agencies to establish conservation teams that have worked to reintroduce and restore populations in several regions.

A swift fox vixen enjoys a rare moment of leisure with her two pups.

Opposite: The closely related kit fox is slightly smaller and has larger ears.

PARKS CANADA

In Canada, it was not until 1978 that the swift fox was federally recognized as extirpated; five years later, thanks to private efforts, governmental protection and co-operative reintroduction programs, it was returned to parts of its former territory in Canada and the US and swift fox numbers began to recover.

In 1999, as a result of the cooperative reintroduction program that began in 1983, Canada upgraded its status to endangered. Pairs of foxes either bred in captivity or captured in the US were released in areas of southern Alberta and Saskatchewan with the goal of establishing a viable Canadian population of swift foxes. Between 1983 and 1996, approximately 942 swift foxes were released, and today there appears to be a small, but self-sustaining, population of around 700 foxes. The animals and their dens are protected on federal, provincial and private lands.

The first American reintroduction program was begun in 1998 and has successfully established a swift fox population on Blackfeet lands in northwestern Montana. The Blackfeet have a strong cultural and historical connection to the swift fox and have welcomed its reestablishment in the home they have shared for thousands of years. This land has not been as touched by commercial agriculture as the rest of the American prairie, and provides excellent habitat for the swift fox. These reintroductions have not met with the resistance that has been expressed towards some other species, because the swift fox poses no threat to agriculture. Overall, they have significantly expanded the swift fox population and range on the northern Great Plains and are beginning to return these delicate foxes to their rightful place in the prairie ecosystem.

The swift fox has long been described as "intelligent looking", and the research of the past 20 years has proven the accuracy of that description. Take the journals of Axel Moehrenschlanger, a Ph.D. candidate who spent three years studying swift foxes in the field. In 1995, he trapped, radio collared and tracked 23 foxes, including a young male he dubbed Rootie.

"It was virtually impossible to find his

dens," Moehrenschlanger wrote on a website called *The Wild Ones.* "When tracked on foot, he would just circle the observers. It seemed ... he was mocking us. He also stole food from live-traps in a manner that, to this day, remains a mystery."

Two years after their first encounter, Rootie seemed to have the world on a string. Happily mated, he and his mate had three healthy pups. Then his mate was taken by a golden eagle. Though she fought the capture and eventually escaped from the eagle's claws, her injuries were fatal and she died, leaving Rootie with their litter.

Most males would either leave the pups, or – in the case of polar bears or lions – kill them. But Rootie "jumped to the task of feeding his three pups," wrote Moehrenschlanger. "As often is the case, two of them died, but one strong fellow we call Lance survived."

The swift fox's slender body and long legs allow it to live up to its name, for it can run up to 50 kilometres or 30 miles an hour, a speed which appears even faster because of the animal's diminutive stature. This permits it to catch rabbits, mice, and other small rodents, though it's an opportunistic eater, and also feeds on birds, insects, reptiles, amphibians and carrion.

COURTESY OF THE US FISH AND WILDLIFE SERVICE

A member of the canid family, the swift fox is related to wolves, coyotes, dogs, and other foxes. Yet foxes share many characteristics with the cat family. Both hunt alone, stalk their prey and pounce on it to pin it to the ground. Both foxes and cats have extremely sensitive whiskers on their muzzles and wrists that provide information about their surroundings. And both have eyes that are superbly well adapted to their nocturnal lifestyles, with vertical pupils that allow them to see in the dark but protect their sensitive eyes from the daytime sun. Cats and foxes also both have very acute hearing that helps them locate prey and be alerted to dangers in the dark. These similarities are not the result of shared lineage, however; the fox and the cat's last common ancestor lived about 40 million years ago. They are the result of convergent evolution, two species developing similar characteristics that best allow them to fill a particular ecological niche.

The swift fox is closely related to another species, the kit fox, which lives in arid regions of the western US and as far south as central Mexico. The two foxes are so similar that some biologists consider them to be one species, but the kit fox has physical traits adapted specifically for its desert environment. It is slightly smaller and slower than the swift fox, and has a proportionally longer tail and larger ears that help it to disperse body heat. The kit fox also has hairy feet and thick fur that insulate it from the heat of the day and cold of the desert night.

A very lucky swift fox can live for eight to 10 years in the wild. But this is rare, for dangers

Swift Fox Fast Facts

Vulpes velox

LENGTH & WEIGHT

Fully grown, swift foxes weigh between two and three kilograms or between 4.5 and 6.5 pounds, and are about 30 centimetres or a foot tall and 80 centimetres or 2.5 feet long, including their bushy, black-tipped tails. Males are slightly larger than females.

HABITAT

At home in mixed grass and short grass prairie, it is most often found in rolling hills where small animals thrive and suitable den habitat is available. Swift foxes use dens throughout the year for shelter and protection, unlike other fox species, which only use them in the spring for giving birth and raising their young. Although swift foxes can dig their own dens, they more frequently take up residence in abandoned ground squirrel or badger holes and modify these as necessary. A family of swift foxes may change dens frequently in search of better food sources or to escape a build up of parasites such as mites, fleas, and ticks. During the day, they rarely venture far from the den, but will emerge to sun themselves nearby. They prefer to hunt and scavenge in the evening and throughout the night.

MATING & BREEDING

Adult swift foxes often live in pairs, although it is uncertain whether or not they mate for life. Mating takes place between late December and early March, becoming later in the year toward the northern extent of the fox's range. Pups are born approximately 50 days later, between March and May. Most litters contain four or five pups, but can be as small as one or as large as eight. Litter size may be influenced by the quantity and quality of resources available to the mother during her pregnancy. Both parents care for the litter and will attempt to defend the pups from intruders. The pups are helpless at birth, and their eyes and ears do not open for another ten to twelve days. They grow quickly, however, and after two months look like smaller versions of their parents. By their first autumn swift fox pups are fully grown and ready to leave the family den.

LIFESPAN

A very smart or lucky swift fox can live for eight to 10 years in the wild and up to 13 years in captivity.

NAMES

Swift foxes were also historically known as kit foxes, but when differences between the two were discovered, the former were renamed for their speed. It's believed that swift and kit foxes may interbreed in areas where their ranges overlap.

abound. Eagles, hawks and great horned owls are among the species that prey upon swift foxes. As adaptable coyote and red fox populations have expanded both their ranges and their numbers, they compete for food resources. Coyotes will also kill swift foxes, but rarely eat them, which suggests that they are seen not as prey but as competition.

However, the greatest threat to the survival of the swift fox, as for many other species, is human activity. Though hunting or trapping the animals for their pelts is illegal in Canada, swift foxes have been inadvertently killed by part-time trappers, and as the population grows in many places, a growing number of swift foxes fall victim to traffic fatalities, in part because they do not seem to have developed the healthy fear of cars that many other species have acquired. In the *Journal of Mammology* in August 2003, Jan Kamler et. al. reported monitoring 42 swift foxes for three years in a landscape of short grass prairie, as well as both dryland and irrigated farming. They found that between half and two-thirds of the study group survived and that the primary cause of death (42 per cent) was vehicle collisions.

Habitat destruction has also been a limiting factor for swift fox populations. Most of the short-grass prairie to which the swift fox has adapted has been converted to ranch and

VIEWING SWIFT FOXES

Because of their propensity to remain in their dens during the day and hunt at night, as well as their wariness and speed, swift foxes are difficult to view in the wild. However, the Cochrane Echological Institute, located in Cochrane, Alberta, west of Calgary, not only raises swift foxes for release to the wild in conjunction with the Blackfeet Reserve in Montana, but also cares for other rare or injured species. For more information, contact www.cei@nucleus.com

farmland. Kamler's study also showed that swift foxes rarely inhabit dry farmland, and are virtually never found in irrigated agricultural fields. The foxes apparently avoid areas with tall, dense vegetation because it limits their vision and movement, giving larger predators a distinct advantage. Cultivated farmland, especially when pesticides are in use, also has fewer insects and other small animals for foxes to eat. Eating insects contaminated with herbicides and pesticides are also known to make swift foxes sick. These factors make the swift fox highly dependent on undisturbed prairie.

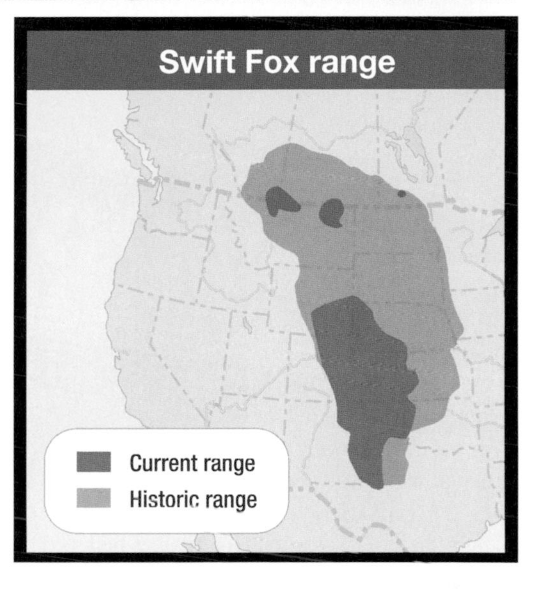

J.J. Audubon's swift fox clearly shows the black markings on its nose and tail, as well as its preference for native prairie.

NATIONAL ARCHIVES CANADA / ACCOUNT NO. 1970-188-1871 / W.H. CLOVERDALE COLLECTION OF CANADIANA C-041776

Black-footed Ferret

Mustela nigripes • Endangered (US, Canada and Mexico)

FEW species have come so close to completely disappearing and yet recovered. Found only on the Great Plains, the black-footed ferret is considered to be among North America's most endangered mammals, largely because black-tailed, white-tailed and Gunnison's prairie dogs, the ferrets' principal prey, were reduced to two per cent of their original abundance during the late 19th and early 20th centuries.

COURTESY OF PAUL MARINARI / US FISH AND WILDLIFE SERVICE

In fact, black-footed ferrets were believed to be extinct until in 1981, when a farm dog named Shep came home with an animal that his owners could not identify. A call to the local taxidermist in Meeteetse, Wyoming, revealed that the animal was an incredibly rare black-footed ferret and a search soon revealed a small population of 129 animals. Because it was so small, this "island" colony undoubtedly caused inbreeding, which perhaps led to lowered immunity. In any event, soon after its discovery, disease swept through the colony, devastating its already precarious population and by 1985, just 18 ferrets remained.

Determined to try to save the species, US Fish and Wildlife Service biologists and Wyoming authorities captured 14 of the remaining animals and launched a breeding campaign. However, breeding ferrets and raising the resulting kits, not to mention reintroducing them into the wild, initially caused considerable frustration. It's likely that the captive raised ferrets were unprepared for the predators, particularly coyotes, they would face, and also true that coyote and fox numbers had soared since the pre-contact period, when prairie wolves and plains grizzlies kept smaller predators in check.

However, over the past three decades methods and release procedures have been refined. In Arizona (the only release site that has Gunnison's prairie dogs), the release method was changed in 2001 to include spring releases of pregnant females, allowing kits to be born in the wild, where it was felt they would be better able to learn survival techniques. And since 2002, ferrets have not only been released at US sites, but also in the state of Chihuahua, Mexico, in 2002, and in Saskatchewan's Grasslands National Park in 2009. Following the Canadian release, the wild population was estimated at more than 500 animals.

However, the recovery has been thwarted by small prairie dog populations and by outbreaks of sylvatic plague, as well as what seems to be a variety of canine distemper. According to the World Wildlife Federation, "Extraordinary restoration efforts are needed to save this species from sliding back to and over the precipice of extinction."

Secretive and playful, North America's only ferret is a member of the Mustelid family, which includes (among many species) skunks, badgers, martens, fishers, wolverines and sea otters (see page 135), as well as the Siberian polecat and the European polecat or domestic ferret, which is widely sold in pet stores. Like most Mustelids, black-footed ferrets are carnivores and not only feed largely on prairie dogs, but live in tunnels built by their main prey.

Ferrets are believed to have evolved in Europe between three and four million years ago. Dispersing across Asia, their weasel-like ancestors likely gave rise to two ancestral species about a million years ago, and these, in turn, entered North America over the Beringian land bridge, moving southeast to the Great Plains during interglacial intervals. Evolving over thousands of years in concert with prairie dogs, they are believed to have eventually covered about 20 per cent of North America's grasslands.

Though they were rarely mentioned in the journals of European adventurers or pioneers, likely because of their subterranean lifestyle (see below), they were occasionally listed in the records of the Missouri and Rocky Mountain Fur Companies. However, they were prized by Native North Americans. The Lakota of the Western plains knew them as *pispiza etopta sapa,* "black-faced ground dog", while the Crow used ferret skins as medicine bundles in sacred ceremonies.

But on the heels of European settlement came deliberate campaigns to destroy North America's prairie dogs (see burrowing owls on

COURTESY OF MIKE LOCKHART / US FISH AND WILDLIFE SERVICE

Kits usually appear above ground in July, at about three months of age, and spend the balance of the summer learning to hunt and avoid predators.

BLACK-FOOTED FERRET RECOVERY PROGRAM / US FISH AND WILDLIFE SERVICE

Above: Born with white hair, kits soon grow a coat nearly the color of the surrounding soil. Those shown above left were raised in captivity, but take on (see above right and below) adult coloring by three months.

page 282), as well as sylvatic plague, which was introduced to North America from Asia about 100 years ago. The combination caused the both prairie dog and ferret populations to plummet. Spread by fleas or direct contact, sylvatic plague is now found in more than 200 mammalian species and on every continent except Australia. It's the only disease known to cause widespread fatalities of black-tailed prairie dogs, and since scientists have not been able to find any evidence of resistance to the disease in the animals' high-density living conditions, the U.S. Geological Survey's National Wildlife Health Center has begun immunizing both ferrets and prairie dogs.

Vaccinating the former in the field has proven remarkably successful, but prairie dogs were more of a challenge. However, studies have shown that they can be successfully immunized orally by feeding them vaccine-laden bait. These strategies, and others, including use of insecticides to kill fleas where the disease has not yet occurred, were all used in Conata Basin in South Dakota's Buffalo Gap National Grassland, after plague was confirmed there in 2008.

Appealing, black-masked creatures with sinewy bodies and perky white ears, ferrets have long bodies, short legs and large front paws and claws. Perfectly suited for digging, they live a mainly subterranean life in pursuit of prairie dogs. Large eyes and ears suggest excellent hearing and sight, but when hunting underground, smell is likely their most crucial sense.

Their short, sleek fur is buff in color, lighter on the underside and nearly white on the forehead, muzzle and throat. However, their legs and feet are dark. Despite the varying colors, ferrets are remarkably well matched to their surroundings and even when peering from their burrows are often nearly impossible to see until they move.

Though they prey mainly on prairie dogs, ferrets are also known to eat ground squirrels, mice, birds and insects.

Captive breeding (and the vaccination regimen) appear to be turning the corner on extinction. Each year, between 350 and 450 kits are born in captivity – more than 6,000 have been raised in total – and approximately half of those are reintroduced into the wild at more and more sites across the Great Plains. In 2009, those reintroductions included Canada for the first time. Grasslands National Park, a wild expanse of mixed grass prairie just over the border with Montana, was selected as the site in which to introduce 34 ferrets into their natural habitat. The recovery plan for the species includes releasing more ferrets in the future. The new residents of Grasslands NP are young ferrets that have been raised in various zoos, including those in Calgary and Toronto. Released at dusk in early October, they were set free near known colonies of prairie dogs.

COURTESY OF THE BLACK-FOOTED FERRET PROJECT / ARIZONA GAME AND FISH DEPARTMENT / US FISH AND WILDLIFE SERVICE

Black-footed Ferret Fast Facts

Mustela nigripes

LENGTH & WEIGHT

Ferrets are usually from 18 to 24 inches (or 46 to 61 centimetres) long with a five- to six-inch (12- to 15-centimetre) tail, they weigh between 1½ and 2½ pounds (slightly less or more than a kilogram). Males are somewhat larger than females.

HABITAT

Historic habitat for the species was the western Great Plains, from southern Alberta and Saskatchewan to northern Mexico. Though well-known among Aboriginal North American peoples, ferrets were not scientifically described until 1851, when naturalist John James Audubon and Reverend John Bachmann found a single specimen along the lower Platte River in in what is now Nebraska.

MATING & BREEDING

Ferrets are sexually mature at a year, but reach their peak reproductive years between three and five. Mating generally takes place in March and, following a gestation period of 41 to 43 days, a litter averaging three or four kits is born in the subterranean den, though single births have been recorded, as have litters of nine or 10. The tiny young are born blind, helpless and covered with thin, white hair. Looked after solely by their mother, they open their eyes at about 35 days then become increasingly active. In July, when they are about three-quarters grown, they appear above ground. Dependant on their mother for food until August, they are then hidden in separate burrows during the day and begin to learn to hunt. By mid-September, they are usually independent.

LIFESPAN

In optimum circumstances, black-footed ferrets can live to be 12 years of age, but the average lifespan in captivity is five to seven years and just three or four years in the wild.

THREATS

In addition to habitat destruction, loss of their main prey and disease, ferrets face such predators as coyotes, great-horned owls, golden eagles, prairie falcons, badgers, bobcats and foxes.

Young ferrets are very playful, leaping, wrestling and jumping backward, open-mouthed – a move that's sometimes called "the ferret dance". They are also very vocal, chortling during mating, hissing with fear and chattering to sound an alarm.

VIEWING BLACK-FOOTED FERRETS

In addition to zoo populations in Calgary and Toronto, the Smithsonian National Zoo's Conservation and Research Center in Front Royal, Virginia, has been breeding ferrets for two decades. In 2009, all 39 kits born earlier in the year were moved to Colorado, some to breeding facilities, and others into a "boot camp" where they live in prairie dog burrows and learn strategies for surviving in the wild. To see more, visit the FONZ (Friends of the National Zoo) website.

Ferrets are also being bred at the National Black-footed Ferret Conservation Center, located north of Fort Collins, Colorado.

Burrowing Owl

Athene cunicularia hypugaea • Endangered (Canada, Minnesota and Iowa); special concern (eight states)

MORE than two million years ago, before North America was buried beneath the first enormous sheets of ice of the Pleistocene era, before American lions and giant ground sloths disappeared from the face of the Earth, and long before humans arrived in this hemisphere, burrowing owls were raising their families in what is now Idaho (and presumably elsewhere on the Great Plains). We know this because fossils of these appealing little owls – slightly larger than their modern descendants – have been found in the Hagerman Fossil Beds National Monument, a site that's world famous today for its fossil horses. Burrowing owls are still found in the national monument, which is located in the Snake River Valley west of Twin Falls, and the park is going to considerable lengths to make them welcome, because in Idaho, as virtually everywhere across Western North America, burrowing owls are in serious decline (see below).

The situation is most dire at the northern and eastern margins of their range, in Canada, as well as in Minnesota and Iowa, where these ground-dwelling owls are listed as endangered. In Manitoba, for example, burrowing owls were once found as far east as Winnipeg, and nested as far north as the Swan River Valley, 600 kilometres or 375 miles from the American border. By the early 1930s, however, these pint-sized raptors were clearly in decline. In 1978, the province's population was estimated at 110 pairs; in 2001, just three pairs were sighted; in fact numbers were extremely low between 1996 and 2006. Then, in 2008, Manitoba Conservaton biologist Ken De Smet sighted eight pairs.

Elsewhere on the Great Plains, where they once were found in the tens or even hundreds of thousands, numbers were also plummeting. And in British Columbia's Okanagan Valley, though an award-winning winery bears the name of this long-legged owl, and despite the Burrowing Owl Conservation Society of BC's attempts to reintroduce the little birds in the 1990s – they are no longer found in the valley. (The conservation society has not given up, however, and still plans to create a breeding centre in the Okanagan.)

There are many reasons for the burrowing owl's steep decline, all of them having to do with human encroachment. Among them are agricultural and urban intrusion, degradation of their chosen habitats – the mixed and short grass prairies and desert landscapes of the Americas – and the widespread use of pesticides. Particularly dangerous is the pesticide carbofuran, which interferes with the transmission of nerve impulses and causes birth defects. Widely sold under the trade name Furadan 480F, it has proven to be particularly lethal to burrowing owls, for grasshoppers are among their favorite foods. Studies have shown that carbofuran has led to a 54 per cent reduction in the number of young and a 50 per cent reduction in the number of pairs that raise one or more chick successfully. According to an Alberta study, this is due at least in part because the use of carbofuran in a given area reduces the number of meadow voles and deer mice by 33 and 40 per cent respectively, significantly reducing the food available for raising nestlings.

Burrowing owls are also killed in collisions with vehicles and, where they live in close proximity to urban areas, such as parts of California, are often harassed or killed by household pets.

Finally, many biologists believe that one of the most important reasons for the shrinking numbers of burrowing owls in North America was the near-annihilation of prairie dogs.

Soon after their arrival on the Great Plains, European settlers began an all-out campaign against burrowing animals, believing they destroyed crops on the eastern prairies and harmed their livestock and even the rangelands farther west. Today, we know better. As repeated studies have found, prairie dogs may deservedly be called the "architects of North America's grasslands", for their feeding on roots and shrubs helps to eliminate invasive woody plants and their tunnelling behavior not only aerates the densely rooted native grasses, but provides homes for many species, including burrowing owls and black-footed ferrets (see page 278). Nevertheless, prairie dogs were despised and exterminated for more than a century, as recently as the 1960s. And it is only now that they are at last being appreciated for their crucial contributions to North America's grasslands and reintroduced into parks and preserves from Canada to Mexico.

In many places, that reinstatement comes almost too late for the burrowing owl. Across Western Canada, bird counts in 2004 found just 795 individual birds. Though this may have meant an actual population of twice that number, it nevertheless marked a devastating decline since the early 1990s, when Canadian biologists estimated there could be as many as 20,000 of these ground-dwelling owls in the four Western provinces. Studies by Saskatchewan's Operation Burrowing Owl program showed a 95 per cent decline between 1988 and 2000, an annual rate of 21.5 per cent.

During the same period in the United States, biologists documented the disappearance of these perky yellow-eyed owls from 89 per cent of sites in Wyoming where they were historically found, while between 1990 and 1996, numbers declined by 58 per cent in

DENNIS FAST www.dennisfast.smugmug.com

Two burrowing owls rest comfortably on their doorstep; one adopts a typical one-legged pose.

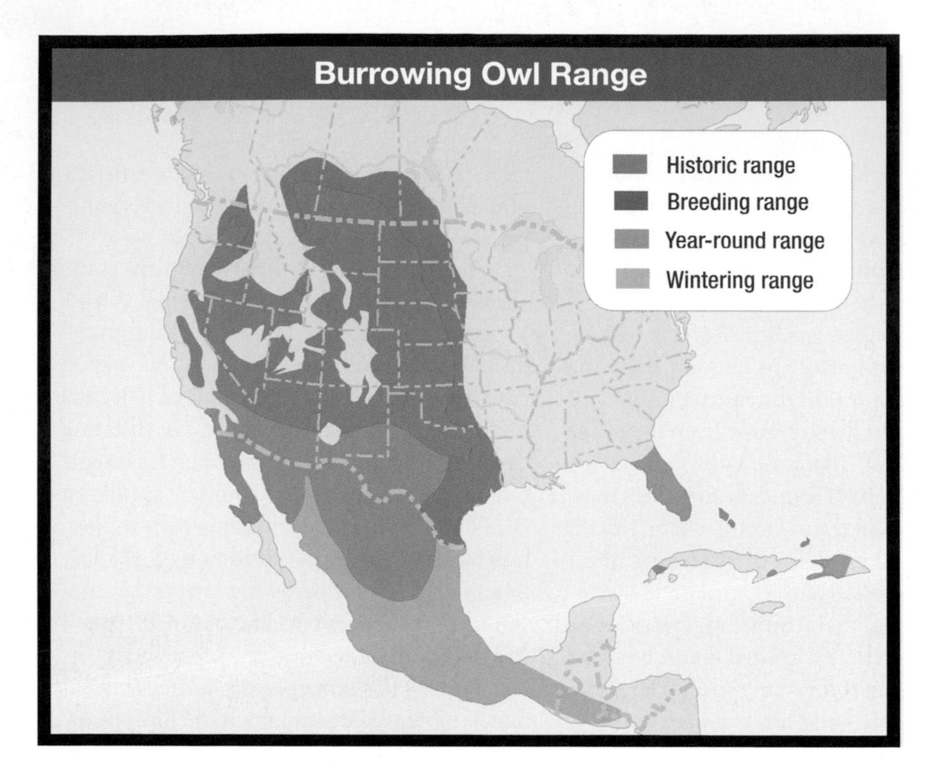

western Nebraska, and, though numbers of non-migratory owls of the Salton Sea region are up, overall burrowing owl numbers have dropped by more than 30 per cent in California.

As the numbers declined, the range of burrowing owls contracted as well, as the map above shows. By the early 1980s, the little owls had disappeared from north-central Washington State and numbers of nesting pairs were so low in Minnesota, Iowa and eastern South Dakota that reintroduction programs were initiated.

In addition to Canada and the United States, western burrowing owls are also found in Mexico, and in the dry grassland and desert regions of Central and South America. In fact, the first report of burrowing owls, published in 1792, was by Giovanni Ignazio Molina, an Italian Jesuit priest who was stationed in Chile at the time.

Though biologists have not yet completely sorted out owl taxonomy, it's clear these magnificent birds have been with us for millions of years. Some paleontologists suggest they evolved more than 100 million years ago, but most believe they are considerable younger.

"Owls," writes the University of Michigan's Danielle Choleviak on the university's website, "are well represented in the fossil record, with several families dating to the Paleocene era, approximately 58 MYA (million years ago)."

Barn owls (members of the Tytonidae family) evolved first, and over their long history gave rise to several very large species, including *Tyto gigantea,* which was larger than the present-day European eagle owl.

The other owl family, Strigidae, includes all other owls and seems to have emerged between 22 and 24 million years ago. It, too, spawned very large species, including the powerful *Ornimegalonyx oteroi,* which lived in Cuba nearly two million years ago and was more than a metre (or between three and four feet) tall. Today, owls live all over the globe, including such relatively inaccessible regions, even for birds, as the Hawaiian Islands.

Burrowing owls have a particularly ancient history, as the Hagerman fossils illustrate. The more robust ancestor found there, which dated to the upper Pliocene (older than 1.8

VIEWING BURROWING OWLS

Southeastern Colorado boasts the highest concentration of burrowing owls in North America, and the Rocky Mountain Arsenal National Wildlife Refuge, just 11 miles NE of downtown Denver, is one of the best places to see these captivating owls, as well as a large prairie dog colony, bison, deer and many other species. For information on programs and the site itself, visit the refuge's excellent website: www.fws.gov/rockymountainarsenal

mya) has also been found in Idaho and Kansas, while younger fossils have been found in South America, very likely indicating that burrowing owls evolved on the Great Plains and spread from that ancestral heartland.

Not surprisingly, perhaps, owls have been associated with many cultures going back into the mists of time. One of the oldest human associations is an ancient cave painting of an "eared" owl found deep in Chauvet Cave, near Ardèche, in southern France, where radiocarbon datings indicate some of the paintings are 30,000 years old. Traced with a finger on the cave wall, the painting illustrates an owl with its head turned 180 degrees, so that it is peering backward over its wings. More recent paintings, as well as owl mummies have been found in Egyptian tombs; remarkably, tests have shown that the owls, as well as dogs and other mummified animals were interred the same way and with the same care used for humans.

In ancient Greek society, as well as modern North America, owls are widely associated with wisdom. As Frances Backhouse writes in *Owls of North America*, "Greek mythology links the goddess of wisdom, Athene, to owls, and this connection is commemorated in the name of the genus to which the burrowing owl belongs."

Perhaps because of their silent flight, often ghostly calls and propensity for night-time hunting, among North America's Aboriginal cultures, as well as in many other parts of the world, owls have long been associated with evil spirits, disease or death. For example, among the Kwakwaka'wakw of northern Vancouver Island and its small coastal islands, the owl is associated with ill fortune and is often considered a messenger of death.

Though owls are widely believed to make stereotypical "whoo-whoo" sounds, in fact their calls range widely, from the western screech-owl's whistles to the breeding barks of the short-eared owl. But no owl has a repertoire like the burrowing owl; its remarkable stock of calls and sounds includes at least 17 vocalizations, from coos and hisses to an almost perfect imitation of a rattlesnake's warning rattle. One of the first sounds burrowing owl chicks are able to make, it not only works on the little owl's natural enemies – badgers, foxes and other predators – but also kept early settlers on the Great Plains at a distance. For a time, some believed that burrowing owls actually shared their subterranean burrows with rattlesnakes.

Unfortunately, almost before settlers had sorted out the source of the sound, they had begun destroying the owl's habitat with intensive agriculture (ultimately accompanied by pesticide use) and an all-out campaign to eradicate prairie dogs. Though burrowing owls – which, like other

DENNIS FAST www.dennisfast.smugmug.com

Burrowing Owl Fast Facts

Mustela nigripes

DESCRIPTION

The burrowing owl is small, long-legged and, according to those lucky enough to have met one, adorable. A mottled sandy brown, with a round, tuftless head and bright yellow eyes, it is most often spotted on the lip of its subterranean burrow or on a nearby fencepost, where it bobs up and down, a motion that amplifies its binocular vision and improves depth perception.

DISTRIBUTION

Originally found from the Canadian prairies to southern Argentina and Chile in grassland and desert landscapes, particularly among colonies of black-tailed prairie dogs, it has been largely extirpated from the northern and eastern prairies. Nests have been found from 200 feet below sea level in Death Valley to 12,000 feet at the Dana Plateau in Yosemite National Park.

HEIGHT & WINGSPAN

"[T]he size of a can of Coke" and weighing half as much, as Peter McMartin of the *Vancouver Sun* put it in April 2008, burrowing owls stand between six and nine inches or 15 to 22 centimetres tall with a wingspan of 22 inches or 55 centimetres.

LIFESPAN

In captivity, the appealing owls have been known to live for 10 years, but three or four years is considered a significant lifespan in the wild. In the past quarter-century, migration to the southern US and Mexico has proven to be a deadly proposition. Though researchers are still not certain why, less than 50 per cent of adult burrowing owls return to their northern breeding grounds each spring, and a mere five to six per cent of young owls return to breed in Canada the year after they are born.

HABITAT

Burrowing owls adopt burrows previously dug by Richardson's ground squirrels, prairie dogs or even badgers, their mortal enemies, for their nests. Prairie dog eradication campaigns in the early 20th century and again in the 1960s, combined with sylvatic plague, are estimated to have killed 98 per cent of the burrowing animals. This significantly contributed to the plummeting numbers of both burrowing owls and black-footed ferrets.

MATING & BREEDING

On the northern Great Plains, burrowing owls arrive in April or early May and mate soon after. A mated pair chooses its nest and may enlarge it with bills or claws. The owls then line the nest with dried cow or horse manure, and occasionally with scrap paper and even human clothing, apparently to mask the scent of the eggs and owlets. When all is ready, the female lays between three and 12 small white eggs, depending on the abundance of food. She incubates them for between three and four weeks, while the male hunts for food or stands guard above. She will often appear at the entrance to the burrow to beg for food. Once the chicks are hatched, the female gradually joins her mate in hunting, particularly at dusk and dawn. The parents stay close to the nest during the day, but travel farther afield at night. After two weeks, the chicks to move to the entrance of the nest where they huddle together (or are moved into neighboring burrows for safety's sake), waiting for the adults to bring them food. They begin to take short flights at four weeks and can fly well two weeks later. Fledging (of an average of three to five young per owl pair) occurs between six and seven weeks of age, after which the young may move to nearby burrows. Burrowing owls on the northern plains migrate south between early September and mid-October, often to southern Oklahoma or northern Texas, or even as far south as northern Mexico. Owls in the southern states are year-round residents.

DIET

The owls feed on grasshoppers, crickets, beetles, lizards, frogs and garter snakes during the day, and mice mainly at night. The birds hunt by running or hopping along the ground, flying from a perch or hovering over tall vegetation. Prey is caught with their feet and then transferred to the beak for carrying and when feeding the female of the chicks. Burrowing owls also cache excess food.

NAMES

In the past, these unique owls were often called prairie dog owls, gopher owls, ground owls, howdy owls and even rattlesnake owls. Despite their English and Latin names,they very rarely dig their own burrows.

owls, do not build their own nests – can also use ground squirrel and badger holes as well as occasional fox dens for nesting, roosting and storing food, none are as perfect as the underground chambers and network of tunnels created by prairie dogs. Not only do prairie dogs keep the plants around their burrows neatly trimmed with their almost incessant grazing and often create lookout mounds with the excavated soil, allowing the tiny owls a clear 360° view and an elevated platform from which to spot predators, but the large colonies also represented an emergency food supply, since burrowing owls are, despite their size, raptors.

Predators are a constant worry for the little owls, which raise their families underground. In addition to badgers (see page 212) and foxes, they must be concerned with skunks, weasels, raccoons, rattlesnakes (the real thing) and bull snakes. Many of these are small enough to access the nest chambers, where eggs and nestlings can be preyed upon. Above ground, and during migration, burrowing owls must deal with coyotes, foxes, great horned and short-eared owls, northern harriers, prairie falcons and many hawk species, as well as domestic dogs and cats.

COURTESY OF JANET NG,
SASKATCHEWAN BURROWING OWL INTERPRETIVE CENTRE

Sherman, an unlucky owl who luckily found friends at the Saskatchewan Burrowing Owl Interpretive Centre, was first hit by a truck and then flew into a barbed-wire fence. He now helps to raise broods of owlets every year in his permanent home in Moose Jaw.

Controlling dogs and cats in regions fortunate enough to have an owl population is just one of many things that citizens can do to bring burrowing owls back from the brink. In addition to education on the benefits of protecting burrowing mammals, others include identifying nesting areas so that they can be protected; banning the use of recreational vehicles in prime habitat, for they not only cause accidental owl deaths but also compact the soil; using non-toxic weed and pesticide control; delaying spraying until chicks are fledged, and – of particular importance – promoting habitat restoration.

Those wishing to get more involved can join programs such as the Saskatchewan Burrowing Owl Interpretive Centre in Moose Jaw; the World Wildlife Fund's Operation Burrowing Owl and BC's Burrowing Owl Conservation Society, which operates a breeding station in Langley and has had considerable success with its "soft-release" programs over the past 17 years. With the help of nearly 100 volunteers, the BC program breeds owls, digs nesting sites in the hundreds and, using an approach society director Mike Mackintosh calls "a shotgun wedding", pairs juvenile birds and releases them into a netted enclosure over the burrows for about two weeks.

"Ideally," he says, "they become compatible and mate." In 2007, the year's release produced 200 fledglings.

Plans are in the works for another breeding station in the Okanagan. In partnership with First Nations, the society hopes to return the birds to the region, despite a very low migratory survival rate.

Greater Sage-grouse

Centrocercus urophasianus • Endangered (Canada); Extripated (Arizona, Kansas, Nebraska, New Mexico and Oklahoma)

ANYONE who has ever watched the mating dance of a greater sage-grouse cock is unlikely to ever forget the experience, for it's nothing short of spectacular. The birds, which are the size of small turkeys and significantly larger than any other species of grouse, are pretty impressive at any time of the year, but during the spring, the displays of the courting cocks are riveting.

About the middle of March in most places in the US, and slightly later in Canada, dominant sage-grouse males begin to establish their territories, strutting grandly about on low ridges or knolls in open clearings in the sagebrush. These are often "ancestral leks", as they're known, which have been in use for centuries. And the ensuing battles are, as they are for most such contests in the wild, to ensure

COURTESY OF GARY KRAMER / US FISH AND WILDLIFE SERVICE

Resplendent with his white feathered chest and fanned tail, a greater sage-grouse cock embarks on his courting display.

the dominant cocks primary breeding rights.

A dominant male and his challenger will occupy the centre of the lek and the two are usually surrounded by yearlings – insurance for the central figures, perhaps, against predators. Wing battles can be fierce when two or more cocks are competing for breeding rights, particularly during the early days of the season. Morning after morning they compete, often until well after sunrise, until at last one cock's dominance is clear.

His right to perpetuate his genes assured, the dominant cock – his pointed tail raised and fanned like a peacock, his white chest feathers puffed almost to bursting and his wings fluffed to enhance his stature – rapidly inflates and deflates the bright yellow air sacs on his throat, producing the distinctive popping that females apparently find so seductive. The noise has been vividly described as "a sort of drawn-out burbling that sounds uncannily like someone gulping underwater."

Then the hens – most have waited until the "lekking" was all but over before arriving – descend on the territory, all heading for the dominant male. According to a study done in Wyoming in 1942, the dominant cock and his primary competitor mated with 74 per cent of the females, making most of the next generation very closely related.

In 1800, an estimated 1.1 million greater sage-grouse were found in 16 western US states and three Canadian provinces. Today, they inhabit just over half of their historic range and have been extirpated from Arizona, Kansas, Nebraska, New Mexico and Oklahoma, as well as British Columbia. In 2005, the National Audubon Society put their total numbers at 142,000. And even in Wyoming, where the sage-covered hills south of Pinedale were once filled with these remarkable birds, numbers are dwindling dramatically, largely because they occupy land that sits atop the nation's prime gas drilling territory.

In 1998, Wyoming's huge Jonah oil field,

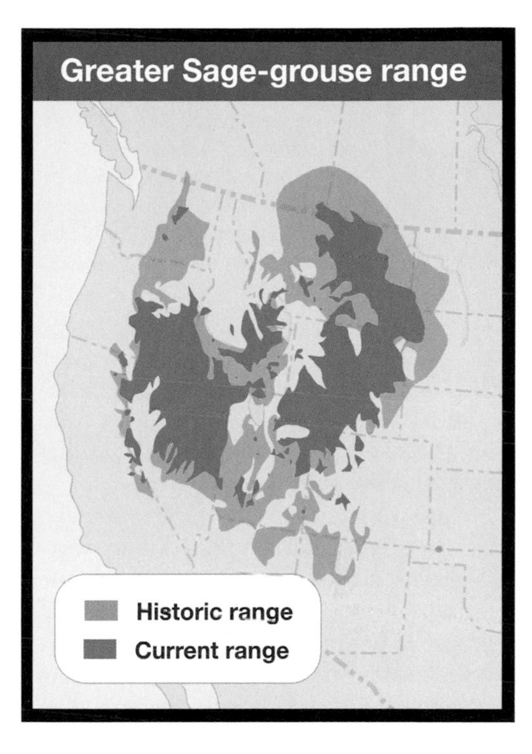

one of the largest on-shore natural gas discoveries in the US with 10.5 trillion cubic feet of gas, had 58 producing wells; six years later, that number had risen to 600, with 3,100 more proposed; when completed, that would average "a gas well every 10 acres", according to an article by Tom Kenworthy in *USA Today.*

Unfortunately, Wyoming's energy development sits right in the middle of some of the nation's best grouse habitat, and astride an important migratory route for pronghorn and mule deer.

Ground-nesting birds found at elevations between 4,000 and 9,000 feet (1250 and 2800 metres), greater sage-grouse are almost completely dependant on the sagebrush ecosystem, and specifically on silver sagebrush, for both food and protection from predators. Not surprisingly, greater sage-grouse are considered an umbrella species for the sagebrush ecosystem, meaning that their conservation

VIEWING GREATER SAGE-GROUSE

Sage-grouse can be found in many places where there are large tracts of sagebrush. In Wyoming, "lek-viewing" is a popular activity in the early spring, so much so that the Wyoming Game & Fish Department publishes "Lek-Viewing Ethics," including arriving at the site at least an hour before sunrise, parking well away from the edge of the lek, staying in your vehicle (which should be turned off) and using spotting scopes and binoculars to watch the birds. Pets, by the way, should be left at home. The website also provides a long list of sites where sage-grouse can be found, as well as the best times to view them. For more information, contact www.gf.state.wy.us/wildlife and click on Sage Grouse Management Information.

In Canada, where the birds number less than 500 in total, they may be found in Grasslands National Park in southwestern Saskatchewan.

tion will ensure high-quality habitat for other species that live only in these systems.

Sagebrush once covered about 155 million acres (or over 62 million hectares) in the western US, but much of it has been lost to cattle grazing, farming and housing developments, and also to fire in the drought-prone Southwest. Climate change may affect the survival of sagebrush and alter natural fire regimes, allowing grasses to thrive in the place of sagebrush. However it happens, history has demonstrated that if the sagebrush habitat is removed, the grouse disappear as well.

In recent years, the West Nile virus has emerged as a direct threat to the survival of individual birds. The risk of transmitting West Nile virus seems be more common in areas of coalbed methane development (CBM), since a byproduct of CBM is large pools of standing water, in which mosquitoes flourish. All of these factors have led to the proposed listing of greater sage-grouse under the Endangered Species Act.

Unfortunately, greater sage-grouse are not only unusually sensitive to human activity, but also difficult to reintroduce, even to areas that seem suitable. These realities have made the task determining whether the birds should be listed as endangered or threatened very difficult. That decision has been twice postponed, in 2008 to study the work of more than two dozen scientists who have been looking at threats to the bird's survival, and a second time, until at least 2010, "leaving in limbo," as Associated Press reporter Matthew Brown wrote in June 2009, "a spate of industries that face sweeping restrictions if the bird is protected." These include, he continued, "livestock grazing, oil and gas drilling and an increasing number of wind turbines."

The delays are not surprising, given the complexity of integrating these various interests with emerging research on global warming and changing weather patterns. Ultimately, all these aspects need to be assessed, in order to set aside the best places on the northern Great Plains for greater sage-grouse to thrive in the future.

COURTESY OF THE NATIONAL PARK SERVICE

Greater Sage-grouse Fast Facts

Centrocercus urophasianus

DESCRIPTION

Almost a third larger than other grouse species, the greater sage-grouse has rounded wings, a long, pointed tail (a characteristic shared by the sharp-tailed grouse, which is much smaller) and legs that are feathered to the toes. Females are a mottled brown, black and white, while males, which are larger, have a white ruff around their neck, white chest feathers, conspicuous yellow fleshy combs above the eyes, and bright yellow air sacs that are inflated during their mating display.

HEIGHT & WEIGHT

The greater sage grouse stands about two feet or 61 centimeters tall and weighs between two and seven pounds (or one and three kilos).

HABITAT

Sage-grouse are almost invariably linked with shrub-steppe landscapes and particularly with silver sagebrush. Also known as white or hoary sagebrush, this erect, freely branching shrub grows up to five feet or 1.5 metres tall and often spreads its lower branches out along the soil. It is here, among the lower leafy branches, that more than 80 per cent (according to some studies) of grouse hens make their nests, hidden from view, yet surrounded by green vegetation, flowers and seeds, as well as insects that thrive in the soil. Studies have repeatedly found that declines in sagebrush habitat lead to declines in sage-grouse numbers; the long-term survival of the species requires at least 65 per cent sagebrush cover.

MATING & BREEDING

Sage-grouse cocks indulge in spectacular mating displays on open mating grounds known as leks. These courtship displays or contests virtually ensure that the dominant and perhaps subdominant cock will mate with the majority of available hens. Hens mature at a year, but not all yearlings breed. Nesting under silver sagebrush, usually within three miles or five kilometres of the lek where the mating occurred, sage-grouse hens lay between six and nine eggs, which are incubated between 25 and 29 days. The survival rate of the chicks – nearly 60 per cent of which are female – is generally low.

In Canada, during two years studied, between 12 and 23 per cent of chicks survived the first eight weeks of life, likely due to predation by raccoons, coyotes, striped skunks, red foxes and even Richardson's ground squirrels, as well as magpies and crows.

DIET

Silver sagebrush constitutes between 47 and 60 per cent of the diet for adults during the summer months and 100 per cent in the winter. Chicks are fed insects during their first weeks, as well as herbaceous plants; both are high in protein.

LIFESPAN

These handsome grouse can live for up to six years. Predators of adults include eagles, owls and hawks.

THREATS

Historically, greater sage-grouse were remarkably free of disease, until they were hit by West Nile virus in 2003. In Wyoming and Montana, the virus claimed 25 per cent of grouse hens that first year. The virus decreased the following years, likely because of cool weather, which reduced the number of disease-bearing mosquitoes. Scientists also believe that small populations, which are found in a number of places throughout its range, result in inbreeding and make sage-grouse more susceptible to parasites and disease.

However, the greatest threat to the species is the destruction of sage-grouse habitat, which is shared by many species in decline, including the swift fox, loggerhead shrike, burrowing owl and many others.

NAMES

The Latin name, *Centrocercus urophasianus*, is derived from the Greek *kentron,* meaning "spiky", *kerkos*, meaning "tall" and *oura phasianos,* meaning the "tail of a pheasant".

COURTESY OF THE NATIONAL PARK SERVICE

Loggerhead Shrike

Lanius ludovicianus exubitorides • Endangered eastern population(Canada and two US states); *Lanius ludovicianus mearnsi* • Endangered San Clemente population; *Lanius ludovicianus migrans* • Threatened prairie population

These pint-sized predators wear a black mask, which not only makes them easy to identify, but is particularly appropriate, for loggerhead shrikes are masters of disguise. Though smaller than robins and classified as songbirds, they behave like hawks and sound, at least at times, like robbery victims.

Shrikes are hunters. Using their sharply hooked beaks, they kill insects – primarily grasshoppers – as well as small rodents, reptiles, amphibians and even other birds, particularly house sparrows. To save food for future use and to aid in consuming comparatively large prey without benefit of the strong talons typical of hawks and other larger birds of prey, or perhaps simply to display their hunting prowess during courtship, shrikes often impale their prey on sharp thorns or on barbed wire. To assist in this, shrikes often nest in near fencelines in open scrubby areas with thorny shrubs such as hawthorn and buffaloberry.

Though both males and females do sing, particularly in the spring when their bubbly courtship song can be heard, shrikes are better known for their shrieking alarm call, which gave them their name. Their vocal repertoire is not limited to singing or shrieking, however; it also includes a variety of peeps, clicks and rattles that are often combined to form a repetitious rhythm that shrikes use like a chorus.

With their remarkable range of behaviors, as well as their wide distribution – true shrikes are found across Europe, Asia and Africa, as well as North America – it shouldn't be surprising that they are a much studied group. Indeed, as the introduction to a report on the 4th International Shrike Symposium in Germany in 2003 held, "Shrikes are so interesting because they combine features of the passerines and raptors. Therefore, it is not surprising that they have given rise to a dedicated group of researchers – shrikeologists – who focus on every possible aspect of shrike biology – shrikeology."

Of particular interest is the evolution of the birds' impaling behavior, which is also seen in Australia's butcherbirds and Africa's bush shrikes. Celebrated Israeli shrikeologist Reuven Yosef believes it likely evolved

NATIONAL ARCHIVES OF CANADA / ACC. NO. 1970-188-2792 / W.H. CLOVERDALE COLLECTION OF CANADIANA / C-040482

Though tiny, loggerhead shrikes are efficient predators that use thorny shrubs to impale their prey.

RON WOLF

from trial-and-error activity, when the prey of ancestral shrikes was inadvertently wedged in a fork or on a projection, allowing the bird to feed at leisure.

From this, impaling techniques have evolved to be used as methods of communication. In Poland, scientists have found that great gray shrikes display the results of their hunting during courtship, giving females a chance to evaluate their prowess. It has even been suggested that male shrikes "buy sex with appropriate food offerings" and that "nuptial gifts ... to extra-pair females [or consorts] are of significantly higher energetic value than those offered to the mate." And, echoing human behavior in some countries or some eras, lesser gray shrike males have been found to punish unfaithful females.

In short, "Many of the shrikes' behavioural traits are assumed to be connected with their food supply and foraging," the symposium report concluded.

In Canada, the migratory prairie population of loggerhead shrikes once nested from British Columbia's Okanagan Valley to south-central Manitoba, and across the northern Great Plains states; they also lived year-round across the central and southern US and through much of Mexico. Today, they no longer nest in BC and numbers are dropping everywhere else.

In Saskatchewan, where most of Canada's prairie loggerheads breed, declines have been most evident in the southeastern corner of the province (and very pronounced in adjacent southwestern Manitoba), as well along the northern edge of their range, in the province's aspen parkland. In Alberta, bird counts conducted in 2008 indicated a 14.9 per cent decline in the number of birds over the previous decade, the greatest declines in any group of birds.

Members of the small and virtually indistinguishable migratory eastern population were once found from southeastern Manitoba east to the Maritimes, as well as across the south-eastern US from Maryland to northeastern Texas. Since the 1960s, however, these birds, too, have declined in number and in Canada their range has dramatically contracted. They were extirpated from New Brunswick and Quebec, though birds have been reintroduced annually into the wild in Quebec, as well as

Quebec, as well as Ontario, since 2002. That same year, at the western end of their range, Manitoba biologist Ken De Smet, estimated that there were just 50 breeding pairs of eastern loggerheads left in Canada, all in Ontario and southeastern Manitoba.

In 2004, Wildlife Preservation Canada had an even gloomier assessment. "It is believed there are only 100 pairs [of eastern shrikes] remaining in North America. In 2004, 36 wild pairs were found in Canada, 27 of those in Ontario," the organization declared in an overview.

Since then, with assistance from many birdwatchers and local residents, Manitoba's Shrike Recovery Action Group has located anywhere between six and eight pairs of nesting eastern loggerheads, most of them in suburban Winnipeg, where they use small trees in front and back yards for nesting and hunting. While this is not characteristic of the broad expanses of pastureland that eastern loggerheads usually prefer, in 2004, those eight Manitoba pairs produced a total of 56 eggs; of those, 42 hatched and all were raised to fledging. This was markedly better than the two previous years.

South of the range of the migratory subspecies, a number of non-migratory shrike subspecies can be found, including the federally endangered San Clemente loggerhead, which is found on the southernmost of the Channel Islands off the west coast of California, and the island loggerheads, which nest on three other Channel Islands (but have been extirpated from two others).

In addition to the problems of habitat loss and pesticide use faced by other North American shrikes, the subspecies on the Channel Islands also face vegetation stripping by feral goats and sheep, as well as domestic cattle, and predation by feral cats and black rats. To assist in the recovery of these island shrikes, efforts are being made to eradicate feral species, reduce the extent of tall exotic grasses (which prevent shrikes from seeing their prey) and create larger and more numerous clearings through the use of controlled burns and mowing.

As is true for many other grasslands species, the decline of loggerhead shrikes has been blamed on agricultural practices that eliminated native grasslands, wetlands and shrubs, as well as the use of pesticides. In fact, grassland birds have shown the greatest declines of any bird species in North America, and loggerhead shrikes are declining faster than any other grassland bird, according to the Manitoba Recovery Action Group. In part this may be because shrikes nest and hunt in shelterbelts along roadsides; as a result, they are often killed in collisions with vehicles.

In an attempt to restore numbers of the prairie subspecies, Nature Saskatchewan has developed a program called Shrubs for Shrikes. Launched in 2003, the campaign began with a mail-out to rural landowners in south-central and southwest Saskatchewan and ads in local newspapers. Both asked for information from anyone who had spotted shrikes and provided a toll-free contact number. Subsequently, a dozen landowners agreed to work with the province in preserving and enhancing shrike habitat by planting nesting shrubs provided by the province.

Despite years of study, and long hours spent trapping and banding shrikes, biologists are still not certain where shrikes from various populations spend the winter, what dangers they are exposed to during migrations or how faithful they are to their nesting sites. In an effort to find out, color banding of birds began in Manitoba in 1987. Using tiny colored bands representative of certain areas and employing different colors in Canada and the US, an effort was made to band the birds before they fledged, in their wintering region and when they nested. While young birds did not necessarily return to the region where they were hatched, proving, wrote Ontario biologist Chris Grooms, "that there is gene flow between

DENNIS FAST www.dennisfast.smugmug.com

VIEWING LOGGERHEAD SHRIKES

The best time to view shrikes is between mid-May and early June, after their eggs have hatched and parents are busy delivering food to the chicks an average of every four minutes. Shrikes are usually seen in open areas of Canada's southern prairies and the US grasslands, where there are nearby shrubs or trees for nesting. The presence of shrikes can often be inferred by impaled prey on barbed wire fences; the birds themselves can sometimes be seen there as well.

Grasslands National Park, Saskatchewan: Loggerhead shrikes are among many endangered prairie species that can be found in the park's two parcels of mixed grass prairie that sit on the border with Montana. Others include burrowing owls, swift foxes, piping plovers and the greater sage-grouse. This is a natural park, where no-trace camping and overnight backpacking are allowed. For further information, access the park's website: www.pc.gc.ca/grasslands

Prairie State Park, Missouri: This tall grass prairie preserve in southwestern Missouri, just east of the Kansas state line, boasts nesting shrikes, as well as greater prairie-chickens. The park, which offers basic camping, preserves almost 4,000 acres or 1620 hectares of tall grass prairie, an ecosystem that once covered about one-third of the state. Less than one per cent remains.

Theodore Roosevelt National Park, North Dakota: Like Grasslands NP to the northwest, this stark and striking landscape is home to hundreds of grassland species that are rare or non-existent in many other places. Prairie loggerhead shrikes are known to nest in the park and northern shrikes are sometimes found in fall and winter. The park offers seasonal camping from May to September in three no-service campgrounds. For more information or to obtain hiking or birding guides, contact www.nps.gov

Badlands National Park, South Dakota: Shrikes nest here, in one of the largest protected mixed grass prairie preserves in the US. Contact www.nps.gov for more information.

Far to the southwest, dwindling numbers of island loggerhead shrikes still nest on Santa Cruz and Santa Rosa Islands in California's Channel Islands National Park. (The park is also home to the critically endangered island fox, and the Channel Islands National Marine Sanctuary is home to endangered blue whales (see page 164). For more information, contact www.nps.gov

COURTESY OF KEVIN L. COLE

Loggerhead Shrike Fast Facts

Lanius ludovicianus exubitorides · Lanius ludovicianus mearnsi Lanius ludovicianus migrans

LENGTH & WEIGHT

Slightly smaller than robins, loggerhead shrikes are eight or nine inches or between 20 and 22 centimetres long, with a wingspan of 13 inches or 33 centimetres. They weigh under two ounces or about 47.5 grams.

HABITAT

Can be found in open to semi-open country, as long as hunting perches are present. They prefer to nest where thorny shrubs or barbed wire fences make it possible to store prey for later consumption.

MATING & BREEDING

Breeding at the age of a year, shrikes are mainly monogamous, but males will occasionally mate with a second female, and females will sometimes leave her mate when the young are fledged and raise a second family with a nearby male. Both male and female find the nest site, and gather the nesting materials, but the female builds the nest alone, creating a well-made structure of thickly woven dead plants, well hidden in thorny trees or shrubs such as hawthorn and buffaloberry. Eggs are laid in late April in the east and mid-May in the west, which the female incubates for 16 days, while being fed by the male. Both parents feed as many as eight tiny, naked, blind chicks so well that by the age of two weeks, they are almost as heavy as their parents. The chicks fledge at between 17 and 20 days of age, but return for a time to be brooded, or warmed, by their mother at night. Migrating south in September, they mingle with non-migratory shrikes living in the southern US.

LIFESPAN

Studies are still ongoing to determine the lifespan of different populations of shrikes.

FEEDING

Shrikes feed primarily on large insects, particularly grasshoppers and crickets, but will also eat small reptiles and small rodents. Small prey is picked up in their beaks, while mice that weigh as much as the bird itself are carried with their feet. Perched in a shrub, shrikes wait for prey to appear, then swoop down to kill it. Vertebrate prey are killed by cutting the spinal cord with a special cutting tooth on the upper beak. Prey are then carried to a place where it can be impaled on a thorn or wedged into a forked branch. This adaptation compensates for a shrike's lack of heavy talons or strong feet, which most birds of prey have.

NAME

Its common names come from the Old English: loggerhead described its disproportionately large head, while shrike meant "shriek", a description of its atonal alarm call. Shrikes are also called "butcher birds" and one newspaper dubbed it "a cute songbird with a mean streak". Collectively, a group of shrikes is known as a "watch".

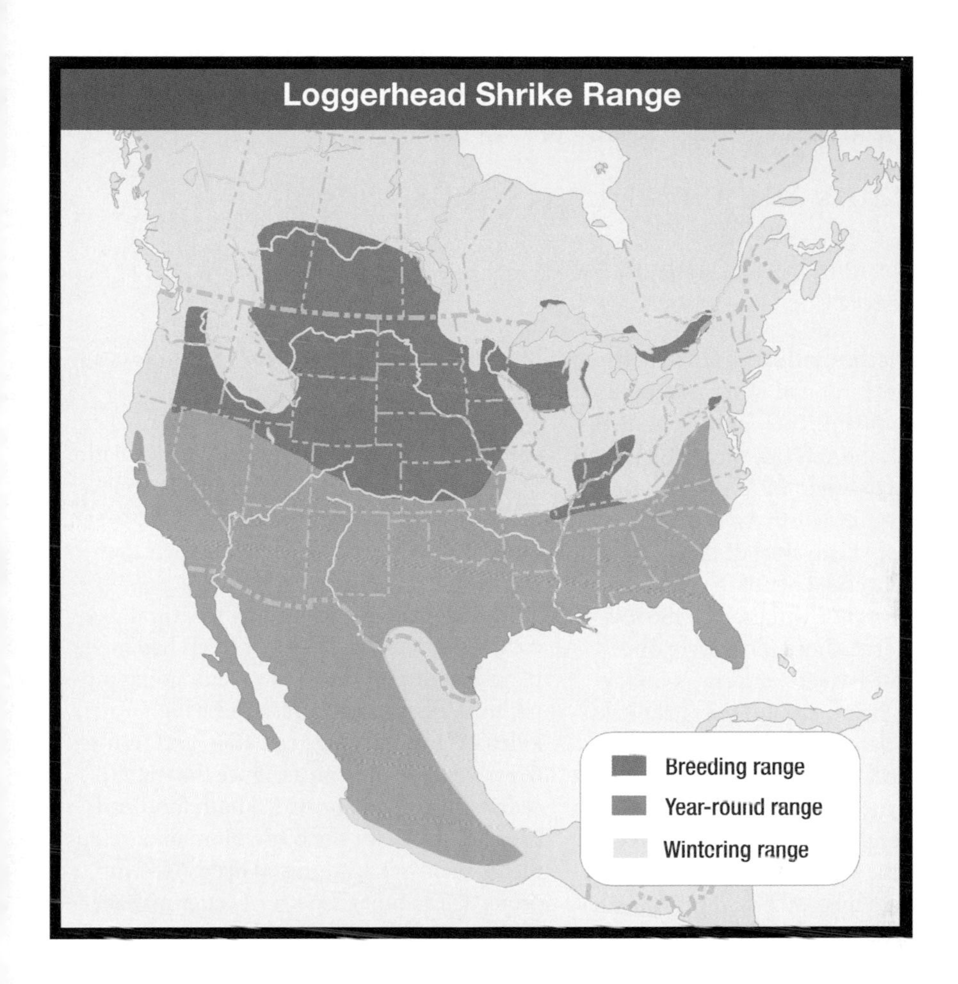

populations," he also believed strongly that a steep decline in the number of young being fledged from nests where chicks were banded, was evidence that the program had significantly altered the number of birds being fledged. It was therefore abandoned in Ontario.

"Bands can kill shrikes by catching on the thorns that are so intimately a part of a shrike's daily life. There are clear, confirmed and documented cases of shrikes being trapped by their bands on thorns of shrubs in the wild in Ontario and in captivity," wrote Grooms in 2009. Grooms was also convinced that, in Ontario at least, predators such as crows and jays learned by watching fieldworkers where to find shrike nests, leading to a drop in young.

De Smet disagrees, believing that other factors, particularly poor weather, have more to do with declines in the survival of nestlings. "During wet years, especially wet, cool stretches, many young (and even entire families) are lost, as adults are caught between brooding their young and feeding them.

"The young die from hypothermia, inadequate food or a combination of both."

Despite their small size, shrikes are striking birds, with a gray crown and back, white underparts, a black tail and wings and a dramatic black eye mask. They are sometimes confused with the northern mockingbird (which lacks the black mask) or the kingbird, (which has a black hood).

Piping Plover
(Great Plains and Great Lakes subspecies)

Charadrius melodus circumcinctus • Endangered (Canada and Eastern US); Threatened (Western US)

ONCE, when beaches and sandbars were the province of shore birds and waterfowl, piping plovers were abundant. Then human sun worshippers arrived, and this tiny bird, which nests just above the high water line on open beaches in three large areas of North America, began what may prove to be its final decline.

The beach lovers were not the first hurdle for piping plovers, nor will they be the last. These sand-colored shore birds were once abundant on lake beaches and rivers on the northern Great Plains, around the Great Lakes and along the Atlantic Coast to Florida. Then, beginning in the late 1800s, unrestricted market hunting for the millinery trade devastated the piping plover population on the Atlantic Coast (see also the short-tailed albatross, page 146). Not only did the feathers adorn women's hats, but the birds were also sent to market for human consumption.

Plover populations recovered somewhat in the 1920s and '30s, following the passage of the 1917 Migratory Birds Convention Act, but after World War II, as beaches all over North America increasingly became the preserve of sun lovers and developers, the decline began again. Atlantic Coast and Great Lakes populations were particularly threatened by summer recreation and cottage, as well as suburban, development. Northern Great Plains populations were also threatened by human disturbance in such high-density summer recreation such as Grand Beach, Manitoba.

However, through most of the Great Plains, where piping plovers are most often found at alkali lakes and wetlands, predators – including crows, coyotes, gulls and grackles – pose the primary threat.

The loss of lakeside beaches was further complicated by the creation and manipulation of reservoirs on many rivers; even when plovers could find a shoreline or sandbar off the beaten track, it was often flooded before their nestlings could mature.

On the flip side of the coin, naturally occurring fluctuations in water levels had long played an important role in maintaining piping plover habitat, for springtime high waters killed off encroaching vegetation and restored the plovers' feeding areas. Once the water receded, the beaches provided both food and breeding space for the birds. Human developments such as irrigation and hydro-electric projects, and other aspects of water management have dramatically altered these natural water cycles in many areas, allowing beaches to become overgrown and unsuitable as nesting areas. And though, in a few cases, human developments have actually created habitat for pipiing plovers, such as at Lake Diefenbaker, a man-made lake on the Saskatchewan River, agricultural and industrial development have greatly reduced the number of available breeding sites. During much of the 20th century, wetlands were drained or the area around them used as pastureland, allowing cattle to trample and destroy nests. And in the piping plover's wintering areas along the southeastern Atlantic and southern Gulf coasts of the United States, residential and commercial development, along with pollution, have all compounded the damage done to plover populations.

Almost a quarter of the world population of piping plovers is found in Saskatchewan, according to the 2006 census.

Human activity not only discourages and damages nests, it often forces adult plovers to use an "injured bird" ruse to lure predators away, leaving their eggs and chicks vulnerable.

As a result, by 1991, when the first comprehensive international census was undertaken, there were 3,509 adult piping plovers in the Great Plains and Great Lakes populations; both populations are members of the subspecies *C. m. circumcinctus.* Of those, 1,437 were in Canada and 2,072 in the US. Five years later, the number in Canada appeared to have grown by 17.5 per cent to 1,688, while the number in the US had slipped by one-fifth to 1,646. Had US birds simply nested in Canada?

However, when the 2001 census was completed, it was clear that the entire subspecies was declining, and that decline was particularly pronounced in Canada. Of the total – 3,025 birds or almost 500 fewer than 10 years before – just 973 were found in Canada. And on the Canadian prairies, the population had declined by 32 per cent between 1991 and 2001. And at Lake of the Woods, which is shared by Ontario and Minnesota and harbors the only population between the Great Plains and Great Lakes, the tiny adult piping plover population declined from 18 in 1991 to just eight birds 10 years later.

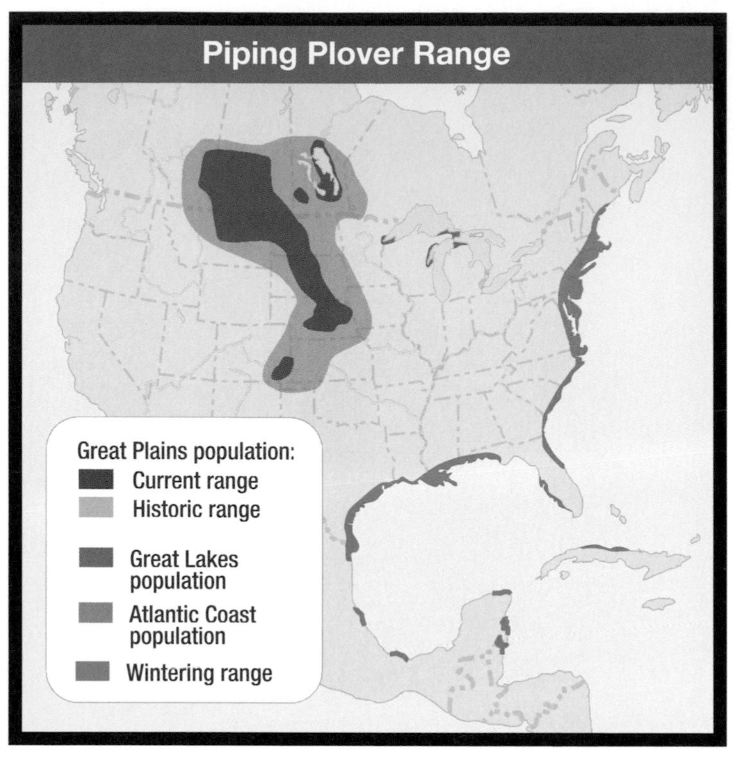

On a brighter note, the endangered US Great Lakes population – piping plovers haven't nested on Canadian territory around the Great Lakes since 1977 – has more than tripled. From 30 birds in 1991, it has grown to 71 nesting pairs in 2009. In part this is due to the efforts of bird keepers at the Detroit Zoo and other zoos across the US, which monitor piping plover nests at the University of Michigan's Biological Station in Pellston each spring. All abandoned eggs are collected for captive rearing; the resulting fledglings are released back into the wild.

Piping plovers have an ancestry that biologists Edward Miller of Newfoundland's Memorial University and Allan Baker of the Royal Ontario Museum believe stretches back into antiquity. "The Charadriiformes [a large and diverse group of shorebirds] arose in the Cretaceous >90 mya …", they wrote in an April 2009 article in *The Auk,* a journal of the University of California Press.

Over the next 60 million years, these primitive shorebirds evolved into many different species, including plovers of various sizes, as well as curlews, avocets and terns. Shorebirds are particularly interesting for investigating rates of evolutionary change, particularly in vocal displays, write Miller and Baker,

DENNIS FAST www.dennisfast.smugmug.com

Piping Plover Fast Facts

Lanius ludovicianus exubitorides · Lanius ludovicianus mearnsi Lanius ludovicianus migrans

LENGTH & WEIGHT
Slightly larger than sparrows, piping plovers are between 15 and 18 centimetres or 5.5 and 7 inches long, and weigh between 43 to 63 grams (1.5 to 2.2 ounces).

DIET
They feed on aquatic and terrestrial invertebrates.

LIFESPAN
In the Atlantic subspecies, few survive beyond nine, though one was known to live to be 14; on the Great Plains, individuals have been known to live to be five years of age.

HABITAT
The Great Plains population nests on sand and gravel bars or shoreline beaches at least 10 metre or 33 feet wide along rivers and around lakes, or on the edges of alkaline wetlands. To watch for predators, they choose sites with little vegetation. Young plovers sometimes use shoreline vegetation to hide from predators or escape nasty weather.

MATING & BREEDING
The Great Plains population winters mainly along the Gulf of Mexico, returning to its northern breeding grounds between April and August. Predominantly monogamous, the males undertake elaborate aerial and ground displays (often called "butterfly flights") to claim their territory and attract a mate and then together, the pair builds a nest by scraping out a shallow hollow in the sand or gravel and lining it with pebbles and bits of seashells. The female generally lays three or four cream or pale brown speckled eggs over a period of seven days, which both parents guard carefully until they hatch in June. Like their larger and more common relative, the killdeer, piping plovers will feign injury to lure predators from their nests. Once the eggs or chicks are out of danger, the parent will itself fly out of harm's way. The chicks emerge fully feathered and can run and feed within hours; their parents "brood" or warm them for several days after hatching; they take wing between 20 and 25 days of age.

MIGRATION
When nests fail (as a result of predation or flooding), piping plovers have been known to fly south as early as late June. Migration generally begins in early August for coastal and island wintering grounds around the Gulf of Mexico (for the Great Plains population). The birds usually fly non-stop to their winter havens.

NAMES
The common and Latin names both come from this little bird's clear-toned and melodic piping sound.

VIEWING PIPING PLOVERS
Saskatchewan and South Dakota have among the highest concentrations of nesting piping plovers. Chaplin, Old Wives and Reed Lakes, large saline lakes located on the TransCanada Highway in south-central Saskatchewan near the town of Chaplin, have collectively been designated a Western Hemisphere Shorebird site of global importance, for they support more than 30 species of migrating and nesting shorebirds. Including among these are 15 per cent of Saskatchewan's piping plovers and 30 per cent of North America's sanderlings. The Chaplin Nature Centre features endangered species, a gift shop, a viewing tower and a picnic area. The best time to visit is between May and July.

Plovers also nest at Big Quill Lake, a pear-shaped alkaline lake in east-central Saskatchewan.

In South Dakota, nesting occurs primarily on natural stretches of the Missouri River (and on engineered sandbars) below the Gavins Point and Fort Randall Dams, although some nesting may occur on tributaries.

In North Dakota, piping plovers breed on barren sandbars on the Yellowstone and Missouri Rivers, as well as around alkaline wetlands.

Hundreds of piping plovers in Saskatchewan have been banded to assist in determining where they nest and winter. The Canadian Wildlife Service office in Saskatoon is asking for the public's help in recording the whereabouts of banded birds. This one would be recorded as "right leg green flag above knee, black band over red band; left leg black band over black band".

DENNIS FAST www.dennisfast.smugmug.com

"for several reasons: their vocalizations are not learned and so are not subject to short-term cultural evolution as occurs in songbirds; vocalizations show little or no geographic variation within species . . ." As a result, these delicate birds "have sounds and sound traits that are tens of millions of years old . . ."

To protect Canada's endangered Great Plains population, attempts are being made to keep predators out in many of the more than 300 sites where they breed across the northern prairies. These projects include building predator fences and constructing fences to keep livestock out.

To keep out human predators, sections of beach that appeal to the birds are often fenced off, even at enormously popular places like Grand Beach on the east side of Lake Winnipeg, which draws tens of thousands to its broad white sand beaches every summer weekend. And these barriers are generally respected, for the delicate little bird with its white undersides, black breeding band, black mark between the eyes and clear-toned "piping" or "organ-like" call is a great favorite.

However, while young chicks are remarkably self-sufficient, foraging just hours after hatching, and while both parents tend their brood assiduously, at least in the first week or two, an average of less than one of the four eggs laid actually fledges, and of those, an average of just under a third survive their first year. Since about three-quarters of adults survive each winter migration, the heavy toll on fledglings makes population growth very difficult.

To stop the decline, other projects – in addition to fencing off nesting areas on busy beaches – have been launched. These include agreements with landowners or ranchers to protect areas where the birds nest, and efforts to maintain water levels at traditional breeding grounds. Other approaches being tested include nest enclosures, which allow adults and chicks to come and go freely but keep out predators, and moving eggs and chicks from areas where they are threatened to safer locations.

When the western prairies received much more rainfall than normal in the summer of 2005, river levels rose dramatically, feeding the lakes around which many piping plovers had built their nests. More than 500 pairs of plovers nest in Saskatchewan, with one of the largest concentrations around Lake Diefenbaker, near Saskatoon. As the lake level rose, a Saskatchewan Watershed Authority team of ecologists raced to move the birds' nests farther up the beach where they wouldn't be flooded. Each nest had to be placed carefully on a dish and moved a few metres at a time, so that the parents could keep track of their eggs. When the water levels continued to rise, it was decided that the nests could not be moved to safety. However, instead of giving up on the 75 nests, most of which contained four eggs, the team collected the eggs and brought them back to their research facility. There, 274 eggs were incubated, and turned by hand every few hours. The result was that about 104 chicks survived. After fledging, the young birds were released in stages onto Chaplin Lake in southern Saskatchewan, which supports a large breeding population.

In the US, piping plovers were once abundant along the Missouri River, which had a natural capacity to create sandbars. However, damming the river greatly limited the creation of sandbars. As numbers of plovers (and endangered least terns) declined, the United States Army Corps of Engineers began to assume the river's historic duty; three sandbars were built in 2004-05, and another, on Lewis and Clark Lake on the border between South Dakota and Nebraska, in 2007.

The results were impressive, as Daniel Catlin found in research for his 2009 Ph.D. at Virginia Polytechnic Institute, for piping plovers chose engineered sandbars over natural ones and their survival rate was significantly higher. And in 2006, both returning adults and juveniles returned to the engineered habitat at a higher rate than to the natural habitat.

Northern Prairie Skink

Plestidon septentrionalis septentrionalis • Endangered (Canada)

LIZARDS are rare in the north and the northern prairie skink is among the rarest. The ranges of only five native lizards extend into Canada and just one of these lives in Manitoba. Though this small creature is found primarily from eastern North Dakota and Minnesota south to central Kansas, two small, isolated populations also live in Manitoba, in the Carberry Sandhills southeast of Brandon and in a tiny site farther west, the Lauder Sandhills, south of Oak Lake. This very limited range makes the northern prairie skink particularly vulnerable to habitat loss, and as a result, it's considered an endangered species in Canada.

This little lizard, which measures between 13 to 20 centimetres or between five and eight inches long, including its tail, is a relic of the last glaciation and the continent's largest glacial lake – just as its habitat is. As Manitoba biologist and naturalist Errol Bredin has written, "At its height, glacial Lake Agassiz covered much of Manitoba and tapered south into North Dakota and Minnesota, terminating in the vicinity of the extreme southeastern corner of North Dakota. As the ... lake receded and the climate moderated, it is believed that the skink moved northward, following the exposed beach lines."

Bredin believes that the skinks moved northward along both the east and west shores of the lake, for two isolated populations have also been found in northwestern Minnesota. One is in the Agassiz Dunes Natural Area just southwest of the town of Fertile, which in turn is southeast of Grand Forks, and another is farther north, in a small tract of sand dunes near Lake Bronson. Just west of Fargo, North Dakota, along the remnant western shoreline of Lake Agassiz, there's another isolated occurrence of skinks. Recent mtDNA studies have shown Bredin to be correct; Manitoba's population of skinks is closely related to those in Minnesota. Studies continue on the Manitoba populations.

The Carberry Sandhills not only mark the shoreline of what was once the continent's (and perhaps the world's) largest glacial lake. This is also the delta of what was, at the end of the last glaciation, an enormous river that carried meltwater all the way from the Rockies – the ancient Assiniboine River. Today's Assiniboine empties into the north-flowing Red at the heart of Winnipeg, but thousands of years ago, this vast river – many times larger than its docile descendent – poured into Lake Agassiz at almost 160 kilometres or 100 miles west.

As it rushed across Alberta and Saskatchewan, it picked up gravel and sand, as well as lighter, finer silt and clay; because the gravel was heavier, when the river reached the delta and its

DOUG COLLICUTT

CONSERVANCY OF MANITOBA

Alert, speedy and disguised to match their surroundings, northern prairie skinks are adept at hiding – particularly the tiny youngsters (inset) with their bright blue tails.

current slowed, it dropped this material first. Its load of fine silt was carried much farther east, into the enormous lake where it settled over south-central Manitoba, creating a huge area of fertile soil. In between, the river dropped its load of sand, creating a vast expanse of sand now known as the Carberry Sandhills (or, more poetically, Spirit Sands).

While all this was going on, the world was dramatically warming; it was this warmth, after all, that melted the mountain glaciers and the continental ice sheets in the first place. This global warming continued as Lake Agassiz slowly receded and left the delta exposed. Over a period of more than 2,000 years, the hot, dry winds of what is known as the Hypsithermal reworked the sand into huge dunes of soft sand. And it was in and among these dunes that the northern prairie skink found its home for the ages.

Alert and quick, with smooth, shiny skin striped with four alternating light olive, dark olive and black lines, northern prairie skinks are hard to spot and even more difficult to catch. Though they live in sandy grassland areas, they require plentiful ground cover for shelter, as well as a nearby water source. Sandy soils are crucial to the skink's survival in northern areas, because it must be able to burrow down below the frost line in order to survive its winter hibernation.

Ground cover is also crucial. Whether it consists of natural objects such as fallen trees or large flat stones, or human detritus such as flat boards, cardboard or tin sheeting, skinks need something under which to "set up housekeeping", as Bredin puts it. "Some of the largest concentrations of skinks I have found are in litter piles and places where old buildings have been torn down, with boards and roofing tin left lying about."

Here, they spend most of their summer days, lying quietly in small excavations that they've constructed and darting out to feed on grasshoppers, crickets, caterpillars and spiders.

When they're out and about, skinks move rapidly, with a serpentine motion; as a result, they're often mistaken for snakes. In fact, lizards are quite closely related to snakes, and, remarkably, more closely related to birds than they are to turtles. According to Paul Hebert, the lead author of an article in the *Encyclopedia of Earth,* all vertebrates evolved from an ancient amphibian ancestor and reptiles diverged from this labyrinthodont ancestor about 300 million years ago, "in large part due to the evolution of shelled, large-yolked eggs in which the embryo has an independent water supply."

VIEWING NORTHERN PRAIRIE SKINKS

Spirit Sands, as the open dunes of the ancient delta of the Assiniboine River are widely known, is one of the best places to see a skink, though skinks go to great lengths not to be seen and, if caught in the open, head for cover with lightening speed. Nevertheless, this remarkable landscape is well worth visiting in its own right, as much for its collection of plants and animals that migrated here thousands of years ago during a period of global warming that makes today's seem mild – pincushion cactus and western hognose snake among them – as for its beauty. Located in the western extension of Spruce Woods Provincial Park, just west of Hwy 5, Spirit Sands is located about 70 kilometres southeast of Brandon Manitoba.

Spruce Woods Provincial Park: Encompassing almost 250 square kilometres or 100 square miles of mixed grass prairie, aspen parkland, white spruce forest and huge sand dunes – Spirit Sands, the last remnants of the delta of the glacial Assiniboine River – this is the confluence of several eco-regions, resulting in uncommon birds, unusual reptiles, including the prairie skink, and rare flowers. The Assiniboine River and underground springs create wetlands and bogs, while the open dunes, with prickly-pear cactus, skeleton weed and sand bluestem, are just a stone's throw from a forest of huge ash, elm, basswood, Manitoba maple and cottonwood. The park boasts a range of campgrounds, many trails and full facilities. Travel west on the TransCanada Hwy from Winnipeg less than two hours to Hwy 5, turn south for 28 km to the park entrance.

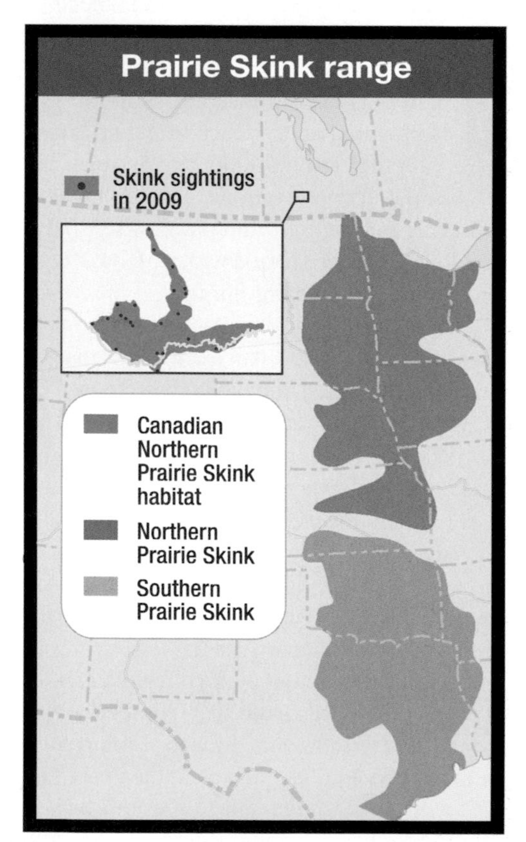

This advance, as well as the development of internal fertilization, allowed reptiles to become terrestrial animals. Radiating out across the land, they became the dominant life form for about 100 million years, during what we now know as the Mesozoic Era, the Age of Reptiles. Amphibians (direct descendants of the ancestral vertebrates) evolved first, and turtles appeared quite early on, prior to the evolution of crocodiles and dinosaurs. And from those, snakes and lizards, and later, birds evolved.

Though birds continue to conceive their young in hard-shelled eggs, some snakes and lizards have further evolved, retaining their eggs (as do mammals) until they have hatched, and giving birth to live young.

The northern prairie skink is not one of those, for it bears its young in eggs, which it broods and defends (see Mating and Breeding, opposite). Nevertheless, it is a member of the most diverse group of living reptiles. Worldwide, there are more than 4,450 species of lizards, ranging from tiny finger-length geckoes to the three-metre- or 10-foot-long Komodo dragon.

Northern Prairie Skink Fast Facts

Plestidon septentrionalis septentrionalis

LENGTH

Adult skinks range from 13 to 20 centimetres or five to eight inches long,

HABITAT

For a skink, the primary factor determining where to live is soil type. The soil must be loose and friable, so that it can dig down beneath the frost level during the winter. In Manitoba, skinks are found almost exclusively on what geologists call the Stockton Loamy Sands, the sands of the ancient Assiniboine Delta. Here, skinks hibernate from September to late April.

A separate subspecies, the southern prairie skink *(P .s. obtusirostris)*, is found in Oklahoma and Texas (see map). Despite its scientific name, which translates as "blunt-nosed northern great skink", *P..s. obtusirostris* is a southern species and not very large.

MATING & BREEDING

When they emerge from hibernation in late April or early May, mature males – those who have completed a second hibernation – develop a bright orange coloring on their jaws and throat and become quite active in preparation for breeding season. In late May or early June, with his tail arched, a male will pursue a female, nudging and lightly biting her. After 10 or 15 minutes, he will grab the skin above her front legs, encircle her and mating begins. About 40 days later, the female lays her eggs in a nest hidden under some form of cover. On average, she produces eight tiny white eggs with thin, elastic shells, which absorb moisture and swell as they develop. For more than a month, she broods the eggs, rarely basking in the sun or feeding, but quietly guarding them, with her body coiled about them and eating those that have spoiled. The eggs hatch in August, releasing hatchlings about five or six centimetres long. Within a few days, the young emerge – easy to spot because of their bright blue tails – dart from the nest and begin foraging for crickets, grasshoppers and spiders. For more information on the prairie skink go to www.NatureNorth.com

DIET

Northern prairie skinks eat small invertebrates such as crickets, grasshoppers, beetles, spiders, caterpillars, which they chase down and eat. They mash their food with their strong jaws before swallowing it.

NAMES

The scientific name, *Septentrionalis* is Latin for "northern" and skink is from the Latin *scincus* or the Greek *skinkos*. The first two parts of the common name – northern prairie – are logical, for they are found on North America's northern prairies.

Skinks have also developed an amazing mechanism for escaping from predators. When pursued, a skink arches its long tail to attract its enemy, but when attacked it can shed a portion of the tail, which continues to leap and vibrate for as long as 15 minutes as the skink heads for cover. Fracturing between vertebrae, the portion of the tail shed is commensurate with the threat. Blood vessels close off quickly once the tail is lost, the flesh soon heals over, and a new tail is grown. This replacement tail is more complete in young skinks than in older lizards, and has a cartilaginous rather than bony support structure.

Following in Errol Bredin's footsteps, Brandon University Biology Professor Pamela Rutherford and her students continue to study these secretive little animals. Their field sampling and research in Manitoba are helping to monitor skink numbers and determine in which areas the population is most at risk.

The most important ways others can help the prairie skink is by preserving its habitat through appropriate land management practices and assisting in raising public awareness about this amazing little lizard and what it needs to survive.

Lake Sturgeon

Aciper fulvescens • Endangered (Canada); Endangered, threatened or special concern (19 US states)

ONCE, during late May and early June, huge fish filled the cold, rushing streams of central and eastern North America. From the Alberta foothills east to New York's Hudson River and south to the Mississippi watershed, they filled the waterways as they swam upstream to spawn. They came in such numbers that to the people who gathered on the shorelines each spring, they seemed to form a bridge of life across the water. These enormous creatures, the largest measuring more than three metres or 10 feet in length, were lake sturgeon, descriptively known on the northern parklands as the "buffalo of the water".

One of North America's nine – and the Northern Hemisphere's 26 – sturgeon species, lake sturgeon are among just three species that have adapted to a purely freshwater existence. And for millennia, they were a source of great bounty for dozens of native North American cultures. Not only did these magnificent fish – some weighing more than 135 kilograms or nearly 300 pounds – provide a bountiful source of food, but their oil was used medicinally, their stomach linings served as drum coverings and the "isinglass" from the swim bladder was used to make glue and paint. Not surprisingly, sturgeon played a major cultural and social role and became the focus of many traditions.

When Europeans arrived, however, they were slow to recognize the value of lake sturgeon. They fumed when the spawning fish threatened to capsize their boats and chafed when their rough skin (for a sturgeon's body is covered with tough skin and rows of bony plates, or scutes, instead of scales, which protect it against attacks by predacious fish) and great size damaged their fishing nets.

Considered a nuisance in many places, those that were caught were fed to pigs, left to rot on the riverbanks or dried and stacked like cord wood on the water's edge as fuel for steamboats.

According to Manitoba's Nelson River Sturgeon Co-Management Board, all this began to change in 1860, when commercial fishermen in Sandusky, Ohio, on Lake Erie, discovered that sturgeon could be sold as a substitute for smoked halibut. Within twenty years, sturgeon fishing had become a major industry on the Great Lakes. Caught by the tens of thousands, their skin was tanned like leather, the isinglass was used to make alcoholic beverages and of course sturgeon eggs, or roe, were harvested to produce caviar. On Lake Erie, where 263,597 kilos or almost 580,000 pounds of sturgeon had been caught in 1885, a decade later, the catch was one-fifth of what it had been, according to a comprehensive 2009 report from the Ministry of Natural Resources.

However, as the numbers dwindled on the Great Lakes, new spawning streams were discovered. Following the burgeoning rail lines, sturgeon fisheries moved west in search of untapped populations.

While this might have allowed another over-fished species to recover reasonably quickly, the life cycle of sturgeon, complicated by industrial and agricultural development, made recovery virtually impossible as one sturgeon fishery after another was devastated across the continent.

Female sturgeon take between 23 and 26 years to reach maturity, and even then spawn only infrequently (see Fast Facts, on page 312). To spawn successfully, they require cold, highly oxygenated water, which became ever less common as industries and agriculture polluted the streams and fast-flowing northern rivers attracted hydro-electric projects. In many regions, the resulting dams prevented sturgeon from reaching their spawning grounds and across the continent, rivers that once swarmed with fish during the weeks of late spring soon contained fewer than a hundred, or in many places, none at all.

The decline of any species is a tragedy, but perhaps particularly so for sturgeon, for they have existed, virtually unchanged, for nearly 70 million years, back to a time when huge dinosaurs walked the land. Even more remarkable, the ancestors of sturgeon stretch back almost 400 million years, to the dawn of vertebrate life on Earth.

Fish were the first vertebrates – animals with backbones – to appear on Earth, appearing about 500 million years ago. (Two of their descendents, lampreys and hagfish, are with us today.) These primitive fish lacked jaws, and had a cartilage rod – a notochord – rather than a backbone. Sturgeon are the only other freshwater fish still living that lack vertebrae or a backbone.

The earliest fish were "shell-skinned" armored with scales and bony plates, but about 400 million years ago, the first of what have been called the "bony fishes" appeared. Though descendents of the earliest of these, lung fish – an ancient group with functional lungs, which ultimately led to terrestrial creatures, including mammals – and their

COURTESY OF ERIC ENGBRETSON / US FISH AND WILDLIFE SERVICE

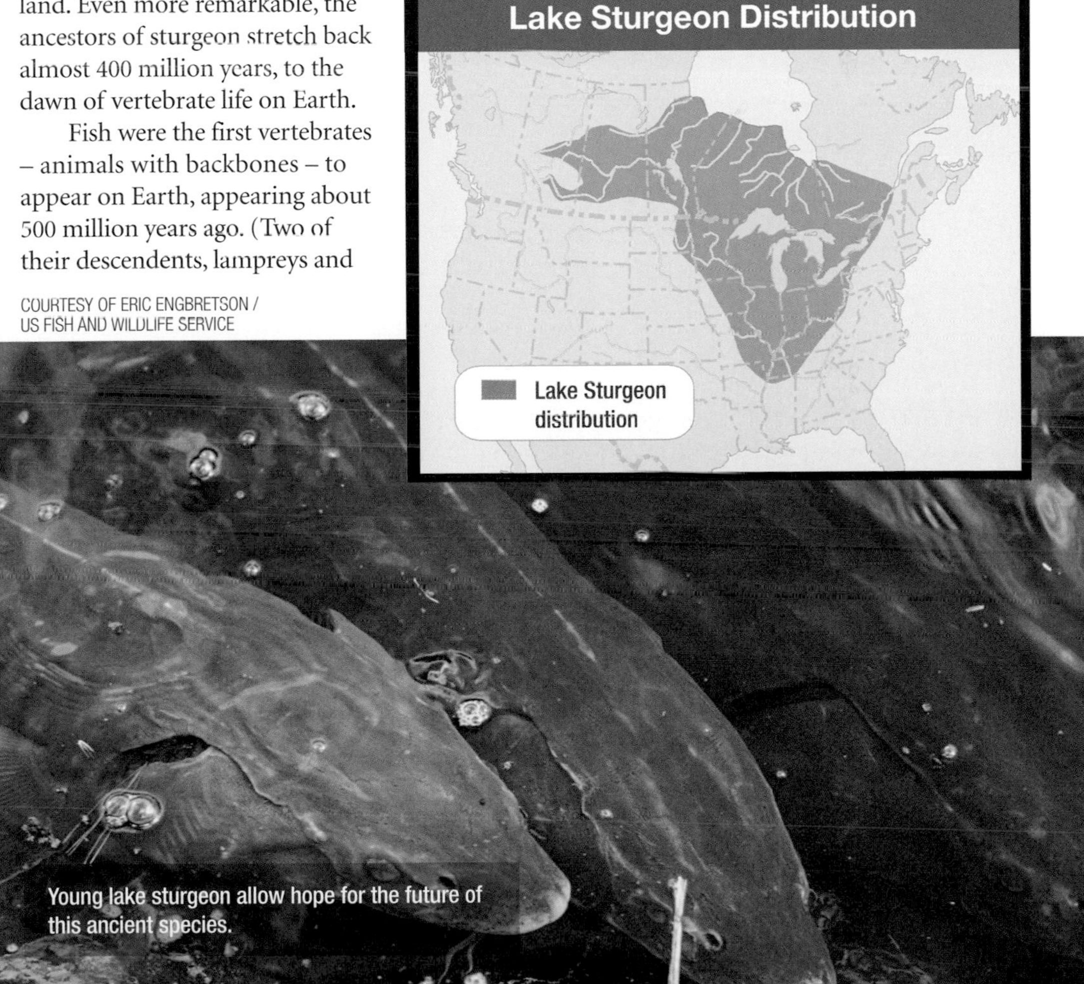

Young lake sturgeon allow hope for the future of this ancient species.

Though their flesh, roe and skin were used, during the late 19th and early 20th centuries, venerable lake sturgeon were often caught as trophy fish.

GULF OF MAINE COD PROJECT / NOAA NATIONAL MARINE SANCTUARIES; COURTESY OF NATIONAL ARCHIVES / NOAA PHOTO LIBRARY / DEPARTMENT OF COMMERCE

lobe-finned successors are rare, they can still be found today. The coelacanth, for example, which was believed to be extinct until it was discovered in the tropical waters off India in 1939, is a lobe-finned fish.

However, it was their successors, the ray-finned fish, which ultimately became the Earth's most successful vertebrates. Today, there are an estimated 30,000 species of bony fish in existence. Among these, sturgeon were the first to appear. For about 175 million years, they were the dominant species; then, during the Jurassic period about 225 million years ago – as *Tyrannosaurus rex* began its rise to North American dominance – they were almost completely replaced by gars, large, slender, thick-scaled predatory fish found in fresh waters of eastern North America, Central America and Cuba, as well as bowfins. These, in turn, largely disappeared when modern fish – known as teleosts – appeared during the Triassic period.

The sturgeon that survived these evolutionary waves were quite different than their ancient ancestors, which had been covered with hard, shiny, interlocking plates. Beneath this armored exterior, prehistoric sturgeon had strong jaws and a bony skeleton. By contrast, their modern descendents have lost most of the heavy scales; they retain only a bony cap over their skull and five rows of bony plates, called scutes, which run the length of their body. Unlike early sturgeon, they have a weak jaw and feed through a round, toothless tube-like mouth.

Most biologists consider the surviving sturgeon species to be degenerative forms of their ancient ancestors. And they differ from modern fish in many ways. In fact, in some ways, including their cartilaginous skeleton and their shark-like tail fin, sturgeon are more like sharks than modern fish.

As indicated earlier, just three species of sturgeon, as well as a very small population of normally anadromous white sturgeon, (see page 244) have made the transition to a purely freshwater existence. Lake sturgeon are by far the largest – at least potentially – of these three. The others are the endangered pallid sturgeon of the middle-Mississippi River and the much smaller shovelnose sturgeon.

The oldest identifiable fossil of a lake sturgeon is 60 million years old and was found in Alberta, indicating that they have been at home in the aspen parklands almost since the end of the Cretaceous period. However, the fish's mid-continent range over the last 10,000 years is almost certainly due to the great glacial lakes – and particularly Lake Agassiz – that dominated central North America at the end of the last glaciation.

Fed by huge glacial torrents from the Rockies to the west, Lake Agassiz's waters fed the lakes and rivers of all of central and eastern North America. Early in its 4,000-year-history, dammed on its north side by the enormous Laurentian ice sheet, it poured south, into the Mississippi Valley. Later, it thundered east, into the Great Lakes, or northwest into the Peace and Mackenzie Rivers. Finally, as the continental ice sheet disappeared, it flowed north into Hudson Bay, as its much smaller descendents, Lake Winnipeg and Lake Manitoba do today. And everywhere, its waters provided a conduit for lake sturgeon (and other species) to return to their ancestral streams.

Over the past quarter-century, scientists, communities and even corporations have recognized that lake sturgeon are in crisis, and face the threat of extirpation in many parts of the continent. In 2006, Canada's lake sturgeon were divided into eight populations; those in Alberta, Saskatchewan and Manitoba are designated as endangered. These designations restrict international trade of sturgeon meat and roe and limit domestic catches. The endangered status of the sturgeon on the Canadian prairies also requires that construction and other development projects undergo much closer scrutiny if they have a potential effect

Lake Sturgeon Fast Facts

Aciper fulvescens

LENGTH & WEIGHT

Growing very slowly, sturgeon take decades to reach maturity. Males may mature in 15 to 17 years, but females generally spawn between the ages of 24 and 26 years of age. Continuing to grow throughout their lives, they can ultimately reach lengths of more than two metres or seven feet and weights of more than 135 kilos or 300 pounds. The largest verified length and weight were from a sturgeon caught in 1943 in Lake Michigan; it was eight feet long and weighed 310 pounds. However, it's possible that larger lake sturgeon were seen in the past.

HABITAT

Lake sturgeon require large areas of open water that are less than 10 metres or about 30 feet deep, to allow them to feed on tiny mollusks, crayfish, worms and insect larvae. To feed, they swim close to the bottom with the ends of their sensitive barbels or feelers dragging lightly over the lake or river bed. When the barbels touch food, the fish's tubular mouth instantly protrudes and the food is sucked in with surrounding silt or sand. The food is then strained and retained, while the bottom materials are expelled through the gills.

MATING & BREEDING

In the spring, adults migrate up rivers to spawn, moving up to 200 kilometres or about 120 miles to areas of rapids. Groups of a dozen or more males congregate near shore, often with the upper parts of their bodies exposed. When a spawning female joins the group, she is flanked by two or more males. The large, adhesive eggs are laid in short bursts, where they stick to the rocky substrate. They are immediately fertilized by the males, whose vibrating fin may be exposed, producing a noise not unlike a drumming grouse. Then the spawning group drifts downstream or out into deeper water, but returns to spawn again, sometimes for a day or more, until the female is spent. Very large females may produce as many as three million black, glutinous eggs, but the average is between 50,000 and 700,000 (an average of 4,000 to 7,000 per pound of fish). Females spawn only once every five to nine years, while males spawn every year or two.

The eggs hatch in between five and eight days, depending on the water temperature and the tiny fry are about the length of a human thumbnail. Growing slowly, it takes about five years for them to reach the length and weight of a pound of butter.

LIFESPAN

The oldest reported age is 154 years.

THREATS

In addition to overfishing, hydro dams, pollution and changes to their spawning habitat, large sturgeon are often killed by propellers, particularly while feeding near shore.

NAMES

In its Latin name, *Acipenser vulvescens, Acipenser* means sturgeon and *fulvescens*, a dull yellow color. Lake sturgeon are also sometimes called "rock sturgeon", and in several European languages, sturgeon means "the stirrer", from the way the fish rummages along the lake bottom silt or sand for food.

VIEWING LAKE STURGEON

Sturgeon can be viewed at historic (and recently recreated) spawning sites in many places across central and eastern North America. Spawning between mid-April and mid-June (depending on when water temperatures reach between 12 and 18° C or 54 to 65° F), adult sturgeon gather at sites such as river rapids and along isolated rocky shores. The fish can sometimes be seen staging displays for one another, rolling underwater and leaping into the air, landing with a great splash.

Sturgeon can also be fished with barbless hooks on a catch-and-release basis on a number of major rivers in central Canada. At least one company in Alberta has made this into a commercial enterprise that draws international clients. However, studies have not yet shown the long-term impact of such activities and it seems clear that sturgeon thrive far better when they are not subjected to human predation.

/ MANITOBA WATER STEWARDSHIP, FISHERIES BRANCH

COIURTESY OF RON CAMPBELL / MANITOBA WATER STEWARDSHIP BRANCH

Today, large lake sturgeon are caught for scientific purposes, rather than for dinner.

on sturgeon or their habitat.

A wide variety of programs have been developed to protect and restore sturgeon populations throughout its traditional range, including some with the cooperation of the industries that have contributed to the species' decline. Among these are field studies involving tagging sturgeon to better understand their habits, life cycles, and populations; stocking tiny fry in rivers where they once abounded; monitoring catches on some river systems and even installing 24-hour patrols in an effort to reduce poaching.

In a unique Canada-US effort, in October 2008, an artificial spawning reef was constructed at the head of Fighting Island in the Detroit River that links Ontario and Michigan. It was an almost instant success, for in May 2009 sturgeon spawned at the site for the first time in 30 years.

Other efforts aim to restock sturgeon runs by collecting eggs and hatching them in aquaria, then releasing the young fish back into rivers and lakes. One example is the Grand Rapids Hatchery in Manitoba. Funded by Manitoba Hydro, it produces sturgeon fingerlings to release into areas where the population has been affected by the province's hydroelectric developments. And in 2008, the Minnesota Department of Natural Resources released 228,000 lake sturgeon fry into two tributaries of the Red River of the North. Half the fish were released in the Red Lake River near Red Lake Falls, the rest stocked the Roseau River near Caribou. The fry were about five days old and about half an inch long.

Other initiatives focus on education and awareness campaigns. The Nelson River Sturgeon Co-Management Board, established in 1993 and including representatives from several Northern Manitoba First Nations, works to raise awareness of issues affecting sturgeon, with an eye on protecting future stocks and the long term sustainability of Aboriginal harvests. Many of the board's outreach programs are aimed at school children in local native communities. It's hoped that these education campaigns will have an impact on society's stewardship choices in the future. These efforts are important, given the sturgeon's life cycle, for it could take up to fifty years to see the results of programs focused on baby sturgeon today.

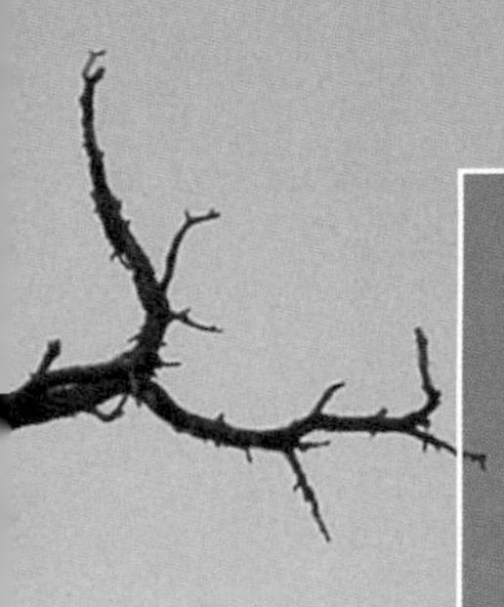

The Western Deserts

Dramatic silhouettes are often found in the western deserts. Here, a stark tree, granite rock formations and spires pierce the sky in Caruthers Canyon, in the New York Mountains of the Mojave Desert.

COURTESY OF RYAN CHRISTENSEN / NATIONAL PARK SERVICE

Snow lingers on Kelso Dunes in the Mohave Desert.

COURTESY OF MATT JATOVSKY / NATIONAL PARK SERVICE

The Western Deserts

North America's deserts run in a broad line down the continent's southwestern interior, and are largely defined by their plant life.

That **plant life** not only determines the species that call each region home, but is, in turn, determined by precipitation. And the amount of rainfall or snowfall each area receives is dependent on its geological history, its elevation and its soil conditions.

Taking some small liberties, this section includes not only the four categories generally accepted as North American deserts – the Great Basin, Mojave, Sonoran and Chihuahua Deserts – but also what is often termed British Columbia's desert (though most of it is found in the United States), the antelope-brush or shrub-steppe ecosystem (see map on 320 and information on page 321)

Internationally, the accepted definition of a desert is a region that receives, on average, less than 10 inches of rain annually. In the shrub-steppe region of BC's South Okanagan and north-central Washington, the average annual rainfall is seven inches. This is sufficient to allow a noticeable covering of perennial grasses, as well as low-lying shrubs, such as sagebrush, antelope-brush and common rab-bit-brush. All three are important nesting or grazing species for a variety of mammals and birds, including greater sage-grouse, big-horn sheep and, in the case of common rabbit-brush, deer and jack rabbits.

Shrublands and shrub-steppe ecosystems can be found in other areas of southwestern BC and the western US, including parts of Idaho, Nevada and Utah, as well as eastern Oregon and California and western Wyoming and Colorado. However, their elevation, as well as their rainfall or, in the case of regions to the north, snowfall, distinguish them from the arid region included here.

A further caveat may be necessary, for the desert classifications outlined here, or their geographic boundaries, are not universally accepted by geologists, biologists or climatologists. And some deserts go by a number of names, which often refer to local subdivisions. Several of these are mentioned below.

The Great Basin Desert

Like the shrub-steppe ecosystem to the north, the Great Basin is a cold desert. And even more than its northern neighbor, it is often snow-covered in winter, largely because much of it lies at elevations of between 4,000 and 6,500

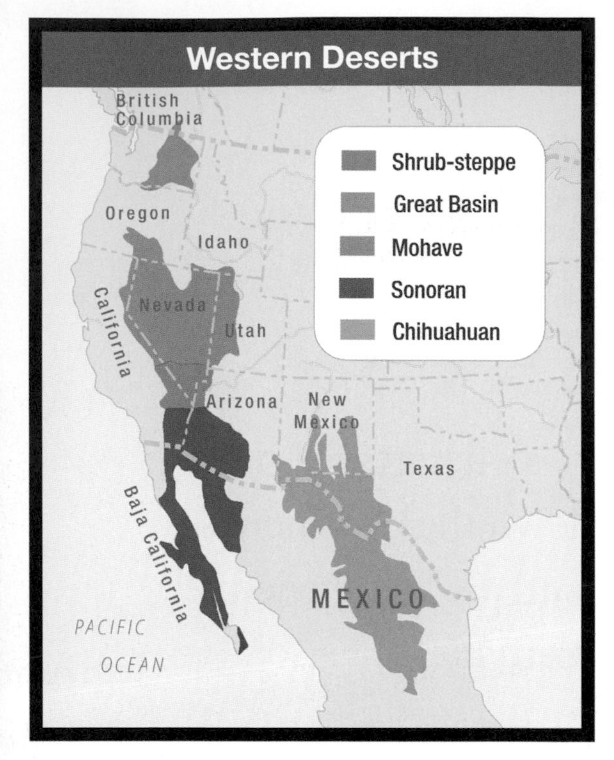

feet (or between 1220 and 1980 metres). (Elevations of BC's southern Okanagan, by contrast, are between 250 and 700 metres or 800 and 2,250 feet.)

Because of its elevation and winter temperatures, vegetation in the Great Basin is largely big sagebrush, blackbrush, Mormon tea and greasewood, with very little yucca and cactus. However, Great Basin National Park also includes forested mountains and valleys, including bristlecone pines more than 4,000 years old, as well as cliffs and peaks of stark rock and beautiful Lehman Cave.

The Mojave Desert

On the southern fringe of the Great Basin, between it and the Sonoran (see below) to the south, is the Mojave Desert. Though it covers a territory of more than 25,000 square miles (64,750 square kilometres), and includes features that are not only nationally, but internationally known – Death Valley, for example, boasts both the lowest point (Badwater, at 282 feet below sea level) and hottest recorded temperature (134° F) in the US – there are those who maintain that the Mojave is not truly a desert, but rather a transition zone.

Despite its Death Valley extremes, the Mojave is generally cooler than the Sonoran Desert to the south, largely because of its altitude. Covering part of southeastern California, and portions of Nevada, Arizona and Utah, elevations here (again, Death Valley excepted) are generally between 3,000 and 6,000 feet. Yet trees are few here, with the exception of the

Joshua tree, a yucca that grows to tree-like proportions. Named by Mormon pioneers, who saw in its uplifted limbs reminders of the biblical figure praying with arms upraised to the heavens, the Joshua tree was much less appreciated by adventurer John Fremont, who called it "the most repulsive tree in the vegetable kingdom".

The Mojave includes four national parks – the aforementioned Death Valley, along with Joshua Tree, Zion and Grand Canyon – and among its many resident species are desert bighorn sheep, pronghorns and kit foxes, as well as iguanas, gila monsters and western diamondback rattlesnakes. And just to the west, high on the slopes of the Sierra Nevada Mountains, endangered Sierra Nevada bighorn sheep (see page 331) are clinging to existence.

The Sonoran Desert

Among the largest and hottest deserts (with the Chihuahua Desert, below) on the continent, the Sonoran sprawls from California and Arizona across the US-Mexico border into Sonora and Baja California. Here, endangered Sonoran pronghorn (see page 336) and jaguars (see page 326) are found, as well as the magnificent "monarch of the desert" and symbol of the American Southwest, the soaring saguaro cactus. Preserved in Saguaro National Park since 1933, this very long-lived cactus – some specimens are more than 150 years of age – grows to a height of 50 feet or more than 15 metres.

COURTESY OF HIGHQUEUE / PUBLIC DOMAIN / www.en.wikipedia.org

Above: The Sonoran Desert, just west of Maricopa, Arizona.

Below: Trees are few in the Mohave Desert.

COURTESY OF THESCHMALLFELLA /CREATIVE COMMONS SHAREALIKE 3.0 / http://creativecommons.org/licenses/by-sa/3.0/

An October view of the Chihuahua Desert, near Van Horn, Texas, looking quite lush and green.

COURTESY OF LEAFLET / PUBLIC DOMAIN / www.en.wikipedia.org

The Chihuahua Desert

Stretching from southeastern Arizona, southern New Mexico and southwestern Texas deep into Mexico, this desert lies astride the Rio Grande River (known as the Rio Bravo in Mexico) and some of its tributaries. It is also higher in elevation than the Sonoran Desert and receives more rainfall than most other desert regions (though still less than 10 inches a year). These factors are crucial to the region's animals and plants, including tiny populations of jaguars and ocelots (see page 348), as well as jaguarundi (see sidebar on page 352).

Though daytime temperatures can be blistering and nights very cold, this is also habitat for dozens of species rarely seen elsewhere, including endangered Mexican wolves (see page 342) and, on the Mexican side of the border, the world's largest colony of black-tailed prairie dogs. In fact, the World Wildlife Fund has said that the Chihuahuan Desert was once the most biologically diverse desert in the world.

In the past half-century – really since Willis Haviland Carrier's "Weathermaker", as the first air conditioners were called, began to be widely used for making hot climates tolerable year-round – North America's desert regions have undergone a population explosion. For example, Las Vegas, which was established in 1911, had a metropolitan population of almost 1.9 million in 2008, while Palm Desert grew from 11,000 in 1980 to over 41,000 in the census of 2000; most of them are aging Baby Boomers.

This growth, with its demand for electricity, and golf courses, has put an enormous strain on the desert region's limited resources, particularly its water resources. This will only worsen as glaciers in the Sierra Nevadas dwindle in size, sending less water down to regions that are already water-starved. Solving these problems, if they indeed are open to solution, will take political will and sacrifice on many levels.

Shrub-steppe Ecosystem

AT the toe of a crumbling slope just west of Osoyoos, on a floodplain of what was once Glacial Lake Oliver, is a small remnant of what the locals call Canada's only desert. Like Nk'Mip, its counterpart on the east side of Osoyoos, as well as the remnants that once were found over much of the valley floors and terraces along the Okanogan and northern Columbia Rivers in Washington State, it is part of the northernmost tip of a dry land system that stretches far to the south, and it is among North America's most endangered environments. Perhaps 10 per cent of this shrub-steppe ecosystem – often called antelope-brush desert in Canada – remains north of the US border, though much is in poor condition due to overgrazing and invasive species. In the US, it is considered critically endangered, with less than two per cent still in its natural state.

Nestled in the rain shadow of the Cascade Ranges, the South Okanagan/Okanogan region experiences summer temperatures well over 30°C or nearly 90°F and an average of 31.7 centimetres (or 12.48 inches) of rain annually. (True deserts receive less than 10 inches.) European settlers, arriving in the valley beginning in the late 1800s, dismissed the dry, shrubby grassland as "wasteland" and wasted little time in converting it to fruit orchards, vineyards, golf courses and suburbs in British Columbia, and to grazing land in neighboring Washington.

Yet this is a bountiful place, long appreciated by the Okanagan or Syilx people of British Columbia and their Okanogan or Colville cousins in Washington. Fully 30 per cent of British Columbia's wildlife species at risk make the southern Okanagan and Similkameen Valleys their home and 22 per cent of all provincially endangered and threatened vertebrates (many of which are also federally listed), as well as hundreds of rare insect and plant species can be found here.

Once, burrowing owls (see page 282) were widely found here. Endangered in Canada and a species of special concern in most of the western states, there are those who plan a second reintroduction attempt, despite an initial failure in the 1990s. Wolverines, which exist in considerable numbers in Canada, but are reduced to between 400 and 500 in the lower 48 states, and bighorn sheep are also found. And, given its huge pre-contact population, which has been proven to have stretched north into Washington's Okanogan Valley, it is very likely that this ecosystem was once home to pronghorn (see page 336).

Like the Great Basin Desert of Idaho, Utah and Nevada, the shrub-steppe ecosystem is a "cold desert", where precipitation falls as snow in the winter and the dominant plant life is not subtropical. Yet dwellers of more southerly deserts would recognize at least some of the vegetation here. Big sagebrush, for example, can be found farther

Antelope-brush is the signature plant of BC's southern Okanagan.

PETER ST. JOHN

south, and is one of the few native species that has been assisted by settlement and the accompanying overgrazing of livestock. Evergreen and remarkably hardy, this aromatic plant, which is also known as big sage, common wormwood and basin sagebrush, grows in vast tracts in the US, covering 470,000 square miles across 11 western states. Nevada has big sagebrush as its state plant.

Growing up to two metres tall, this branching gray-green shrub was used by the Okanagan people in a wide variety of ways: the wood was burned for fuel; the seeds were eaten and the leaves, which contain camphor, were used to make teas to treat colds and coughs, and to soak sore feet. Branches were widely used as a fumigant and to the northeast, the Nlaka'pamux used the bark, which naturally shreds, to weave mats, bags and cloaks.

Antelope-brush, the fragrant shrub for which Canada's portion of the ecosystem is named, is often found growing with big sagebrush in dry, sandy grasslands soil. It's easy to spot in late spring with its shaggy, upright branches and bright yellow flowers and provides crucial food for deer and bighorn sheep.

Antelope-brush is also known as antelope-bush, bitter brush and greasewood. The last of these refers to the plant's pitchy quality that, according to ethnobotanist Nancy Turner of the University of Victoria, "makes it good for producing a hot fire quickly". Bundles of branches served the Okanagan as a dependable and portable fire starter that was often used in winter or when travelling.

The seeds, which are shaped like tiny spindles, were widely eaten by chipmunks, ground squirrels, great basin pocket mice and deer mice; seeds that had been cached for the winter in underground burrows were often overlooked, resulting in clumps of antelope-brush seedlings in the spring.

Another, smaller shrub found here (as well as other dry, low-elevation regions and dry, open ponderosa pine and Douglas-fir forests) is rabbit-brush. Its gray, velvety leaves were used for a variety of purposes by Aboriginal women during their childbearing years: as sanitary napkins and to make a tea to ease cramps. True to its name, its leaves and flexible stems are a favored food of jackrabbits, as well as deer and bighorn sheep.

This arid environment also sustains milkweed, the Monarch butterfly host, giant wildrye, a tall grass with smooth, stiff gray-green leaves that were widely used to decorate baskets and line steam pits and food caches, and bluebunch wheatgrass, a native grass that was "excellent forage for both domestic stock and wildlife," according to the trio of authors of *Plants of the Southern Interior: British Columbia and the Inland Northwest.* Unfortunately, they write, it "is susceptible to damage or local extinction from overgrazing in spring."

Saskatoons (also known as serviceberries, and as "real berries" by the Secwepemc people) grow in many places in the southern Okanagan and were collected in great numbers, then dried and traded west to coastal peoples.

On the glacial or kame terraces and valley slopes, ponderosa pines grew to heights of 30 metres or more, forming beautiful open forests that provided winter range for deer, elk and bighorn sheep, seeds for many small animals and crucial nesting and foraging areas for many birds, including the endangered white-headed woodpecker. The wood of the ponderosa pine (which is also known as yellow pine, western yellow pine, bull pine and rock pine), proved valuable for use in construction and very little of this magnificent old-growth forest remains. A glimpse of it can be seen at Nk'Mip, just outside Osoyoos, BC.

Though many believe them to be creatures of the high mountains, bighorn sheep have long thrived here, inhabiting the dry canyons and high plateaus. As they have since icy waters filled this canyon at the end of the last glaciation, mule deer hide in plain view among the scattered ponderosa pines.

Pondorosa pines and big sagebrush cover a swath of land at Nk'Mip, just east of Osoyoos, providing a glimpse of what this bountiful ecosystem was once like.

PETER ST. JOHN

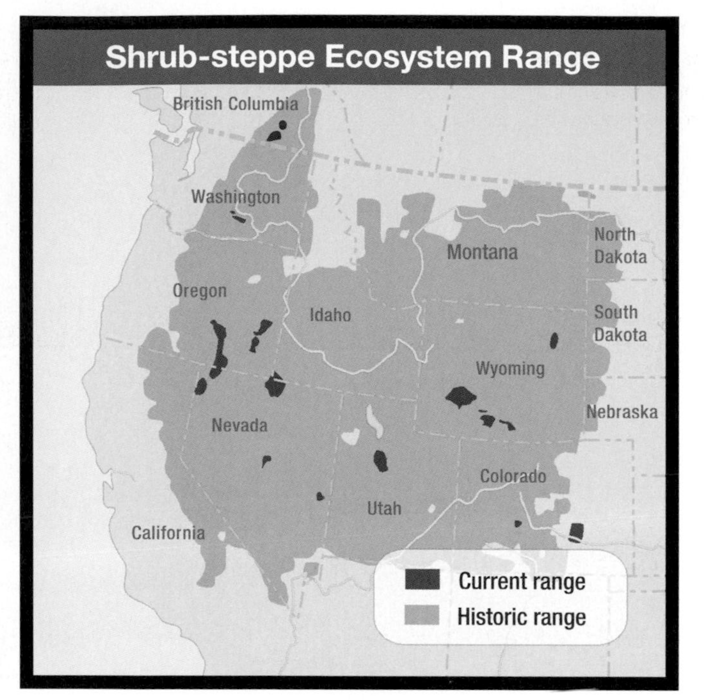

Bears, including grizzlies, hunt in the ravines and tributary valleys to the west and east on the Canadian side of the border, and a tiny population – likely less than two dozen bears – remains in the mountains and valleys of Washington's North Cascades grizzly bear recovery ecosystem. This large region stretches from Mount Baker to the western edge of the Okanogan Valley and south, like a teardrop, almost to Interstate 90, encompassing almost all of North Cascades National Park, Glacier Peak Wilderness and Wenatchee National Forest. Many would like to see the great bears return to the mountains and forests they inhabited for millennia.

Howls heard in the forests of the central Okanogan in the summer of 2008 and photographs taken a year later by a US Forest Service remote camera are also testament to the return of a pack of gray wolves to Washington State after an absence of nearly 70 years.

"Genetically, these new arrivals appear to have worked their way south from British Columbia," wrote Craig Welch in *The Seattle Times* in June 2009, "where wolves have been less likely to key in on livestock as prey, focusing instead on small black-tailed deer, even salmon, and other marine species. Rocky Mountain wolves tend to feed on much larger deer and elk – or sheep and cattle.

"Yet the transition to living with these new predators has been rocky for some."

The wolves – a small pack of perhaps five – have been sighted most often near Twisp, a ranching community in the Methow River Valley west of Okanogan. The ranchers have been in the Okanogan region since the late 1850s and despite the relatively peaceful relationship that exists between their northern ranching neighbors and wolves in Canada, they apparently see the reappearance of these members of what is normally an apex species in a healthy ecosystem as a threat.

In February 2009, a bloody wolf pelt was found stuffed inside a FedEx box bound for Canada, according to Welch, triggering an investigation of an Okanogan County rancher by state and federal agents. Wolves are considered an endangered species in Washington under state law and in the western two-thirds of the state under federal law.

Overhead, bald and golden eagles soar and the entire valley serves as a crucial "corridor" for migrating birds, animals and insects travelling between northern Canada and the central and south US.

Given the desert climate, the lakes and tributaries of the Okanagan/Okanogan River, as well as their adjacent woodlands or riparian areas are of utmost importance to the valley's wildlife.

The valley is also home to a small and

Pines thrive on a high, dry ridge.

COURTESY OF CREATIVE COMMONS SHAREALKIE 3.0 / http:creativecommons.org/licenses/by-sa/3.0

VIEWING THE SHRUB-STEPPE ECOSYSTEM

To protect a portion of this endangered arid ecosystem, as well as to engage in research and promote public education, a number of private and public sites have been set aside in both Canada and the US. Among them are:

1) The Osoyoos Desert Centre just northwest of Osoyoos, established by the Osoyoos Desert Society in 2000. With an information centre, elevated boardwalk trails, kiosks, signage, self-guiding brochures and guided tours, the centre hosts a variety of activities and research. The Osoyoos Desert Society also offers habitat restoration projects, native landscaping workshops and lecture series.

2) Nk'Mip Desert Cultural Centre, the centrepiece of the spectacular desert reserve of the Osoyoos people and their visionary chief, Clarence Louie. Located just east of Osoyoos, Nk'Mip is a luxury resort with an award-winning winery, a wine-tasting facility that puts many in Europe to shame, a full-service spa, a lakefront campground and Sonora Dunas, a nine-hole desert golf course. But Nk'Mip's centrepiece is its architecturally dramatic cultural centre, which offers programming, tours, exhibits and a gift shop, as well as three kilometres of broad, well-groomed trails wind through the sagebrush grasslands and among huge ponderosa pines, some of which are several hundred years old. The trails lead to a reconstructed Okanagan village with a sweatlodge and two large kekulis, or pithouses, which are cool and quiet when the temperature soars and cozy during the winter months.

3) Davis Canyon Natural Area, on Hwy 97A about four miles southwest of Chelan, Washington, which in turn is at the southeast end of Lake Chelan in Okanogan County. David Canyon, a 293-acre valley, is considered one of Washington's best remaining examples of the shrub-steppe ecosystem, with a rainbow of spring wildflowers, including lupines, mariposa lily, death camas, prairie-smoke and larkspur. It was designated a National Natural Landmark in 1986.

4) Hanford Reach National Monument is one of many areas in this ecosystem that are now protected. On the last non-tidal, free-flowing stretch of the Columbia River, it offers jetboat and kayak tours where visitors might see endangered burrowing owls or threatened loggerhead shrikes, as well as bald and golden eagles, pelicans, blue herons, many types of waterfowl as well as elk, deer, coyotes and porcupines. Spawning steelhead and endangered chinook salmon (see page 229), as well as more than 40 other species of fish are found in the river. Hanford Reach also stands apart thanks to its history, for plutonium reactors stand along the river, remnants of World War II and the Cold War. Plutonium from one of them fuelled "Fat Man", the atomic bomb that was dropped on Nagasaki, Japan, on August 9, 1945. The reactors are now being dismantled, and the lands and waters cleaned.

typically shy population of northern Pacific rattlesnakes, the only venomous snake in British Columbia and considered vulnerable in the province.

While BC's northern Okanagan Valley has mainly succumbed to development and vineyards, in Washington's Okanogan, overgrazing has caused the demise of native plants in many places, particularly the bluebunch wheatgrass that was often found in combination with big sagebrush. In its place, a host of introduced non-native species have dramatically changed the landscape. Perhaps the worst of these is cheatgrass, which, according to the U.S. Fish and Wildlife Service, "is so pervasive that it isn't even listed as a weed anymore".

Other "exotics", as introduced species are often termed, include Russian, diffuse, tumble and spotted knapweed; purple loosestrife, and both Canada and Scotch thistle. For a complete list of noxious weeds, go to www.fws.gov/hanfordreach/weeds.hmtl

Jaguar

Panthera onca • Endangered (US); Red List (IUCN)

THE Old World has the tiger, the leopard, and the lion as their "big cats" – those of the *panthera* genus, characterized by their ability to roar. The New World has just one big cat – the jaguar. Though spotted like the leopard, the jaguar is typically larger and of sturdier build. Glamorous, powerful, elusive and lethal, it has figured strongly in the culture and mythology of the Aboriginal cultures of Mexico and Central and South America, where its range has long been most concentrated.

Fossil records indicate that during the last glaciation jaguars, then larger and longer-legged than the modern species, were found in much of the southern half of the United States. And until the mid-19th century, thcy were distributed over the southwestern US as far north as the Grand Canyon – in the mountains of eastern Arizona, southwestern New Mexico, and in southern and eastern Texas. They were even found, some experts believe, in southeastern California and western Louisiana. Today, however, a jaguar sighting in the US is very rare; Listed as endangered there, its numbers are declining almost everywhere throughout the hemisphere.

Jaguars can live in varied landscapes, ranging across tropical, subtropical and dry grasslands, but they prefer to live in proximity to water – by rivers, streams, lakes or swamps, or in dense rainforest that provides cover for stalking prey. They carve out large ranges for themselves; a female's range is usually 10 to 15 square miles (or 26 to 40 square kilometres). Males often require twice that, depending upon prey density, habitat composition, and human exploitation. With the exception of the Amazon rainforest, which remains a vital sanctuary, jaguar ranges almost everywhere are shrinking and fragmenting, most rapidly in drier regions, such as the Argentinean pampas, the arid grasslands of Mexico, and the US Southwest. Over the past century, as a result of deforestation, growing human populations, and excessive and illegal hunting, the jaguar's northern range has receded 600 miles or 1,000 kilometres southward, while its southern range has contracted twice as far northward. The Wildlife Conservation Society estimates that the species has lost 37 percent of its historic range, with its status unknown in an additional 18 percent.

Knowledge about the jaguar's historic US habitats is limited. Records suggest that river valleys were important as travel corridors and as places to forage and den, but in the Southwest such habitats have been radically altered over the last century in the wake of agricultural expansion, settlement and urban development. With increased human water consumption, ground water tables have dropped, drying up many of the area's remaining streams.

By the late 20th century, the jaguar was thought to be extirpated in Arizona and New Mexico as a resident species, with records of occasional sightings attributed to transient individuals from Mexico, where the closest jaguar population lies about 200 kilometres south of border in the Sierra Madre Mountains of Mexico's

COURTESY OF THE ARIZONA FISH AND GAME DEPARTMENT / US FISH AND WILDLIFE SERVICE

Opposite: Macho B, caught and radio-collared in 2009 at an elderly 15 or 16 years of age, was found to be suffering from kidney failure.

This page: Macho B and another male were repeatedly filmed by trail cameras in southeastern Arizona and southwestern New Mexico between 1996 and 2009.

COURTESY OF JOHN AND KAREN HOLLINGSWORTH / US FISH AND WILDLIFE SERVICE INSET: COURTESY OF GARY M. STOLZ / US FISH AND WILDLIFE SERVICE

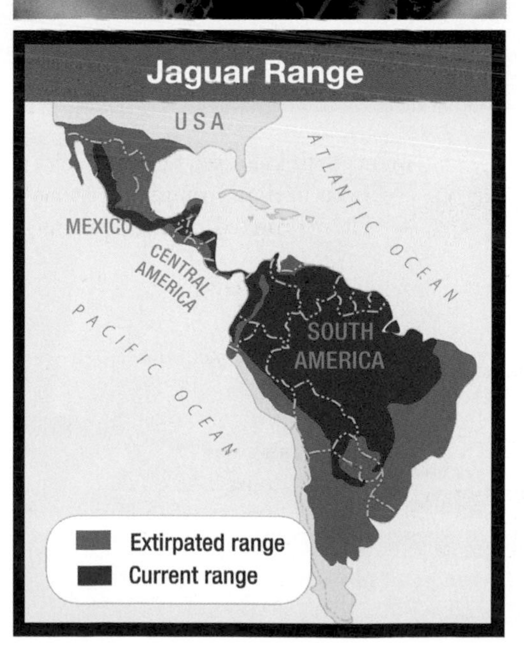

Sonora state. However, there is some speculation that the cats may at least occasionally be present in both countries. In 1996, on two separate occasions, a jaguar was photographed in southeastern Arizona. In the wake of these sightings, heat-triggered surveillance cameras were established along recognized animal trails through the mountains of southwestern New Mexico and adjacent southeastern Arizona. The monitoring confirmed the presence of two adult males, each photographed dozens of times between 2001 and 2009, and possibly a third unidentified jaguar. In February 2009, a 15-year-old male, nicknamed Macho B, was caught, radio-collared and released southwest of Tucson. However, a month later, he was recaptured and euthanized when he was found to have kidney failure.

Experts are divided over the meaning of the presence of these cats. Some believe that they are transient individuals who have travelled north, displaced by dominant males from the breeding population in Sonora, Mexico. Others theorize that a small breeding population may exist. Males, they reason, would not remain unless females were accessible and females have not been photographed by trail cameras because their home ranges are more restricted.

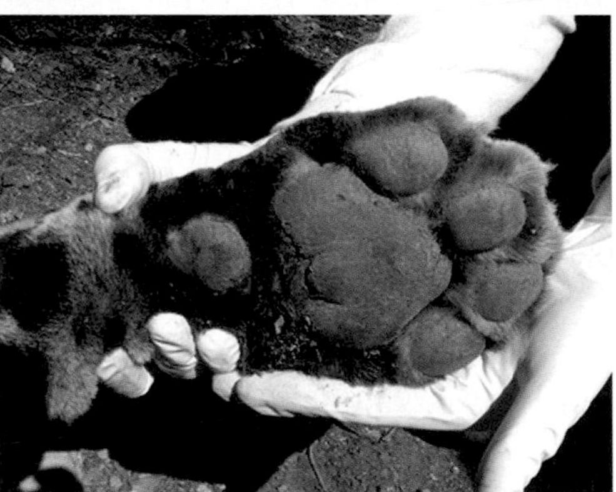

COURTESY OF THE ARIZONA FISH AND GAME DEPARTMENT / US FISH AND
WILDLIFE SERVICE

Top: Black jaguars make up less than six per cent of the population; bottom photos: a sedated cat has its paws studied and measured, a further means of identification, based on the size of its prints.

Jaguar Fast Facts

Panthera onca

DESCRIPTION

Compact and well muscled, the jaguar is shorter than its spotted cousin, the leopard, and has a stockier build. They vary considerably in size, between 100 and 250 pounds (45 and 113 kilograms), with males typically 20 to 30 per cent larger and heavier than females. Lengths vary from 5 to 8.5 feet (1.5 to 2.6 metres), from the nose to tail. The largest jaguars are found in Brazil and Venezuela. Jaguars at the northern end of their range, in Mexico and the US, tend to be smaller. Jaguars have fur ranging from tawny yellow to reddish brown marked with dark rosettes that act as camouflage in jungle habitats. Whereas the rosettes on leopards are plain, the rosettes on jaguars have black spots in the middle, and, as with human fingerprints, these patterns can be used for individual identification. There are also black jaguars, due to a condition known as "melanism," though they represent less than six per cent of the population and are more common in South America. (A spot pattern is visible under close examination.) Jaguars also have broader foreheads and wider jaws than leopards and are considered to have the strongest bite-force of any member of the cat family. Unlike other big cats, jaguars kill prey by a crushing bite to the skull and are capable of breaking open the shells of tortoises.

LIFESPAN

Jaguars live an estimated 12 to 15 years in the wild, and in captivity as many as 23 years.

DISTRIBUTION

In North America, north of Mexico, jaguars are only likely to be found in southeastern Arizona and southwestern New Mexico.

DIET

Jaguars are carnivores, but feed on many species. Their preference is for large prey such as deer, tapirs, peccaries, and domestic livestock, but they will also eat smaller species such as fish, frogs, mice, snakes and turtles.

BEHAVIOUR

Jaguars, like most cats, are solitary animals that roam large land areas likely to contain sufficient numbers of prey species to sustain them. Only courting pairs and mothers with cubs constitute community. They hunt primarily at dawn and dusk and stalk their prey mostly on the ground, though they are excellent climbers and can ambush their prey from trees or rock ledges. Jaguars generally prefer avoidance behaviour to fighting, and broadcast their presence and warn others with a roar that resembles a deep, hoarse cough, as well as scent marking and scraping.

MATING & BREEDING

Jaguar females reach sexual maturity after two years and jaguar males after four years. Because they are widely distributed through different climate zones, there is no specific heat season that applies to all. Births may occur any time of the year in tropical zones, but in northern areas they are more likely take place in the spring. The gestation period lasts four months. One to four cubs may be born, with two to three the average, each weighing between 21 and 32 ounces or 600 and 900 grams. A litter may include a black cub among the spotted. Born blind, they open their eyes at two weeks and remain sequestered with their mother for more than two months. After a month, their diet begins to include some meat along with their mother's milk. By three months they are eating meat almost exclusively. At six months, they begin to hunt with their mother; by two, they begin to hunt on their own.

THREATS

Habitat loss is the main threat to jaguars, but poaching is still a threat. Forests cleared for logging, farming, and urban development fragment and destroy habitat, reduce prey species, and facilitate illegal hunters. Historically, as commercial hunting and trapping declined, jaguars were threatened by cattle ranchers, who will often kill cats that prey on their livestock.

NAMES

The name jaguar derives from *yaguára*, a word in the South American Tupi-Guarani language that denoted any larger beast of prey.

VIEWING JAGUARS

Given the cat's tiny and unconfirmed representation in the United States, its secretive nature and its nocturnal habits, viewing opportunities are extremely remote.

THE JAGUAR, 1846, LIBRARY AND ARCHIVES CANADA / C-041812

RECOVERY PROGRAMS

Given the animal's inaccessibility and secretive behaviour, calculating global jaguar numbers is difficult, with estimates ranging between 10,000 and 20,000 through Mexico, Central America and South America. Estimated numbers in the United States, however, are fewer than the fingers on one hand. In the US, conservation agreements prevent the killing of jaguars on more than 800,000 acres of national forests, wildlife reserves and private ranches, but the return of the jaguar to the US in any significant numbers will depend on the success of cooperative efforts with Mexico, where a small breeding population exists in adjacent Sonora state. Among the organizations dedicated to bringing back the jaguar is the Arizona-based Northern Jaguar Project, founded in 2002, which has partnered with Mexican nonprofit conservation organization Naturalia to purchase and manage two adjacent ranches totalling 70 square miles (181 square kilometres). Located about 200 kilometres (125 miles) from the Arizona and New Mexico borders, the Northern Jaguar Reserve is a safe-haven sanctuary. This and the surrounding lands form one the largest unbroken expanses of wildlife habitat in northern Mexico and link with protected areas elsewhere in Sonora and in Arizona and New Mexico. The objective, promoted by leading jaguar expert Alan Rabinowitz, formerly of the Wildlife Conservation Society, now head of the Panthera Foundation, which is dedicated to protecting the world's wild cats, is to create a network of interconnected corridors and refuges extending from the United States into South America. It is called *Paseo del Jaguar* – Path of the Jaguar. However, a new man-made element threatens to impede jaguar recovery in the US – the completion of the United States–Mexico border fence. Though it is intended to discourage illegal immigration and drug smuggling, it is expected to curb animal migration, including that of jaguars, as well.

Sierra Nevada Bighorn Sheep

Ovis canadensis sierrae • Endangered (US)

THEY live their lives in the skies, at elevations of up to 13,000 feet (or more than 4,000 metres) along the alpine crests of California's Sierra Nevada. Bounded on the west by the Central Valley and on the east by the Great Basin, this "snowy mountain range", as Juan Rodriquez Cabrillo called it when he spotted the snow-capped peaks from the sea off San Francisco in 1542, is considered a part of the Pacific mountain system. But from the perspective of the rare bighorn sheep that make these mountains their home, its open areas among the mountain crests and eastern slopes are crucial to whether they continue to slowly increase in numbers, as they have for the past 15 years, or slide toward extinction, as they seemed to be doing between 1985 and 1995.

Before the mid-1800s and the California Gold Rush, herds of this bighorn subspecies occupied at least 16 areas from Sonora Pass, just north of Yosemite National Park, south along the Sierra Nevada to Olancha Peak, east of Sequoia National Monument. Though the pre-contact population is unknown, scientists believe it was well above 1,000. And in a story that's all too familiar, the decline in numbers coincided with the arrival of Europeans, as market hunters targeted the stocky animals for meat to supply mining towns and trophy hunters went after their magnificent horns.

The introduction of domestic sheep and goats also played havoc. In the 1870s, large numbers of bighorn died as a result of scabies, a mite-borne contagious skin disease that causes animals to loose their hair, which was transmitted by domestic sheep. Others are believed to have succumbed to pneumonia, also contracted from domestic herds.

By 1900, nine Sierra Nevada bighorn herds remained, and a half-century later, it was estimated that those had been reduced to five (though only three were actually sighted). By the 1970s, sheep were found in just two areas, though one of those – on Mount Baxter northwest of Independence – was later found to consist of two separate populations.

Concerns that the subspecies would die out completely led to intensive field studies in the late 1970s and annual monitoring of the herds through the mid-1980s. The smallest of the three populations, located on Mount Williamson, southwest of Independence, seemed static at about 30 bighorn, while the Mount Baxter and Sawmill Creek herds, totalling over 220, were slowly increasing. Beginning in 1979, the two larger herds were used to produce stock to

A young bighorn RON WOLF

reestablish populations in areas that once were home to bighorn. Over a period of eight years, 103 animals were moved to Wheeler Ridge, Mount Langley, Lee Vining Canyon as well as the southern Warner Mountains in northeastern California. The last of these herds died out quickly following contact with domestic sheep, but the others continue, the first two expanding from the outset and the herd at Lee Valley declining at first, due to a combination of poor weather and cougar predation, before finally beginning to grow after one cougar (or mountain lion, as they are sometimes known) was killed in each of three consecutive winters.

Cougar predation had been an increasing problem for Sierra Nevada bighorns, perhaps because a smaller herd size allows fewer "watch sheep", particularly when the herds were using low-elevation ranges during the winter and early spring. As a result, beginning in the 1980s, one herd after another stopped descending to the warmer winter ranges – where food was more abundant and the weather more clement – until in the last half of the 1980s, none of the populations were using the low-elevation ranges, preferring to stay high on the mountain cliffs, where they had a better chance of spotting predators. Though this was clearly based on the herd's negative experiences in the winter ranges, it had several consequences, among them exposure to extreme cold and wind during the winter months; a higher death rate from avalanches and deep snow, as well as poorly nourished animals, particularly in the later winter and early spring.

The result, exacerbated by large populations of mule deer, and drought, was population crashes among all the herds, so that by 1995 experts estimated that only about 100 Sierra Nevada bighorn remained, and of those, only 17 were females in their lamb-bearing years.

Responding remarkably quickly to this crisis, the subspecies was listed as endangered in January 2000, and in the years since, threats have been assessed, recovery plans made and in 2007, a recovery plan finalized. As of 2009, according to the Sierra Nevada Bighorn Sheep Foundation, herds have grown to number between 350 and 400 in five locations. However, that growth had slowed in recent years, in large part because suitable wintering locations were often compromised by domestic sheep grazing nearby or by the recreational use by off-road vehicles, including snowmobiles.

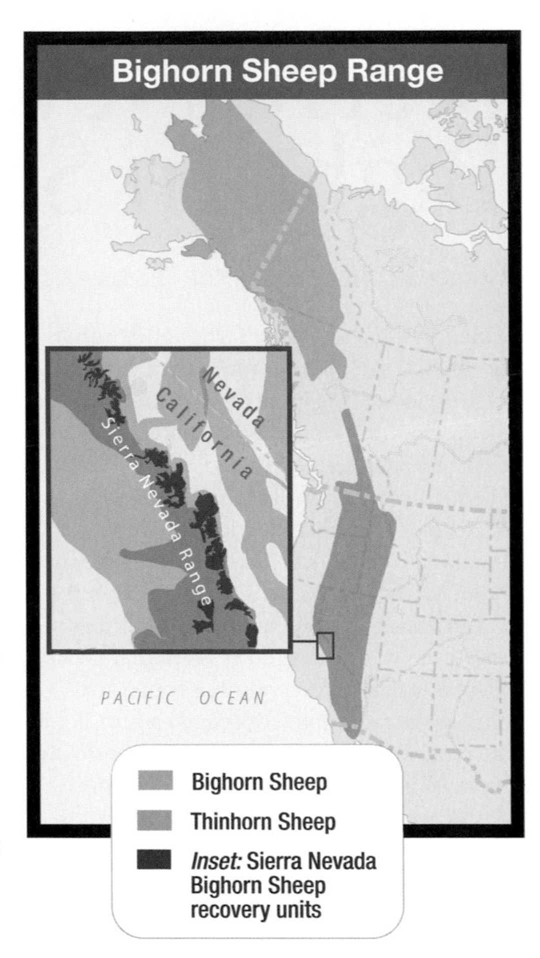

To allow the wild population to continue to grow, in 2008 the US Fish and Wildlife Service designated more than 400,000 acres (or 162,000 hectares) of land in the eastern Sierra Nevada as critical habitat, land deemed crucial for the survival and continued growth of the bighorns. To ensure that they don't foul watersheds, compete for food or drive herds away, domestic sheep are being removed from these areas of critical habitat, and off-road vehicles may be limited.

Ungulates, or hoofed mammals, are the most successful group of living land mammals, with about 180 species worldwide. The primitive ancestors of modern ungulates were the Condylarths, a group of mammals that arose during the Paleocene epoch, beginning about 65 million years ago. Eventually true ungulates evolved, diverging in the early Eocene into two branches: Perissodactyls with an odd number of toes and the Artiodactyls with an even number of toes. Three families of Perissodactyls – horses, rhinoceroses and tapirs – remain in existence today. Their modern forms first evolved in North America, as did camels, and migrated back to Asia and Europe, where they survived while their American precursors gradually died off. The Perissodactyls were initially predominant, but when the climate changed and grasses became more common around 20 million years ago, the Artiodactyls began to diversify and specialize as grazers. Modern Artiodactyls fall into nine families, including deer, pronghorn and horned ungulates.

True sheep first appeared in Europe during a period of global cooling (though prior to the Pleistocene or ice age in which we live) about 2.5 million years ago. Very successful as the Earth cooled, they proliferated throughout Europe and Asia and are believed to have crossed Beringia, the Bering land bridge, from Siberia at some point during the middle Pleistocene. As the glaciers waxed and waned, they spread down the mountainous spines of western North America, as far south as Baja California and Mexico's northern Sierra Madre.

According to the US Fish and Wildlife Recovery Plan for Sierra Nevada bighorns, which uses material from a 1993 study by Rob Ramey, then curator of vertebrate zoology at the Denver Museum of Nature and Science, "Divergence from their closest Asian ancestor (Siberian snow sheep; *Ovis nivicola*) occurred about 600,000 years ago."

In North America, the wild sheep diverged again, according to Canadian scientist and naturalist E.C. Pielou, into two subspecies of what are called "thinhorn" sheep – the all-white Dall's sheep and the gray-backed stone sheep – which found a home in northern Canada and Alaska, as well as several subspecies of bighorn sheep, which range from southern Canada to Mexico.

Originally, bighorn sheep were divided into Rocky Mountain bighorns (which are often seen in Canada's mountain parks) and California bighorns. However, Ramey's studies into the mitachondrial DNA of herds in the Sierra Nevada show that they are distinct from those in northwestern Nevada, which are now classified as desert bighorns (*Ovis canadensis nelsoni*).

With stocky bodies and relatively short legs, bighorn sheep are remarkably agile on precipitous rocky slopes (something even young lambs learn by imitating their elders), but not particularly fleet of foot. As a result, they depend on their keen eyesight and open, steep and rocky terrain to avoid predators. This works well during the summer months, when warm temperatures melt the 20

DENNIS FAST
www.dennisfast.smugmug.com

Sierra Nevada Bighorn Sheep Fast Facts

Ovis canadensis sierrae

HEIGHT & WEIGHT
Standing between three and three-and-a-half feet (or 90 to 105 centimetres) at the shoulder, and weighing up to 160 pounds or 72 kilograms, Sierra Nevada bighorns are smaller than other subspecies. Males are larger than females and both sexes are sturdy and well-muscled. Both have permanent horns; on old rams the horns can be very large, spiralling back and outward, often in a full circle. Females have small, erect horns.

LIFESPAN
Males can live to be between nine and 12 years; females from 10 to 14 years.

DIET
Bighorn sheep are remarkably flexible in the food they consume, thanks to a large rumen (or first stomach, where grasses, sedges, shrubs or rushes can be predigested) and a large reticulum, (the second stomach, which is lined with a ridged and honeycombed membrane). The amount of snowfall seems to dictate both quality and quantity of their upper elevation diet during the summer months, while winter's first soaking, with more than an inch of precipitation, is needed to ensure that growth will be bountiful on lower elevation meadows.

DESCRIPTION
Their smooth coat consists of guard hairs and dense fleece, varying from dark brown to grayish and pale tan depending on the region. The belly, rump patch, back of legs, muzzle and eye patch are all white. The brown horns are found in both sexes, but are much larger in males. In females, the horns are slender and sabre-like. In males, the horns are massive. In old (seven- or eight-year-old) males, the horns may begin a second curl.

BEHAVIOUR
Bighorn sheep feed during the day, largely to allow them to watch for predators, and spend the nights on rocky slopes. How far they venture from steep escape slopes is largely dependent on how open the terrain is (which determines their ability to see danger), as well as the weather and season, and the presence of young.

MATING & BREEDING
Males and females live in separate groups for most of the year. During the summer months, females generally stick to high elevations, while the males graze somewhat lower. Males join the females during the fall breeding season, and indicate their interest in an ovulating ewe with a series of nudges and bumps. Following a gestation period of 174 days, usually one but occasionally two lambs are born between late April and early July. Both sexes mature sexually at two, but females may not breed until four, while males, which are polygamous, have to compete for the right to breed.

NAMES
Formerly known as California bighorn sheep, DNA testing has demonstrated that the populations in the Sierra Nevadas are distinct and have therefore been renamed.

to 80 inches (or 51 to 200 centimetres) of snow that falls each winter and nourishes the plants at ever higher altitudes.

However, if Sierra Nevada bighorns have learned that they can not avoid cougars, their main predators, at lower levels where they might find food in the sagebrush-bitterbrush scrub in the winter and early spring, the result is very hungry herds, poor yearling survival, losses of entire herds in avalanches and fewer and weaker lambs in the spring.

Bighorn sheep are not picky eaters; they are able to digest a wide variety of grasses, sedges and rushes, so biologists do not believe it was grazing choices that prevented the herds from descending to low-elevation ranges during the winter. Instead, apparently lacking strength in numbers, they learned from experience that winter ranges meant danger. Larger herds allow more eyes to watch for the danger that is certainly there; in the past 25 years, of 147 bighorn deaths recorded in the Sierra

Desert bighorns look into the camera at California's Mojave National Preserve.

VIEWING SIERRA NEVADA BIGHORN SHEEP

Hikers along the Pacific Crest National Scenic Trail may see these wary wild sheep when traversing dizzy heights and deep canyons of the Central California section. In addition to bighorns, marmots, coyotes, deer and black bears also make the mountains their home. Desert bighorns are more easily viewed throughout the southwest.

COURTESY OF THE NATIONAL PARK SERVICE

Nevada, 54.5 per cent were killed by predators.

Studies have also shown that bighorn sheep in the Sierra Nevada may be lacking in phosphor- ous, which is low in alpine soils in Yosemite NP, perhaps because of leaching from snowmelt. Bighorn sheep in the Sierra Nevada seem to be aware of this, for biologists have found that they consistently select forage areas with higher phosphorous content, even at the expense of high protein levels.

Diseases that spread from domestic sheep populations are a challenge to prevent, for domestic sheep stray, even when well watched, and bighorn males have been known to move into domestic herds, likely looking for mates. The solution is to create buffer zones of up to nine miles or 13.5 kilometres between the two species. Pack goats, sometimes used in the back country of the Sierra Nevadas, are another possible source of contagion, though horses, mules and llamas do not seem to be.

In recent years, tule elk have been introduced to regions of the Sierra Nevadas used by bighorns. But despite their propensity, and that of native deer, to graze on the winter ranges historically used by wild sheep, there does not seem to be competition for food, likely because sheep eat plants the elk and deer disdain.

In spite of concerns about off-road vehicles, bighorns tolerate the presence of humans in other areas, including the Canadian Rockies, where they graze contentedly at the side of major throughways as cars and trucks drive slowly by, their occupants snapping photos. Perhaps, if the Sierra Nevada bighorn population continues to grow – and there's room in the mountains for 1,000 animals, according to the experts – visitors to eastern California might have the same privilege.

As conservationist John Muir wrote in *Mountains of California* in 1894, "The wild sheep ranks highest among the animal mountaineers of the Sierra. Possessed of keen sight and scent, and strong limbs, he dwells secure amid the loftiest summits, leaping unscathed from crag to crag, up and down the fronts of giddy precipices, crossing foaming torrents and slopes of frozen snow, exposed to the wildest storms, yet maintaining a brave, warm life, and developing from generation to generation in perfect strength and beauty."

Sonoran Pronghorn

Antilocapra americana Sonoriensis • Critically imperilled (US and Mexico)

PRONGHORNS are indisputably North America's most ancient indigenous land mammals. And some of their more remarkable characteristics bear witness to millions of years of evading predators. Their speed, for example, exceeds all other animals in the Americas and is only equalled (and then over short distances) by African cheetahs. This has led zoologists to believe that pronghorns evolved their running ability to escape from *Miracinonyx trumani,* one of at least two species of American cheetah, a big – and very fast – cat of the American grasslands that evolved more than two million years ago and disappeared about the end of the last glaciation.

But if they are built for speed over long distances, with an oversized trachea, huge lungs and a large, powerful heart, if they are blessed with superb eyesight and an acute sense of smell, if they survived the many glaciations over the past 1.8 million years, and the coming of humans to the Americas, they were unprepared for the great killing spree that accompanied the arrival of Europeans on the Western grasslands.

It's not certain how many pronghorns greeted the first European adventurers. Estimates of 30 million or even twice that have been suggested. What is known is that herds of hundreds ranged from Alberta's western foothills east to Manitoba's Pembina Valley and south through the mixed-grass and shortgrass prairies into the Great Basin. Able to thrive on tough shrubs and even thorny cactus, they also grazed north into the shrub-steppes of Washington, according to a study by R. Lee Lyman of the University of Missouri-Columbia, published in *Northwest Science* in 2007, and south to the grasslands of California and the deserts of Arizona, New Mexico and northwestern Mexico.

But these beautiful, wary and dazzlingly fleet animals – in September 1804, Meriwether Lewis wrote in his expedition diary, "... I beheld the rapidity of their flight along the ridge before me it appeared reather the rappid flight of birds than the motion of quadrupeds" – could not outrun the guns of the newcomers. Beginning with the fur traders of the 1820s, and increasing as the first rail lines stretched across the West, market hunters killed millions of pronghorn, shipping enormous quantities of rich, lean meat to the East and West coasts.

As Lisa Hutchins wrote in an article entitled "Prairie Racer: The Pronghorn Antelope" in 1999, "In Denver during the 1860s, just twenty-five cents could buy three or four entire pronghorn carcasses – a single coin for hundreds of pounds of meat. Despite this, pronghorns continued to be shot by the hundreds or even thousands and [like the bison] the carcasses simply left on the prairie to rot."

Curious and unaccustomed to the wiles of humans, pronghorn would often approach a flapping handkerchief decoy, or find themselves trapped and unable to escape behind a barbed-wire fence. Though they can leap almost 20 feet, or more than six metres, at a bound, they will not jump fences and have been known to starve, huddled in a fenced corner. Worse, despite Yellowstone National Park's status as the world's first national park, thousands of pronghorn were killed there annually during the 1870s.

Farmers and ranchers, wrongly assuming that pronghorn would take forage from their cattle, joined in the slaughter and by the early 1900s, the vast herds had disappeared. By 1920,

Ancient pronghorns very likely developed their speed to avoid such fleet and deadly predators as the now-extinct North American cheetah.

MICHAEL ROTHMAN

it was estimated that between 13,000 and 25,000 pronghorn remained across North America. At last, conservation-minded organizations, along with state, provincial and federal programs began to curtail hunting and provide protection.

The Dirty Thirties assisted, in a way, for the prolonged drought and rock bottom prices drove settlers from their land and greatly reduced livestock herds, allowing, according to the *2006 Pronghorn Management Guide* and A.E. Neilson's "A Brief History of Antelope in Idaho", "sizeable areas of cultivated land to revert to native vegetation. State, provincial, federal, and private organizations now began regulating the harvests of pronghorn, which were ... being reintroduced to unoccupied historic rangelands. Only in a ... few areas was damage to vegetation by drought and livestock foraging so severe that pronghorn were unable to survive."

By 1983, pronghorn numbers continent-wide had rebounded to more than a million, and in recent years, pronghorn populations have fluctuated between 600,000 and 800,000 animals, depending on northern winter conditions, drought in the southern states and the number of hunting licenses awarded.

However, if the main herds of pronghorn were doing well, the more delicate subspecies of the southwestern deserts – *A. a. sonoriensis*, the Sonoran pronghorn – was clinging to existence. In 1923, a survey of Sonoran pronghorn, one of four pronghorn subspecies, estimated that perhaps 75 animals could be found in the US: a few lived on the Tohono O'odham Nation south of Phoenix, perhaps five inhabited California's Sonoran Desert and the rest were found in on what is now Organ Pipe Cactus National Monument. Small herds were also in Sonora, Mexico, and the Mexican peninsula of Baja California.

As concern mounted over the pronghorn and other desert species in the 1930s, a four-million-acre federal desert wildlife reserve was proposed, but Arizona ranchers opposed the plan. In the end, less than two million acres were set aside, including what are today Organ Pipe Cactus NM and Cabeza Prieta and Kofa National Wildlife Refuges, all in southwestern Arizona. Between the two national refuges are the Yuma Proving Ground and the Barry M. Goldwater Range.

Despite these efforts, by 1950 Sonora pronghorn, which had once ranged over 35,000 square miles in the US and Mexico, had been extirpated from California, and had disappeared from the Kofa NWR north of the Gila River. By 1967, when they were listed as endangered, the Sonoran pronghorn occupied less than a tenth of their original range.

In the 1980s and 1990s, the US Fish and Wildlife Service began to study these delicate and elusive animals. Over the next decade, the scientists discovered that they migrate seasonally, particularly when forage is poor; they make little use of standing water, often actively avoiding it when it seems artificially placed; they rarely cross highways, railway tracks and, initially at least, responded poorly to capture and release projects, particularly when chased prior to capture. Following capture for the purpose of radio-collaring in 1994, for example, five of 22 adults died within the following month, leading scientists to believe that at least some of them might have suffered from capture myopathy, a physiological condition caused by intense fear or stress, leaving the animal dehabilitated. Changing the capture and handling processes over the succeeding years seemed to address the problem. Concerned that air force training flights might also cause stress, the military began avoiding areas of Cabeza Prieta NWR most used by the pronghorn.

Despite all that had been learned,

COURTESY OF JIM HENDRICK / CABEZA PRIETA NATIONAL WILDLIFE REFUGE

COURTESY OF THE ARIZONA GAME AND FISH DEPARTMENT

Sonoran Pronghorn Fast Facts

Antilocapra americana Sonoriensis

DISTRIBUTION
Able to withstand temperatures from 45°C (or 113°F) to -45°C (-49°F), pronghorn are adapted to almost every grassland ecosystem, from prairies to brushlands and deserts.

LIFESPAN
The average for both sexes is between seven and 10 years, though pronghorn as old as 15 have been found in the wild.

MATING & BREEDING
Males come of age at a year, but rarely mate then, for dominant males establish small harems during the rut in late summer. Males with a territory that includes a water source and physical features that allow him to corner does will do better than those without. Breeding for the first time at about a year-and-a-half, pronghorn does deliver twins about 60 per cent of the time, following a gestation of eight months, longer than most ungulates. All does in a given population will give birth within a few days, and the mothers will eat the afterbirths to prevent detection by predators. Each fawn averages seven pounds (or just over three kilos); one tiny twin is lost almost half the time. Within five days of birth, fawns can outrun a human, and though does cluster together during the two months when their fawns suckle, a significant percentage of the young are nevertheless taken by coyotes, cougars, bobcats and golden eagles. Does care for their young for a year or 18 months to fend for themselves.

DIET
Depending on the ecoregion in which they live, pronghorn herds feed on grasses and sagebrush; Sonoran pronghorn feed heavily on cacti and chain fruit cholla, a treelike plant with fleshy, green hanging fruit; both provide fluid when water is scarce. They have been described as "dainty" feeders, dining sparingly on a wide variety of plants.

NAMES
Their Latin name, *Antilocapra americana*, means the "American antelope goat", though they are neither antelopes nor goats. Pronghorn have also been called antelope, prong bucks and pronghorn antelope.

COURTESY OF THE ARIZONA FISH AND WILDLIFE DEPARTMENT

Opposite: A big-eyed doe gazes at the camera, while inset, Arizona Game and Fish employees tag a youngster.

A tiny fawn trails its mother; within days of birth it can outrun a human.

VIEWING SONORAN PRONGHORN

Visitors to Organ Pipe Cactus National Monument, where a semi-captive breeding program is ongoing, may see one or more pronghorn, since numbers are once again slowly growing. For more information, contact www.nps.gov

Pronghorns are the only animals in the world with branched horns, rather than antlers.

COURTESY OF THE BUREAU OF LAND MANAGEMENT

however, the number of wild Sonoran pronghorn continued to slide, from a US population of more than 200 adults in 1994 to just 99 in December 2000. Despite the decrease, in 2001 it seemed things were taking a turn for the better, for thanks to widely distributed rains, 78 fawns were recorded, the highest fawn-to-doe ratio ever recorded and of the 78, 50 survived.

However, the record year was followed by a terrible year of drought. Between mid-August 2001 and early September 2002, less than three-quarters of an inch of rain fell. As they normally do, most of the does produced twins, but this time none survived. By December 2002, biologists estimated that perhaps 21 animals remained in the US population (with an estimated 280 in the two, quite separate, Mexican populations).

The drought was too much for the pronghorn, added as it was to many other threats, including predation by coyotes, bobcats and cougars, the desiccation of and lack of access to regional rivers, including the Gila River, and disease. Blood samples taken in 2000 from five pronghorn indicated that all tested positive for several diseases, including bluetongue, a viral disease carried by sheep and cattle and transmitted by biting midges. Those severely affected are feverish, with swollen faces and tongues, and may have open sores on their feet. Not all those with the virus develop these symptoms, but those that do often sicken quickly and can die within a week. In cattle and sheep, the virus can be transmitted to a fetus or cause weak or deformed offspring. Bluetongue apparently originated in Africa and spread to Asia, Europe and now the Americas. Vaccination plans for cattle and sheep have been developed and, since 2008, are being carried out in Europe.

In 2003, fearing that Sonoran pronghorn were about to be extirpated in the US, the federal government stepped in with an aggressive program of water and forage enhancements, seasonal area closures in parts of the wildlife reserve and Organ Pipe Cactus National Monument, and a semi-captive breeding program. In partnership with the Air Force, Marine Corps, Mexican government, two Arizona hunting clubs, zoo veterinarians and University of Arizona, three major recovery projects were developed, including one that proved that if they are thirsty enough, pronghorn will drink from artificial water sources.

Four years later, these measures seemed to be having an effect. As of October 2009, the captive population in the 640-acre (or 260 hectare) breeding facility, which is fenced to

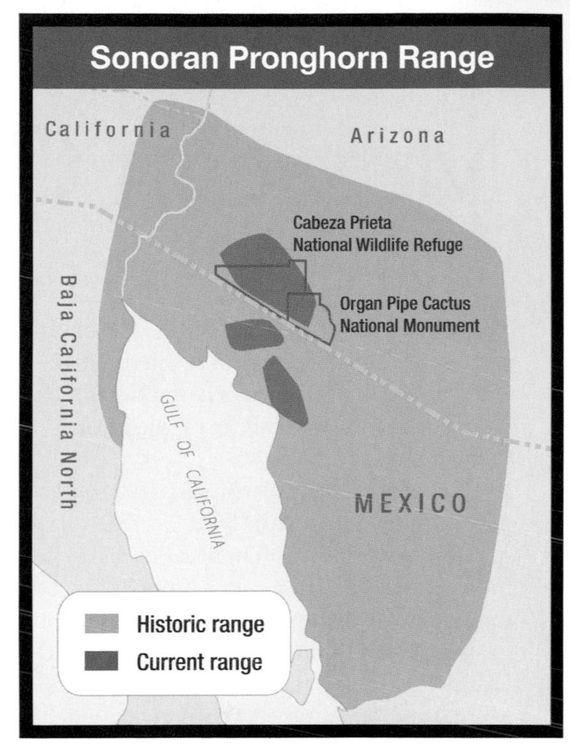

keep out predators and irrigated to increase water and plant growth, was 71, with a total of 21 released into the wild since 2006, totalling a population estimated to be about 68, according to 2008 aerial surveys.

Like their prairie cousins, Sonoran pronghorn are members of a unique North American species – *Antilocapra americana,* the "American antelope goat" – though they are neither antelopes nor goats and are not genetically linked to any other animal in the world. Over the past 20 million years, there may have been as many as 12 or 13 pronghorn genera, including four-horned and six-horned species, and it was perhaps a combination of speed and luck that allowed the pronghorn we know today to survive. The plains of the past were populated by a remarkable assemblage of predators, including dire wolves, plundering dogs, dholes and protocyons – both pack-hunting dogs capable of hunting medium-sized prey – giant short-faced bears, sabre-toothed cats; North American lions and the cheetahs mentioned earlier, as well as jaguars (which were once considerably more widespread in North America than they are today – see page 326).

As indicated earlier, Sonoran pronghorn are one of four pronghorn subspecies. The others are *A. a. americana*, the pronghorn of the Great Plains; *A. a. mexicana*, the Mexican pronghorn and the endangered *A. a. peninsularis*, the pronghorn of Mexico's Baja California peninsula.

Long-legged and small-bodied, pronghorn have woolly undercoats, covered with straight, course guard hairs varying in color from gold to tan. They have large white areas of hair on the rump, cheeks and underparts and bands of white on the throat. Sonoran pronghorn are paler in color than their northern relatives.

Pronghorn are unlike other North American ungulates in several ways: both sexes have permanent horns, rather than antlers. However, unlike the horns of cattle, these bone blades are branched in males and covered with skin – rather like those of giraffes – that develops into a sheath of keratin, which is shed annually.

Pronghorn are built for running, with fewer leg bones than elk; no dew claws, as found in elk, deer and moose, and two well-cushioned sole pads. Their speed comes from a very fast recovery on each stride and long leaping strides. However, the same physique that gives them speed and endurance prevents them from jumping, creating a fear of fences.

Pronghorns have a number of gaits, including an extreme "pushed-back" position in which front and back legs actually cross in the air, rather like a greyhound, and a leaping gait called a pronk. At top speed, pronghorn have been regularly clocked at more than 60 miles or 100 kilometres an hour; and they can keep up a fast pace for hours.

Large, bulging eyes and superb eyesight allow them to spot predators more than three miles or five kilometres away, while an early warning system sends danger signals to other pronghorn nearly a mile away.

Mexican Gray Wolf

Canis lupus baileyi • Endangered (US and Mexico)

THIS delicate wolf with its distinctive coat – the famed *el lobo* of the southwestern US – is the rarest wolf subspecies in the North America and was virtually extirpated from the Southwest by the 1970s.

Though in many places across the continent, dominant predators lived for millennia in relative harmony with indigenous peoples, the arrival of Europeans often brought them to the brink of extinction. However, the situation for the Mexican wolf was somewhat different. Even after the arrival of the Spanish in New Mexico and Texas, Mexican wolves continued for almost three centuries to hunt the elk and deer that abounded in the pine, oak and juniper forests and low mountains and valleys of southwestern New Mexico, central and southeastern Arizona, southwestern Texas and central Mexico.

Things changed with the opening of the Santa Fe Trail in 1821, which inaugurated more than a half-century of turmoil that included Civil War and Apache independence battles, the mining boom of the 1870s and 1880s and the settlement period that followed hard on its heels. Not only did all the newcomers target the ungulates that constituted the majority of the wolves' prey, but as the elk and deer were replaced by cattle, the wolves turned in desperation to stalking calves, cows and steers and ranchers accelerated their campaign to eliminate the wolf packs.

It's unlikely the ranchers were motivated only by stock losses. Just as they were elsewhere across the US, wolves were often killed simply because they were wolves, for European settlers carried with them fears that were largely rooted in "Old World myth and folklore", to quote U.S. Fish and Wildlife Service biologist S.H. Fritts. The extermination program, which involved federal, state and private groups, used every method available and was nothing if not thorough. Wolves (as well as bears, cougars, coyotes, foxes and even eagles – the nation's national symbol) were shot, trapped and poisoned until, by the mid-1900s, they were believed to have been extirpated from the US.

Pushed far south into Mexico's Sierra Madres, Mexican wolves appeared only occasionally in the border counties of Arizona, New Mexico and Texas. In the 1980s, the Mexican authorities reported that perhaps 30 Mexican wolves remained and by the end of that decade, they were believed to be extinct in the wild.

However, in 1976, three years after gray wolves (including the Mexican subspecies) had been listed as endangered under US law, the Fish and Wildlife Service hired Roy McBride, a former federal trapper who had once been involved in the eradication program, to determine whether there were in fact Mexican wolves still in the wild in Mexico. Over the next three years, he reported between 30 and 50 wolves were still living and in 1978, McBride was hired to trap and bring in alive as many Mexican wolves as he could.

It took two years, but in the end, he caught four males and one pregnant female. As Theresa Delene Beeland wrote in her 2008 Masters thesis at the University of Florida, "Most experts agree the last five [wild] Mexican wolves alive in the world were likely in the hands of the Fish and Wildlife Service and that the [Mexcan] gray wolf had been largely exterminated before much was known about its natural history or biology."

In fact, the current theories on the evolution of wolves in North America are complex.

Pacing in a pre-release pen, this Mexican wolf is ready for the U.S. Fish and Wildlife Service's reintroduction program.

COURTESY OF THE ARIZONA GAME AND FISH DEPARTMENT / US FISH AND WILDLIFE SERVICE

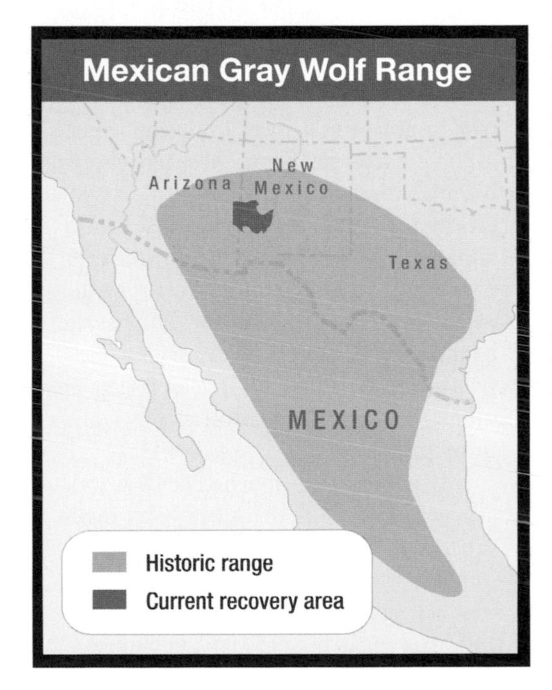

Among them were findings published in the *Canadian Journal of Zoology* in 2003, concerning mtDNA research conducted on two wolf skins from Eastern North America (including the last eastern timber wolf purportedly killed in New York State in the 1890s and another wolf killed in Maine in the 1880s), gray wolves from elsewhere in North America, as well as red wolves and coyotes. The researchers found that eastern timber wolves (also known as eastern Canadian wolves) are very closely related to the red wolves once found in Florida and the American south, and more closely related to coyotes than they are to gray wolves.

Based on these findings, Trent University biologist P.J. Wilson and his colleagues wrote, "We show the absence of gray wolf mtDNA in these wolves [the century-old skins from New York and Maine]. They both contain New World mtDNA, supporting previous findings of a North American evolution of the eastern timber wolf (originally classified as *Canis lupus lycaon*) and red wolf (*Canis rufus*), independently of the gray wolf [*Canis lupus*], which originated in Eurasia."

The scientists suggested that there was a branching sometime between one and two million years ago from a common North American ancestor of gray wolves, eastern North American (or timber) wolves and coyotes. One of the branches migrated to Eurasia, presumably over the Bering land bridge, and there gave rise to the gray wolf. The other branch remained in North America and sometime between 150,000 and 300,000 years ago branched into the ancestors of eastern North American wolves and coyotes.

The descendants of the gray wolf then returned to the Americas, likely during a glacial period between 150,000 and 120,000 years ago, but possibly later, during the early Wisconsin glaciation about 50,000 BP.

What is clear from fossils in California's La Brea Tar Pits is that gray wolves, which spread throughout western North America, must have co-existed, for a time, with the sturdier and now extinct dire wolf.

So, where do Mexican gray wolves fit into this complicated scenario? *In Wolves: Behavior, Ecology and Conservation,* edited by L. David Mech, senior research scientist with the U.S. Geological Survey and Luigi Boitani, a professor at the University of Rome, and published in 2003, the following was written of Mexican wolves: "Two of the captive Mexican wolf populations display a single divergent mtDNA haplotype, found nowhere else, that is more closely related to a subset of Old World haplotypes than to any New World haplotype, suggesting that these Mexican wolves share a more recent ancestry with wolves from the Old World … Further the basal position of the Mexican wolf haplotype … suggests that it is a relict form stemming from an early invasion of gray wolves from Asia."

COURTESY OF THE US FISH AND WILDLIFE SERVICE

COURTESY OF THE WILD CANID SURVIVAL AND RESEARCH CENTRE

A female, top, and male wolf. Opposite bottom: a wolf pup howls; opposite top: very young pups, named the Saddle Pack 2004 litter, born under the Species Survival Program in Sevilleta, New Mexico.

Wolf biologists now had DNA as well as physical evidence that this was a unique subspecies. Mexican wolves constitute the smallest gray wolf subspecies, reaching an overall length of about five feet or 1.5 metres and weighing between 50 and 80 pounds (23 to 36 kilograms), about the size of a German shepherd and three-quarters the size of gray wolves. Unlike gray wolves, their grizzled coats are mottled or patchy in color, mixed with shades of buff, gray, rust and black.

Using the five wolves that Roy McBride had trapped, the U.S. Fish and Wildlife Service and, after 1985, a consortium called the Mexican Wolf Captive Management

Committee, raised a captive population of 107 wolves by 1995. After 1993, the American Zoo Association's Species Survival Plan took over in the US and the federal wildlife agency took responsibility in Mexico.

Responding to concerns about the wolves' very narrow gene pool, wolves from two captive populations – the Ghost Ranch lineage, descended from two wolves taken from the wild in 1959 and 1961, and the Aragón lineage, descended from three wolves from Mexico's Chapultepec Zoo in the 1970s, all of which were proven to be pure Mexican wolves – were added to the breeding line.

In 1997, the US federal government formally approved a plan to reintroduce Mexican

COURTESY OF THE US FISH AND WILDLIFE SERVICE

wolves into the White Mountain region in east-central Arizona and west-central New Mexico. Chosen after years of studying potential sites, the region, called the Blue Range Wolf Recovery Area encompasses nearly 4.5 million acres of mountains, forests and grasslands and includes Arizona's Apache National Forest and New Mexico's Gila National Forest. The region has both plentiful water and recovered and relatively abundant populations of elk and white-tailed deer, as well as bison, mule deer, pronghorn, bighorn sheep and collared peccary, the only place in North America where all these species overlap.

COURTESY OF THE US FISH AND WILDLIFE SERVICE

In March 1998, eleven captive-raised wolves in three family groups were released from acclimatization pens and in 2002, the first wild-born litter of a wild-born parent was born. In 2008, the wild population was estimated at just 52 wolves, including just two breeding pairs, down in numbers from 2003. This has occurred, in part, because wolves have been removed to prevent conflicts with ranchers, and because there have been few serious consequences for those who have targetted and killed Mexican wolves. There are concerns about this situation, for the two breeding pairs have produced small litters, with pups of low body weight. Another 300 wolves are in breeding facilities in New Mexico and elsewhere.

To facilitate the 1998 release, Mexican gray wolves were described under the Endangered Species Act as an "experimental, non-essential" population. In 2009, Tucson, Arizona's Centre for Biological Diversity petitioned the US government to again list Mexican gray wolves under the Endangered Species Act, which would allow government agencies to act more aggressively on behalf of the wolves.

COURTESY OF DAVID TOLEDO

COURTESY OF STEVE DOBROTT

As it has in other regions, including Montana, where wolves from Yellowstone have migrated into a number of areas in the state, and Washington, where wolves are repopulating the northern forests on their own, the reintroduction has not been without controversy. Though studies in Montana have shown that wolves are responsible for the predation of less than a tenth of a per cent of the cattle actually killed in the state, in 2009, 120 prize rams were killed by a pack of wolves, and in the fall, with an estimated wolf population of 1,300, the state began licensing hunters. A wolf licence was $19 for local hunters and $350 for those from out of state; more than 7,000 were purchased in the first two weeks of the six-month season.

The move has many critics, including Minnesota's Wolf Trust, which reported that between 1979 and 2001, Minnesota's 2,500 wolves were verified to have killed 1,200 cattle, 879 sheep, 173 dogs and 1,251 fowl – about one-half an animal (including fowl) per wolf per year. In Alberta, which has an estimated 4,000 wolves (including about 1,500 in close proximity to livestock), an average of fewer than 250 cattle are lost annually to wolves.

Joe Miele, president of The Committee to Abolish Sport Hunting in Las Cruces, New Mexico, wrote in an editorial to Montana papers, "Wolves are not a problem in Montana. With ranchers often 'crying wolf', it may be surprising to many to learn that government run poisoning campaigns aimed at killing off natural predators are more deadly to cattle and sheep than wolves are. Are we to relive the early 1970s when hunters killed wolves to near extinction, necessitating their placement on the Endangered Species List?

"With wildlife watchers outnumbering hunters by a margin of two-to-one and contributing $325 million more to the local economy than hunters do (according to the U.S. Fish and Wildlife Service), it behooves the state to protect its wolves from hunters, trappers and ranchers so they can be enjoyed by the majority of Montana's non-hunting residents."

Healthy wolf packs also benefit other species, culling old or sick animals and creating populations that are more alert, more robust and healthier.

Nevertheless, old fears die hard. Though livestock owners are compensated when it is proven their cattle or sheep are killed by wolves, and studies have shown that disease, birthing problems and severe weather, as well as other predators – including coyotes, bears, eagles and even dogs – are eight times more likely to kill cattle (particularly calves) and four times more likely to kill sheep (particularly lambs), it is nevertheless wolves that are often blamed and targeted.

Mexican Gray Wolf Fast Facts

Canis lupus baileyi

LENGTH & WEIGHT
Adults weigh between 50 and 80 pounds (23 to 36 kilos) and are an average of five feet (or about 1.5 metres) in length, the same size as some European wolves. Males are generally larger than females.

LIFESPAN
Under optimum circumstances, wolves can live for 15 years, however accidents, predation by humans and cougars, and injuries sustained by prey such as elk often shorten their lives.

HABITAT
Historically, Mexican wolves lived in the mountains, forests and semi-arid grasslands from Mexico's Sierra Madre north to southwestern Texas, western New Mexico and eastern Arizona, and possibly into southern Colorado and Utah. Today, the reintroduced packs live in Arizona's Apache National Forest and New Mexico's Gila National Forest. Depending on food supply, a wolf pack's territory may encompass several hundred square miles.

DIET
Traditionally, scientists believe Mexican wolves preyed largely on deer, particularly the collared peccary and Coues white-tailed deer, both among North America's smallest ungulates which, some believe, might account for their small size. Today, however, studies have shown that 70 per cent of their prey is elk, with deer and other, smaller species, as well as fish, making up the balance. Bringing down a large ungulate takes pack cooperation, revolving around the chase. Packs are often unsuccessful and feed only a couple of times a week, eating large quantities when food is available.

BEHAVIOUR
Like other gray wolves, Mexican wolves are social animals that live in family groups called packs with a dominant or alpha pair, their pups and related yearlings. Their intricate communication system includes scent marking, body postures, whining, barking, growling and howling.

MATING & BREEDING
Generally monogamous, wolves are fertile by age two, but only the dominant pair will mate and the female will give birth about once a year after a 63-day gestation period. Litters average five to seven pups, which are born blind and helpless in dens in broken, sloping country near a good water source. The dens may be natural holes enlarged for the purpose and often have several entrances and a panoramic view. Breeding usually occurs in February, the pups are born in April and emerge from the den about four weeks after birth. The entire pack cooperates in feeding the pups and their mother, and while they are nursing, bring pieces of prey and eventually whole carcasses to the den. By three months, the pups begin to accompany the pack on hunts and are able to hunt alone (though the pack continues to provide food) by December. Once they reach sexual maturity, subordinate wolves often leave the pack to create their own family groups.

NAMES
Ironically, the Mexican wolf's Latin designation – *Canis lupus baileyi* – honors Vernon Baily, a trapper who worked for the US government in the early 1900s. In Mexico it's known as *el lobo.*

VIEWING MEXICAN GRAY WOLVES
Numbers are still so small and the territory they occupy so large that viewing wild Mexican wolves takes real determination. As indicated below, they are usually found at elevations above 4,000 feet, though they occasionally travel to valley grasslands. Captive Mexican wolves can be seen in the Mountain Woodland Exhibit at the Sonoran Desert Museum outside Tucson, Arizona.

Viewing the larger population of gray wolves is rather easier. Members of the Yellowstone National Park population are regularly seen, as are gray wolves in Canada's mountain parks, Banff, Jasper, Yoho, Kootenay and Waterton, which shares a boundary with Montana's Glacier National Park.

Ocelot

Leopardus pardalis • Endangered

WHERE the ocelot is found in arid areas, such as southern Texas, its tawny hide, backdrop to a miscellany of dark spots, stripes and smudges, takes on a creamy golden yellow hue. Little wonder this wild cat was long an irresistible prize to hunters serving the fashion for exotic furs. At one time, an estimated 200,000 ocelots through its range – Texas to Argentina – were killed each year for their beautiful hide. Since 1989, the Convention on International Trade in Endangered Species has prohibited trade in skins and live animals, and few would risk contempt by openly wearing an ocelot coat today – all of which signals a general improvement in the ocelot's fortunes.

Classified as a "vulnerable" endangered species from 1972 to 1996 by the International Union for the Conservation of Nature and Natural Resources, the ocelot is now rated "least concern" by the body, the world's main authority on the conservation status of species. However, there is little evidence of significant improvement in the dappled cat's fortunes in its most northerly range, along the brush country of the lower Rio Grande Valley of south Texas. In the United States, the ocelot has been classified as endangered since 1982. With fewer than 50 believed resident, the ocelot is now among the rarest wild cats in the country.

Today, it isn't hunting the ocelot that takes a toll on its US population. It's habitat loss. In the pre-contact period, ocelots ranged from Texas into Louisiana, Arkansas and Arizona, and by some estimations may have extended over the entire southeastern United States as far north as Ohio. That range in the historic period has diminished to the most southerly tip of Texas. To forage, rest, and establish dens, ocelots require dense vegetative cover, which the thick, almost impassable chaparral of south Texas provided for millennia, until the advance of human settlement. When Spanish explorers arrived in the Rio Grande Valley in the 16th century, they found a landscape markedly different than today's. An intersection of tropical and temperate climates, the coastal savannahs and interior brushlands were dense with wildlife. Those days are gone. Since the 1920s, more than 95 per cent of the scrubland in the lower Rio Grande Valley has been cleared for agriculture and urban development. Experts estimate only about one percent of the area could be defined today as optimal habitat for ocelots; much of that has been fragmented into isolated patches. Isolated communities lead to inbreeding and loss of genetic diversity, which makes ocelots more vulnerable to disease and genetic defects. It also leads to ocelots travelling great distances to find sufficient

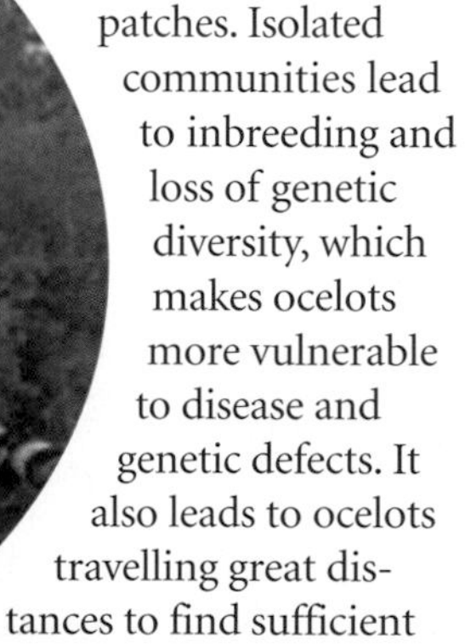

COURTESY OF NOVA MCKENTLEY / US FISH AND WILDLIFE SERVICE

Though at last protected from fur trappers and hunters, ocelots paid an enormous price for their beautifully marked coats.

Opposite: Laguna Atascosa NWR veterinarian Jody Mays examines an ocelot.

COURTESY OF TOM SMYLIE / THE US FISH AND WILDLIFE SERVICE

RECOVERY PROGRAMS

The challenge is to restore and expand native brush habitat and link existing breeding populations with habitat corridors on both public and private land in south Texas. To that end, several organizations, including the US Fish and Wildlife Service, The Nature Conservancy, and Environmental Defense, are working with private landowners, offering incentives to take land out of food production and restore native vegetation. The Fish and Wildlife Service is also working with the Texas Department of Transportation to install wildlife underpasses below new roads, which would help reduce the number of ocelots killed by traffic.

habitat to survive, often crossing several roads in the process. The animals may either starve or be run over by a car, which is now the chief cause of ocelot mortality in the US.

The continued survival of the ocelot population in the US is uncertain. Three national wildlife refuges in south Texas – Lower Rio Grande Valley, Santa Ana, and Laguna Atascosa – combining 137,000 acres (or 555 square kilometres) – support good ocelot habitat, as do privately owned parks such as The Nature Conservancy's Lennox Foundation Southmost Preserve and the Sabal Palm Audubon Center.

Ocelot Fast Facts

Leopardus pardalis

LENGTH & WEIGHT

Including its black-banded tail, a male can be three feet, nine inches (about 1.14 metres) long, with females 20 per cent smaller. About twice as heavy as a domestic cat, they weight between 15 and 30 pounds (7 and 14 kilos).

DESCRIPTION

Slender and elegant, the ocelot's underlying coloration varies with its habitat, with the base color of its fur ranging from a creamy yellow in arid areas to a darker yellow-brown in forested habitats. Its black spots and darker rosettes lend it a distinctive appearance, as do the two black lines on either side of its face.

LIFESPAN

Ocelots live an estimated eight to 13 years in the wild.

DIET

Carnivores, ocelots feed on a variety of prey, with small mammals such as rodents and rabbits making up much of their diet. They will also eat reptiles, amphibians, fish and birds, and have been known to eat domestic poultry or young pigs or lambs, if allowed the opportunity. Almost all their prey is much smaller than the ocelot itself.

BEHAVIOUR

Like most cats, ocelots are solitary, usually only meeting to mate, although ocelots of the same sex will sometimes share a spot in a tree or in dense brush during the day when at rest. Like most cats, they are active at night and territorial. Males may maintain an area as large as 12 square miles (30 square kilometres) overlapping that of several females who maintain territories half the size. Though they engage in occasional territorial disputes, ocelots prefer to avoid belligerent encounters, including those with human beings.

MATING & BREEDING

Ocelots become sexually mature after three years. There is no particular mating season in tropical areas, or in southern Texas. Females have been documented denning year round, with slight peaks in spring and fall. The gestation period is between 70 and 80 days, with one to three kittens the result. The kittens remain in their mother's territory until they are about a year old.

THREATS

Habitat loss is the main threat to ocelots. Large-scale clearing of brush in south Texas for cropland conversion and urbanization through the 20th century reduced wildlife habitat by 95 per cent. Vehicular traffic is the leading cause of ocelot mortality, as ocelots are often forced to cross roads to reach desirable habitats.

NAME

The name "ocelot" derives from *ocelotl*, a word in the Nahuatl language of central Mexico that refers to the jaguar, not the ocelot.

VIEWING OCELOTS

Given the cat's small numbers in the United States, its secretive nature, its nocturnal habits, and its preference for scrub that a human could only enter by crawling, viewing opportunities are extremely remote. Only good fortune would likely afford a visitor a glimpse at or near one of the national wildlife refuges or private sanctuaries in south Texas.

OCELOTO, CASTORE / LIBRARY AND ARCHIVES CANADA 2558

Together, they form an embryonic ribbon of wildlife habitat along the Rio Grande River, but much of the land between and surrounding these preserves remains in private hands and can't be directly managed to accommodate the needs of ocelots or other threatened or endangered species. The hope for ocelots in the Rio Grande Valley, however, lies in part with more ranchers orienting their businesses to increasingly profitable game species, such as ducks or deer, rather than cattle, which obliges them to restore native habitat. Hope for ocelot survival also depends on collaboration with conservation groups on the Mexican side of the Rio Grande, in Tamaulipas state, where ocelots are similarly endangered. A cross-border corridor would help ensure a greater number of available mates to maintain genetic diversity. Thwarting this strategy, however, is the United States–Mexico border fence, currently under construction through prime riparian habitat along the Rio Grande River, which will block animals from accessing protected areas on either side of the border.

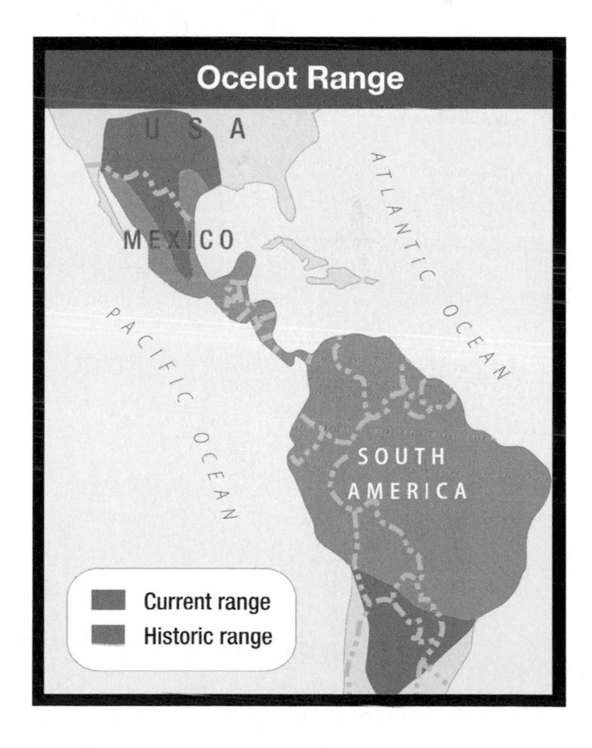

Jaguarundi

Herpailurus yagouaroundi • **Endangered**

EVEN rarer and more elusive than the ocelot, but within the same geographic range in Texas, is the jaguarundi, an unspotted cat with either black, reddish-brown, or brownish gray fur. Small enough to be mistaken for a feral cat – they range between six and 20 pounds or about three and nine kilos – jaguarundi were historically spared by the fur trade. But, like the ocelot, they have suffered from the loss of dense brush in the lower Rio Grande Valley. Though jaguarundi share many of the characteristics of the ocelot, less is known about their biology and habits. The size of the jaguarundi population in the US, too, has proven difficult to pin down. However, recovery efforts aimed at the ocelot, including conservation of remaining habitat in south Texas and the creation of corridors connecting habitats within the United States and cross-border into Mexico, are expected to also benefit jaguarundi populations.

COURTESY OF GARY HALVORSEN / US FISH AND WILDLIFE SERVICE

Above: A jaguarundi sits at the opening to its cave. Bottom images: left, the McKittrick Canyon Valley, and right, Guadalupe Mountains National Park in Texas are both jaguarundi territory.

BOTTOM IMAGES: COURTESY OF LEAFLET / CREATIVE COMMONS, NO RESTRICTIONS / EN.WIKIPEDIA.ORG

Glossary

anadromous fish: fish that spend part of their lives in salt water and part in sea water

baleen: finged plates of keratin that hang from the upper jaws of baleen whales. Often known as "whale bone", it is a gray or black material much like a fingernail. In the 18th and 19th centuries, it was used for corset stays, shirt hoops, buggy whips and umbrella ribs.

benthic: organisms living on sea or lake bottoms

Beringia: the area of northwestern Alaska and the Bering Strait that was exposed when ocean levels dropped dramatically during Pleistocene glaciations. Despite its northerly position, much of Beringia was ice-free during these glacial periods.

BP: Before Present, a system of geological dating in which the age of artifacts, cultures or events are given in years prior to the present. The year 1950 was chosen to mark the present.

Celsius: a temperature scale that registers the freezing point of water as 0° C and the boiling point as 100° C under normal atmospheric pressure. Named after Anders Celsius (1701 1744), a Swedish astronomer who devised the scale.

cetaceans: belonging to the order Cetacea, which includes aquatic mammals like whales and propoises.

copepods: small marine and freshwater crustaceans

COSEWIC: the Committee on the Status of Endangered Wildlife in Canada; a body that researches at-risk species and makes recommendation to the federal government for inclusion under its Species at Risk Act (SARA)

Cretaceous: the third and last period of the Mesozoic era, which lasted 74 million years and was marked by the final flowering of the dinosaurs, and the early development of mammals and flowering plants.

DNA: deoxyribonucleic acid, the chief constituent of chromosomes. Able to replicate itself, it is responsible for transmitting genetic information in the form of genes from parent to offspring.

Dorset period: a cultural period from about 3000 to 1200 BP involving Arctic maritime people who lived off the resources of the sea in northeastern Canada. They are known for their beautiful ivory carvings, bone needles and awls, as well as hand pulled sleds, microblades and harpoons, which were used to hunt seal, walrus and beluga whales.

echolocate: the ability to bounce high-frequency sound off surrounding objects, including schools of fish, allowing whales, dolphins and bats, among others to "see" their surroundings

Eocene: the second-oldest of the five epochs of the Tertiary period, marked by the rise of mammals

extirpated: gone from a particular region

Fahrenheit: a temperature scale that registers the boiling point of water at 32° F and the boiling point of water at 212°. Named for Gabriel Fahrenheit (1686-1736), a German physicist who lived in Holland.

genus: a category used in taxonomy, which is below 'family' and above 'species'. A genus tends to contain species that have characteristics in common. The genus forms the first part of a 'binomial' Latin species name; the second part is the specific name.

isinglass: a form of gelatine obtained from the air bladder of sturgeon and other fish, which was used as an adhesive or as a clarifying agent for wine

IUCN: the International Union for the Conservation of Nature and Natural Resources. Founded in 1948 and based in Switzerland, it publishes a Red List of species threatened with extinction every four years.

Holocene epoch: the present period of the Quaternary period (our glacial age), beginning at the end of the last glaciation about 11,000 BP

Jurassic: the second period of the Mesozoic or dinosaur era, following the Triassic and preceding the Cretaceous, between about 208 and 140 mya

krill: small marine crustaceans that constitute the main food of baleen whales

Linnean or Linnaean: named for Karl Linnaeas, also known as Karl Linné (and after 1761 von Linné), a Swedish biologist who created the system of classification for plants and animals that is used today, describing them by genus and species

ma or mya: scientific shorthand for millions of years or millions of years ago

melon: the rounded forehead on a beluga whale

Mesozoic: the geologic era after the Paleozoic and before the Cenozoic, between about 245 million and 65 million years ago

Miocene period: the fourth epoch of the Tertiary period in the Cenozoic era, from about 24 and 5.3 mya

Mitacondrial DNA (mtDNA): DNA found outside the nuclei of cells and passed to the next generation only by the mother

notochord: a cartilage rod that serves as a backbone in some very primitive species of fish

odontocetes: whales that have a single blowhole

orogeny: the process of mountain building

Pangaea: "all lands"; one of a series of supercontinents that formed about 280 million years ago.

PCBs: polychlorinated biphenals

plankton: tiny plant and animal organisms that float or drift in great numbers in fresh or sea water

pinnipeds: seals and sea lions

Pleistocene: the last major ice age, which began about 1.8 million years ago and is ongoing

Polynyas: a Russian term for open areas of water surrounded by ice

POPs: persistent organic pollutants

Protocetid: ancestors of whales

Magnificent and elusive, cougars (or mountain lions as they are sometimes called) are important predators across Western North America. In the East, where they were once believed to be extirpated (in the US) or endangered (Canada), they are making a comeback.

SHARON CHESTER

refugia: in climatological terms, areas that were not ice covered during the last glaciation

reintroduction: an attempt to reestablish a native species in an area where it previously occurred

riparian: riverside

suture zone: the area where a terrane (see below) has attached to a new crustal plate

takin: relatively small, plump bovine animals that live in Asia

tectonic: pertaining to, causing or resulting from the structural deformation of the Earth's crust

terranes: a fragment of the Earth's crust which has broken off one tectonic plate and attached itself – acretted or sutured itself — to crust lying on another plate. The crustal fragment preserves its own distinctive geological history, which is different from that of the surrounding areas. Terranes are therefore sometimes termed "exotic" landmasses. The area where a terrane is attached to a new crustal plate – termed the suture zone – can usually be identified as a fault.

Tertiary period: the first period of the Cenozoic era, from the end of the Cretaceous about 65 mya to the Quaternary, characterized by the rise of apes and modern flora

Tethys Sea: a large sea that opened between Gondwana (the southern continent) and Laurasia (the northern continent) as Pangaea, the Earth's last supercontinent, split in two during the Jurassic. The Mediterranean Sea is the last large remnant of the Tethys Sea.

Thule: named for Thule, Greenland, this refers to a cultural period in the Canadian Arctic from 1000 to 400 BP; peopled by the ancestors of today's Inuit

tonne: a metric ton

ungulate: having hoofs or belonging to a group of mammals having hoofs.

WHA: wildlife habitat area

Wisconsin glaciation: the last and one of the largest incursions of ice during our current ice age. It had two distinct phases, the Early Wisconsin, from about 75,000 to 65,000 BP and the Late Wisconsin, from about 23,000 to 10,000 BP

EON	ERA	PERIODS	EPOCH
Phanerozoic	Cenozoic Era (65 mya to today)	Quaternary (1.8 mya to today)	Holocene (11,000 BP to today)
			Pleistocene (1.8 mya to 11,000)
		Tertiary (65 mya to 1.8 mya)	Piocene (5 to 1.8 mya)
			Miocene (23 to 5 mya)
			Oligocene (38 to 23 mya)
			Eocene (54 to 37 mya)
		extinction of dinosaurs	Paleocene (65 to 54 mya)
	Mezozoic Era (245 to 65 mya)	*reptiles & flowering plants* Cretaceous (146 to 65 mya)	Divided as: Upper Middle Lower
		Jurassic (208 to 146 mya) *first birds & mammals*	
		Triassic (245 to 208 mya)	
	Paleozoic Era (544 to 245 mya)	Permian (286 to 245 mya) *amphibians*	
		Carboniferous (360 to 286 mya): Pennsylvanian (325 to 286 mya)	*large primitive trees*
		Carboniferous (360 to 286 mya): Mississippian (360 to 325 mya)	
		Devonian (410 to 360 mya)	*fishes, first amphibians*
		Silurian (440 to 410 mya)	*first land plants*
		Ordovician (500 to 440 mya)	*invertebrates: first fish*
	first shells, trilobites dominant	Cambrian (544 to 500 mya); Tommotian (530 to 527 mya)	
Precambrian (4,600 to 544 mya) *first multi-celled organisms*	Proterozoic Era (2500 to 544 mya)	Vendian (650 to 544 mya) or Ediacaran	No Epochs
		Neoproterozoic (900 to 544 mya) Mesoproterozoic (1600 to 900 mya) Paleoproterozoic (2500 to 1600 mya)	
	Archaean (3800 to 2500 mya)		
	Hadean (4500 to 3800 mya) *Approx age of oldest rocks*		

This view of a bay off Buttle Lake was taken in Strathcona Provincial Park on central Vancouver Island. Snow-covered Mount Myra is in the right background.

JACK MOST www.themostinphotography

The Authors

Heather Beattie holds three degrees, a B.A. from the University of Winnipeg, and two M.A. degrees, one from York University in Toronto and the other from the University of Manitoba in Winnipeg. She works as an archivist at the Hudson's Bay Company Archives, located at the Archives of Manitoba.

Barbara Huck is an award-winning journalist, author and publisher, who has authored or co-authored several national bestsellers, including *In Search of Ancient Alberta* (with Doug Whiteway); *Exploring the Fur Trade Routes of North America* and *In Search of Ancient British Columbia*, as well as *The Land Where The Sky Begins.*

Nature & History Titles by Heartland

In Search of Ancient Alberta

Pelicans to Polar Bears: Watching Wildlife in Manitoba

Mistehay Sakahegan: The Great Lake

Thanadelthur / Blackships: Young Heroes of North America, Volume I

Exploring the Fur Trade Routes of North America

Wapusk: White Bear of the North

Sacagawea: The Making of a Legend: Young Heroes of North America, Volume II

Magical, Mysterious Lake of the Woods

Dancing Backwards: A Social History of Canadian Women in Politics

In Search of Ancient British Columbia

The Land Where the Sky Begins

Heartland Associates
PO Box 103 RPO Corydon
Winnipeg, Manitoba Canada R3M 3S3
Tel (204) 475-7720 Fax (204) 453-3615
Email: hrtland@mts.net

To learn more about these and many other titles by Heartand,
please visit www.hrtlandbooks.com